PRINCIPLES OF POLYMER ENGINEERING RHEOLOGY

PRINCIPLES OF POLYMER ENGINEERING RHEOLOGY

JAMES LINDSAY WHITE

Polymer Engineering Center
Department of Polymer Engineering
The University of Arkon
Arkon, Ohio

A WILEY-INTERSCIENCE PUBLICATION

John Wiley & Sons, Inc.

NEW YORK / CHICHESTER / BRISBANE / TORONTO / SINGAPORE

Copyright © 1990 by John Wiley & Sons, Inc.

All rights reserved. Published simultaneously in Canada.

Reproduction or translation of any part of this work
beyond that permitted by Section 107 or 108 of the
1976 United States Copyright Act without the permission
of the copyright owner is unlawful. Requests for
permission or further information should be addressed to
the Permissions Department, John Wiley & Sons, Inc.

Library of Congress Cataloging-in-Publication Data:

White, James Lindsay, 1938–
   Principles of polymer engineering rheology.

   Bibliography: p.
   1. Polymers and polymerization—Rheology.   I. Title.

QD381.9.R48W48   1988   668.9   88-33830
ISBN 0-471-85362-3

Printed in the United States of America

10 9 8 7 6 5 4 3 2 1

PREFACE

The polymer industry is an area of great energy and dynamism. New materials with new combinations of properties continue to come to the marketplace. Engineering parts ranging from sophisticated extrusion profiles and injection moldings to tires are undergoing continuous development. Engineering plastics continue to make rapid headway against metals in aircraft components. The engineer requires a detailed understanding of the flow behavior of these materials in order to design the dies and molds required to form tomorrow's structural components.

The flow or rheological behavior of polymer melt systems is very complex and is still not fully understood. Such systems include not only the traditional pure melts, but solutions and compounds containing anisotropic (fiber or disc) or colloidal particles. Pure polymer melt systems themselves consist not only of flexible chain materials, which include polyolefins, but rigid chain substances that form ordered structures, that is, liquid crystals. New and complex materials continue to reach the market.

It is the purpose of this book to provide the basic bakcground needed by engineers to (1) determine experimentally and interpret the rheological behavior of polymer melts and (2) apply it to analyze flow in processing operations. To accomplish this we must critically treat three different areas or disciplines. We develop the fundamentals of continuum mechanics and show how it may be applied to devise methods for measurement of rheological properties, formulation of three-dimensional stress–deformation relationships, and analysis of flow in processing operations. We also discuss the structure of polymers and interpret the rheological behavior in terms of structure. Finally, we discuss polymer fabrication technology and seek to analyze and design these processes.

The book is divided into three parts. Chapters I–III represent the background in polymer science, continuum mechanics, and polymer technology needed to understand and appreciate the need for rheological characterization. Chapters IV–VI deal with the experimental foundations of modern rheology and rheo-optics and the interpretation of experimental data. Chapters VII and VIII deal with constitutive equations relating stress to deformation history in non-Newtonian fluids and their applications. Chapter VIII concerns specific applications to polymer processing. Each chapter begins with a critical historical survey of the subject matter and then proceeds to develop the material in a pedagogical manner.

This book is written for first-year graduate students in polymer engineering. It could be used equally well by chemical or mechanical engineering graduate students.

JAMES LINDSAY WHITE

Arkon, Ohio
January 1990

CONTENTS

PRINCIPLES OF
POLYMER ENGINEERING
RHEOLOGY

BACKGROUND

I

INTRODUCTION TO POLYMERS

A. INTRODUCTION

Rheology treats the deformation and flow of materials. To properly design and utilize instruments, the rheologist must know the material with which he or she deals, whether the material is homogeneous or multiphase, and its operating temperatures and chemical stability. Often the rheologist must use his/her talents to interpret experiments in terms of structure.

    In this chapter we discuss the nature of polymer materials, the various classes of possible structural variation, and the nature and states of matter of the major polymer of commercial interest.

B. COMMERCIAL POLYMERS AND DEVELOPMENT OF THE MACROMOLECULAR HYPOTHESIS

Polymers have been used in the service of mankind since the beginning of recorded history. Leather, wood, wool, linen, and cotton are polymeric substances of plant and animal origin. Textile structures are polymeric composites and the textile industry is a polymer industry, though they are rarely considered as such. It was only in the period after 1820, with the development of a rubber industry based upon coagulated rubber latex from South America (*38, 42, 95, 96*), that concern was exhibited over the structure of polymeric substances. Early entrepreneurs such as Thomas Hancock (*42, 95, 96*) and Charles Goodyear (*38*) exhibited concern over the composition and molecular structure of their products. Perhaps the first careful study of a polymer was made at the behest of Hancock by Michael Faraday (*31, 42*), who carried out an analysis of the chemical composition of natural rubber and its latex.

In the half century stretching roughly from 1838 to 1890, entrepreneurs discovered they could chemically modify many industrial substances which we now consider as polymeric to produce more useful products. Nathaniel Hayward (*44*), Goodyear (*37, 38*), Hancock (*41, 42*), and Alexander Parkes (*42, 67*) developed procedures for modifying (vulcanizing) rubber with sulfur and heat. Christian Schonbein (*76*) shortly thereafter nitrated cellulose and in succeeding years the acetates (*45*) and xanthates (*9, 29, 45*) of cellulose were developed. These developments, coupled with the discovery and application of gutta percha, led to the plastics industry. Charles Hancock (*40*) and others developed gutta percha, which we describe today as a 'plastic'. From about 1845, it was widely used as electrical insulation. Largely through the efforts of Alexander Parkes (*50, 68*) and the Hyatt brothers, John Wesley and Isaiah Smith (*47, 48, 97*), in the period 1865–1880, cellulose nitrate was developed as a plastic. As *celluloid* it became very widely used. In the decade 1884–1894, a man-made fiber industry based on cellulose nitrate came into being largely through the efforts of Chardonnet (*27, 45*). This later led to regenerated cellulose or rayon fibers based on cellulose xanthate (*9, 29, 45*).

The realization of the polymeric nature of rubber and the cellulosics awaited the development of quantitative measurements of molecular weight for substances that could not be vaporized. This came with the development of colligative property measurements in the 1880s and 1890s. In 1889, Gladstone and Hibbert (*35*) found, using freezing-point depression, that rubber had a molecular weight of 12,000. Amylodextrin formed in the hydrolytic degradation of starch was noted by Brown and Morris (*12*) at about the same time to have a molecular weight of about 30,000. In 1914 Caspari (*24*), extrapolating osmotic pressure data to infinite dilution, determined molecular weights of 100,000 for rubber and 40,000 for gutta percha. Most contemporary chemists refused to accept these high molecular weights and argued that they were due to associations of small molecules. Polymers were confused with colloidal suspensions. In some cases, it was argued that these materials were ring structures associated together (*70*).

A new era in polymer science came in 1920 when Hermann Staudinger (*81*) clearly elucidated the modern theory of the structure of polymers, arguing that materials such as rubber and polystyrene were linear long chains with repeating units, indeed the same repeating units we accept today. During the next decade and more, Staudinger (*81–85*) carried on a heroic struggle to establish this macromolecular hypothesis. Through extensive molecular weight measurements and critical investigations of chain structure, he gradually won over his opponents or forced them to retire from the field. He showed that if rubber is hydrogenated or converted to other derivatives, it always maintained its "colloidal" characteristics. He clearly showed the distinction between polymers and associated colloids whose characteristics depend on the suspending medium. Staudinger's efforts climaxed with his 1932 monograph *Die hochmolekularen organischen Verbindungen* (*83*).

In the late 1920s and 1930s, a younger generation of convinced adherents of Staudinger began the rational synthesis of new macromolecules. The most spectacular, carefully documented effort of this type was carried out by Wallace H. Carothers (*1, 18–22*) in the period 1929–1936. This led to polychloroprene, aliphatic polyamides (nylons), aliphatic polyesters, and other polymers (*10, 22*).

Commercial development of truly synthetic polymers dates to the beginning of the century, more than a decade before Staudinger's efforts and a generation before their acceptance. Leo Baekeland (*6, 26*) commercialized phenol-formaldehyde resins in about 1909. The first commercial developments of synthetic rubber were carried out at about the same time at Farbenfabriken Bayer in Germany under the leadership of Fritz Hofmann (*39, 88, 92,94*). The 1920s saw the commercialization of polystyrene and polyvinyl chloride by the German I. G. Farbenindustrie a combine of Farbenfabriken Bayer, BASF and Farbwerke Hoechst (*50*). The same firm developed butadiene–styrene copolymer (Buna-S) and butadiene–acrylonitrile (Buna-N) synthetic rubber (*88*). The efforts of Carothers and his coworkers led to the commercialization of polychloroprene synthetic rubber (*22, 28*) and nylon synthetic fiber (*10, 19*) by E. I. duPont de Nemours in the United States.

By 1940, the development of polymer science and the polymer industry was well under way. The activities of scientists and industry during World War II accelerated it into the modern era, with the contributions of the Rubber Reserve Program (*93*) in the Unites States being most significant. The I. G. Farbenindustrie developed nylon-6 and polyurethanes in the same period. Kurashiki Rayon (Kuraray) developed polyvinyl alcohol in Japan in the same period and applied it to fibers.

C. POLYMER STRUCTURE

1. General Remarks

Polymers are in general high-molecular-weight, long-chain molecules. In the simplest case, a polymer consists of a single repeating unit:

$$A A A A A A A A A \cdots \qquad \text{(1-C-I)}$$

Many of the most important polymers are of this type.

The most important type of linear polymers are vinyl polymers, with the repeating structure unit (*32*)

$$\underset{R}{\underbrace{CH_2 - \overset{*}{C}H}} \qquad \text{(1-C-II)}$$

They include polyethylene (R = H), polypropylene (R = —CH$_3$), polystyrene (R = phenyl, 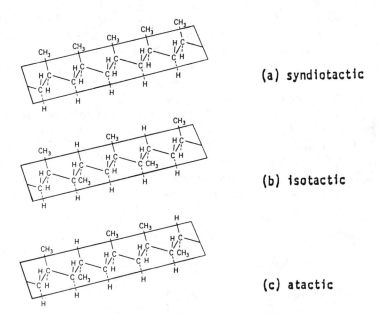), polyvinyl chloride (R = Cl). They have been shown to possess primarily head-to-tail structures, as indicated in Figure 1-C-1.

The carbon atom with the asterisk is asymmetric, thus leading to the possibility of geometric or stereoisomerism. Such polymers were first synthesized and characterized by Natta and his coworkers (*62–66*) in the mid-1950s. If all the asymmetric carbons possess the same configuration, the polymer is said to be *isotactic* (*62, 64, 66*). If the configurations of the asymmetric carbons alternate, the polymer is said to be *syndiotactic*. If there is not regular order, the term *atactic* is used (Figure 1-C-1). One may quantify the intermediate level of tactic structures, but such a development would be out of place here.

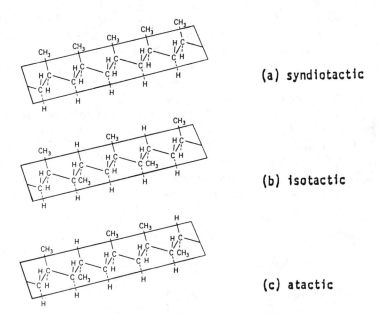

(a) syndiotactic

(b) isotactic

(c) atactic

FIGURE 1-C-1. Tacticity in polypropylene.

Related to the vinyl polymers are structures of the form

$$\begin{matrix} & R_1 \\ & | \\ -CH_2-C- \\ & | \\ & R_2 \end{matrix}$$

(1-C-III)

When R_1 and R_2 are the same, the polymer is referred to as *vinylidene*. When $R_1 = R_2 = Cl$, we have polyvinylidene chloride. There are no tacticity distinctions in vinylidene polymers. However, when R_1 is not the same as R_2, the situation is the same as in vinyl polymers. Generally polymers with the structure of formula (1-C-III) are also tied together in a head-to-tail manner.

Regularities in geometric configurations allow a polymer to crystallize. Commercial polypropylene is primarily isotactic, while commercial polystyrene is atactic. Thus polypropylene readily crystallizes, while polystyrene cannot crystallize and vitrifies to form a glass. Vinyl polymers are used commercially primarily as plastics, though some polypropylene is melt spun to form fiber.

Another important class of polymers is the *polydienes*, which are formed from the monomer

$$\begin{matrix} CH_2{=}C{-}CH{=}CH_2 \\ | \\ X \end{matrix}$$

(1-C-IVa)

and generally consist of a mixture of the structural units

$$-[CH_2C{=}CH{-}CH_2]- \qquad -[CH_2{-}CH]- \qquad -[CH_2{-}\overset{\displaystyle X}{\underset{\displaystyle CH}{C}}]-$$

(1-C-IVb)

| 1,4-addition | 3,4 addition | 1,2 addition |

The double bond may be *cis* or *trans*. The commercial dienes are used as elastomers and generally have a predominantly 1,4 addition form (*88, 93*). Isotactic and syndiotactic structures for 1,2 and 3,4 addition polydienes are possible.

Mention should also be made of condensation polymers (*18, 32*), including, notably, polyamides (*19, 21, 52*), which have the amide $-[NHC]-$ linkage with $\overset{\|}{O}$

$$-[NH{-}R{-}NH{-}\overset{O}{\overset{\|}{C}}{-}R'{-}\overset{O}{\overset{\|}{C}}]-$$

(1-C-Va)

or

$$\{NH-R-\overset{\overset{\textstyle O}{\|}}{C}\}$$

(1-C-Vb)

and polyesters, which have the $\{OC\}$ linkage:

$$\{O-X-O-\overset{\overset{\textstyle O}{\|}}{C}-X'-\overset{\overset{\textstyle O}{\|}}{C}\}$$

(1-C-VIa)

or

$$\{O-X-\overset{\overset{\textstyle O}{\|}}{C}\}$$

(1-C-VIb)

Both polyamides and polyesters are important as synthetic fibers, fibers, and plastics. When R in formula 1-C-Vb is an aliphatic polymer chain, the material is called a nylon; when R is $\{CH_2\}_5$, the material is called nylon-6. A similar distinctions exists for formula 1-C-Va where one speaks of nylon-mn polymers. If R is $\{CH_2\}_6$ and R' is $\{CH_2\}_4$, we have nylon-66. Other important polycondensates are polycarbonates

$$\{O-Y-O-\overset{\overset{\textstyle O}{\|}}{C}\}$$

(1-C-VII)

and polysulfides

$$\{Z-S\}$$

(1-C-VIII)

Certain natural products should be cited also, notably polypeptides, which are a special class of polyamides:

$$\{NH-\underset{\underset{\textstyle R}{|}}{CH}-\overset{\overset{\textstyle O}{\|}}{C}\}$$

(1-C-IX)

which include the proteins. The substituents R vary considerably. In silk, an important protein material, R is largely H and CH_3. Other proteins have more complex distributions of substituents. Synthetic polypeptides, especially poly(γ-benzyl glutamate) have been widely studied.

A second important biological macromolecule is cellulose, whose structural unit is a saccharide ring:

Cellulose esters and ethers are important commercial products.

Polymer chains may be branched as well as linear, that is, they may possess structures of the form

```
—A A A A A A A A A A A A A A A A A A A—
   A      A        A   A   A        A
   A      A        A   A   A        A
                       A   A
              A A A        A
                   A       A
                   A       A
                   A       A          (1-C-XI)
```

The branches may be either short or long and may themselves have branches.

In addition, polymer chains may be cross-linked into three-dimensional structures:

```
A A A A A A A A A A A A
     A           A
     A           A
     A           A
     A           A
     A           A
A A A A A A A A A A A A
          A        A
          A        A
          A        A
          A        A
A A A A A A A A A A A       (1-C-XII)
```

This causes the polymer system to be unable to flow under the action of stresses and to respond as a solid rather than a fluid. *Thermosetting* resins and vulcanized rubber are three-dimensional network structures.

Another class of polymers is *copolymers*, which contain more than one structural unit. The arrangement of the units in a copolymer chain may vary considerably. The units may be arranged in random

$$A A B A B A B B A B A A A A B B A \qquad \text{(1-C-XIIIa)}$$

alternating

$$A B A B A B A B A B A B A B A B A \qquad \text{(1-C-XIIIb)}$$

or block structures such as

$$AB: \quad \cdots A A A A A A \cdot B B B B B B \cdots \qquad \text{(1-C-XIIIc)}$$

or

$$ABA: \quad A''AAA---AA \cdot BBB---BBB \cdot AA \cdots AAAA'$$

(1-C-XIIId)

All types of intermediate structures between (1-C-XIIIa)–(1-C-XIIIb) are possible. The random, A B, and A B A type block copolymers are produced commercially, notably with styrene and butadiene or styrene and isoprene segments.

The copolymers indicated above are linear. It is, of course, possible to have branched copolymers with random, alternating, block, or intermediate structures. Of special interest are the graft copolymers with structure

$$
\begin{array}{ll}
\text{A A A A A A A A A A A A} \\
\quad \text{B} \qquad\qquad \text{B} \\
\quad \text{B} \qquad\qquad \text{B} \\
\quad \text{B} \qquad\qquad \text{B} \\
\quad \text{B} \qquad\qquad \text{B}
\end{array}
\qquad \text{(1-C-XIV)}
$$

where B chains grow out of an A backbone.

Generally, block and graft copolymers are thermodynamically incompatible with themselves and phase separate with the phase morphologies dependent upon the conditions of separation, deformation, and temperature history (43, 75, 90).

2. Molecular Weight and Molecular-Weight Distribution

Commercial polymers generally contain a distribution of molecular weights (17, 32, 57, 77, 86). This distribution is frequently specified in terms of average molecular weights, which are ratios of moments μ_i of the distribution. We may define the *number, weight, z, z + 1, and s average molecular weights* M_n, M_w, M_z, M_{z+1}, and M_s by

$$M_n = \sum N_i M_i / \sum N_i = \mu_1/\mu_0 \qquad \text{(1-C-1)}$$

$$M_w = \sum N_i M_i^2 / \sum N_i M_i = \mu_2/\mu_1 \qquad \text{(1-C-2)}$$

$$M_z = \sum N_i M_i^3 / \sum N_i M_i^2 = \mu_3/\mu_2 \qquad \text{(1-C-3)}$$

$$M_{z+1} = \sum N_i M_i^4 / \sum N_i M_i^3 = \mu_4/\mu_3 \qquad \text{(1-C-4)}$$

$$M_s = \sum N_i M_i^s / \sum N_i M_i^{s-1} = \mu_s/\mu_{s-1} \qquad \text{(1-C-5)}$$

where N_i represents the number of molecules of species of molecular weight M_i. The ratio M_w/M_n is frequently used as a polydispersity index.

Methods of measuring molecular weight and its distribution have been summarized in various monographs. Early work to 1952 is concisely reviewed by Flory (*32*), and we also reference more recent monographs by Tanford (*86*), Cantow (*17*), and Morawetz (*57*). The major methods of absolute determination of molecular weights are colligative properties (osmotic pressure, freezing point depression) and light scattering. The most commonly used "relative" method of characterizing molecular weight is dilute solution (intrinsic) viscosity. This is defined by

$$\lim_{c \to 0} \frac{\eta - \eta_0}{\eta_0 c} = [\eta] = K M^\alpha \qquad (1\text{-C-}6)$$

Here η is solution viscosity, η_0 is solvent viscosity, and K and α are empirical constants. For polydisperse systems, this leads to a viscosity average molecular weight

$$M_\eta = \left[\frac{\sum N_i M_i^{1+\alpha}}{\sum N_i M_i} \right]^{1/\alpha} \qquad (1\text{-C-}7)$$

Molecular-weight distributions are most usually estimated with chromatographs operating on a molecular sieve principle (gel permeation chromatography, GPC) (*17*)

The absolute levels of molecular weight of commercial polymers vary considerably, but are in any case in excess of 10,000. Below molecular weights of 5,000–50,000 depending on polymer type, the material has the mechanical properties of a wax. This roughly corresponds to about 1000 carbon atoms in an aliphatic carbon–carbon backbone. Only above this range of values do polymers achieve the unique combination of properties which makes them of commercial importance.

Commercial polymers such as polyesters and polyamides tend to be in the lower range of molecular weights, with values of the weight average molecular weight M_w of 15,000–40,000, while vinyl polymers and polydienes generally have values of M_w in excess of 100,000.

The breadths of molecular-weight distributions vary considerably among polymers. Anionic polymerization is capable of producing very narrow distribution polymers. The standards for gel permeation chromatographs developed by this technique have values of M_w/M_n less than 1.1. Commercial polybutadienes and butadiene–styrene copolymers produced by this mechanism often have M_w/M_n of 1.5–2 (*53*). The kinetics of condensation polymerization generally results in polyesters and polyamides having values of M_w/M_n close to 2 (*32, 52, 77*). Commercial polystyrenes usually have values of M_w/M_n in the range of 2.5–4 (*53*), and polypropylenes of 5–10. Polyethylenes are frequently in the range 20–30. Emulsion-polymerized butadiene–styrene copolymers appear to have dispersity indices of order 10 (*53*).

3. Compositional Heterogeneity

In copolymer systems, one must be concerned not only with heterogeneity in molecular weights, but also in composition (*49, 53*). Thus, a nominal 50/50 copolymer of A and B, because of the mechanism of its synthesis, could be a blend of molecules with compositions ranging from 25/75 to 75/25. These species would be expected to have different transition temperatures and chemical and physical properties, and perhaps even be thermodynamically incompatible.

Thin-layer chromatography (*43, 53*) has proven a most useful procedure in studying compositional heterogeneity characteristics.

D. STATES OF MATTER AND PRIMARY TRANSITION TEMPERATURE

Polymers exist in amorphous and crystalline states. The primary transition temperature determines when a polymer transforms from a solid crystal or glass to a fluid state capable of processing. In some cases this is a first-order transition, a crystalline melting point T_m. In other cases it is a *glass transition temperature* T_g. All polymers seem to exhibit a T_g, but this is masked in crystallizing systems. The T_g appears to correspond to a (low) level of molecular motion in the backbone of the chain. At lower temperatures, the polymer chains behave in a sluggish manner; equilibrium is difficult to achieve and the polymer takes on a high tensile modulus. Some polymers may be quenched below their T_g, where they will not crystallize. Values of T_m and T_g for important commercial polymers are listed in Table 1-D-1. Many complex minor transitions seem to exist, which are called α, β, and so on (*11*).

Many polymers exhibit crystalline structures at low temperatures. These structures may be represented in terms of Bravais crystal lattices (*73*). Generally, lattices at low levels of symmetry, notably triclinic, monoclinic, and ortho-rhombic, are found in polymer systems. The crystal structures of the major commercial polymers are listed in Table 1-D-1, together with references to the literature. These crystal structures were determined by x-ray diffraction and are readily detected by this method.

Generally, in the bulk state polymers are only partially crystalline, for example, 15–80%. However, single crystals have been prepared by precipitation from dilute solution (*33, 51*). These possess thin (80–200 Å) lamellar structures but are unbounded in lateral directions. The chains are largely normal to the surfaces of the lamellae. This has led to the hypothesis of chain folding in both single crystals and bulk polymers (*33*).

Not all polymers crystallize. The process of crystallization is limited, as indicated earlier, to regular structures such as tactic hydrocarbons and regular linear polycondensates. Random copolymers do not crystallize. Among import-ant homopolymers that crystallize are polyethylene, isotactic polypropylene, and nylon-66. Polymers that vitrify to form glasses include atactic polystyrene and

TABLE 1-D-1   Crystal Structures, Transition Temperatures, and Operating Temperatures for Important Commercial Polymers

Polymer	Crystal structure	T_m (melting temperature (°C)	T_g (glass transition) temperature (°C)	Normal melt processing temperature range (°C)
High-density polyethylene (HDPE)	Orthorhombic (13)	140	−100	160–240
Low-density polyethylene (LDPE)	Orthorhombic (13)	110	−100	160–240
Isotactic polypropylene (PP)	Monoclinic (64)	165	−15	180–240
Polyvinyl chloride (PVC)	Orthorhombic (65)	240	80	170–190
Polystyrene (PS)	—	—	100	180–240
cis-Polyisoprene (NR)	Orthorhombic/monoclinic (13, 58)	35	−70	90–110
Butadiene–styrene copolymer (23.5% styrene) (SBR)	—	—	−55	90–110
Polyethylene terephthalate (PET)	Triclinic (30)	265	70	275–290
Nylon-66	Triclinic	265	40	275–290
Nylone-6	Polymorphic monoclinic (46) is main form	220	40	230–260

atactic poly(methyl methacrylate). Poly(ethylene terephthalate) and commercial polycarbonate [poly(bisphenol A carbonate)] crystallize very slowly and usually form glasses. However, under extreme conditions they crystallize. Nylon-6 is peculiar in that it usually supercools to a glass, absorbs moisture, which lowers its T_g below room temperature, and then crystallizes (7).

In practice, the value of T_m is frequently suppressed by finite cooling rates. The crystallization rates at T_m are very small and achieve a maximum at temperatures approximately halfway between T_m and T_g. At T_g and below the rates are zero. Thus, isotactic polypropylene ($T_m = 165°C$, $T_g = -15°C$) frequently crystallizes in the range 90–110°C during processing operations (55, 70). Polyethylene terephthalate, as mentioned earlier, usually crystallizes so slowly that during many processing operations it cools through the T_m-T_g range and forms a glass.

Crystallization rates are greatly enhanced by the action of applied stresses. Thus, cis-1,4-polyisoprene, usually amorphous at 20°C, crystallizes on stretching (34, 89). Crystallization rates in polydienes and polyolefins are well known to be increased orders of magnitude by applied stresses (34, 60, 79, 89). Poly(ethylene terephthalate) crystallizes in melt-spinning operations when subjected to high stresses, speeds, and rates of drawdown (78).

In typical polymer processing operations, polymers are subjected to both supercooling and stress-induced crystallization. The results for each material depend upon the individual material characteristics.

E. MAJOR COMMERCIAL POLYMERS

1. Plastics (and Fibers)

Plastics form the largest class of commercially available polymers. They possess their primary transition temperature above room temperature. If they are crystalline it is T_m, and if they are amorphous it is T_g. More than half of the volume of polymeric products sold may be categorized as plastics. Several important materials used as plastics (i.e., as film, moldings, and extrusion products) are also fabricated into flexible filamentous products, .that is, as fibers. In this section we will briefly summarizes the characteristics of these materials.

The largest-volume commercial plastic is polyethylene.

$$\overline{}\!\!\leftarrow\!\! CH_2 - CH_2 \!\!\rightarrow\!\!\overline{} \qquad \text{(1-E-I)}$$

Polyethylenes are largely crystalline with their chains in a zig-zag conformation. Many polyethylenes produced by different polymerization processes are being manufactured and utilized. They may be divided into three categories. Linear or *high-density polyethylene* (HDPE) possesses largely linear chains with few branch points. The earliest polyethylene produced commercially and continuing to be important is a long-chain branched polymer known as *low-density polyethylene* (LDPE). More recently, copolymers of ethylene with small amounts of butene-1 or octene-1 are being marketed as low-density polyethylenes with linear chains or *linear low-density polyethylene* (L-LDPE). Density is a measure of crystallinity in polyethylene (as wll as in other polymers). High density means high crystallinity, as would be expected for linear chains. Branching disrupts crystallinity, whether the branches involve long or short chains. Thus, the low-density polyethylenes would be expected to have branching. From considerations of its crystal lattice, polyethylene would be expected to have a density of close to 1.00. Densities of HDPE are in the range of 0.95–0.97. Low-density polyethylenes have densities in the range 0.91–0.92. Generally, HDPE melts at about 135°C and LDPE/L-LDPE at about 110°C. They are melt processed at 160–240°C. The volume of LDPE produced exceeds that of other polyethylenes and is generally used in film as well as molded products and insulation; HDPE is mostly applied for blow-molded bottles, injection-molded products, and pipe.

Perhaps the second most important commercial plastic is polyvinyl chloride (PVC) (*61*)

$$\text{---}CH_2\text{---}CH\text{---}$$
$$|$$
$$Cl$$

(1-E-II)

a partially crystalline plastic ($\sim 15\%$) which melts at about 240°C and has a glass transition temperature of about 80°C. Becuase of its chemical instability (Section 1-F-3), PVC is melt processed in the range 170–200°C and thus below its crystalline melting point. Polyvinyl chloride is an extremely versatile material. Rigid PVC, which has few additives beyond stabilizers is used for pipe and extrusion applications. The majority of PVC is "plasticized" by adding compatible high-boiling liquids to varying extents. This is then used to produce flexible film, sheet, and upholstery. Dioctyl phthalate is a favorite plasticizer.

Another important plastic is isotactic polypropylene (PP)

$$\text{---}CH_2\text{---}CH\text{---}$$
$$|$$
$$CH_3$$

(1-E-III)

This material has a crystalline melting temperature of 165°C and a glass transition temperature of $-15°C$. The chains crystallize in a helical conformation, giving it a relatively low density. Polypropylene is melt processed in the range 180–240°C. At higher temperature it exhibits severe chemical degradation (Section 1-F-3). Polypropylene is widely applied in making film, injection molded products, and fibers (including nonwoven fabrics).

The styrenes form a most important class of commercial polymers. These are based on polystyrene (PS) (*2, 74*),

$$\text{---}CH_2\text{---}CH\text{---}$$
$$|$$

(1-E-IV)

a glass with T_g at 100°C. The larger-volume PS products are copolymers/blends of varying types, the most important of which is high-impact polystyrene (HIPS), which contains a few percent of cross-linked polybutadiene (2, 79). To these may be added the acrylonitrile–butadiene–styrene (ABS) resins, which contain a styrene acrylonitrile copolymer matrix and a largely polybutadiene disperse phase.

Nylon-6 (52)

$$\mathrm{+NH-(CH_2)_5-\overset{\displaystyle O}{\overset{\displaystyle \|}{C}}+} \tag{1-E-V}$$

plays an important role as a commercial plastic and synthetic fiber. It has a crystalline melting point of about 215°C. Nylon-66 (52)

$$\mathrm{+NH-(CH_2)_6-NH-\overset{\displaystyle O}{\overset{\displaystyle \|}{C}}-(CH_2)_4-\overset{\displaystyle O}{\overset{\displaystyle \|}{C}}+} \tag{1-E-VI}$$

possesses similar characteristics, but has a crystalline melting point of 265°C.

Poly(ethylene terephthalate) (PET)

$$\mathrm{+\overset{\displaystyle O}{\overset{\displaystyle \|}{C}}-\!\!\bigcirc\!\!-\overset{\displaystyle O}{\overset{\displaystyle \|}{C}}-O-CH_2-CH_2-O+} \tag{1-E-VII}$$

has a crystalline melting point of 260°C and a glass transition temperature of 70°C. It crystallizes slowly from the melt and thus extrudates and moldings are generally amorphous. They are subsequently crystallized on stretching. Thus PET is widely used for fibers, biaxially oriented films, and bottles.

Polycarbonate (PC) (based on bisphenol A)

$$\mathrm{+O-\!\!\bigcirc\!\!-\overset{\displaystyle CH_3}{\underset{\displaystyle CH_3}{C}}-\!\!\bigcirc\!\!-O-\overset{\displaystyle O}{\overset{\displaystyle \|}{C}}+} \tag{1-E-VIII}$$

has a glass transition temperature of 140°C and crystallizes only very slowly. It is a widely used engineering thermoplastic. Its products are invariably in the glassy state.

Poly-p-phenylene sulfide (PPS)

$$\mathrm{+\!\!\bigcirc\!\!-S+} \tag{1-E-IX}$$

has a T_g of 85°C and a crystalline melting temperature of 285°C. It crystallizes very slowly and resembles PET in its characteristics. Its use on engineering plastic is rapidly expanding.

2. Elastomers

Elastomers possess glass transition temperatures below room temperature (8, 59, 93). Rubber technology began with natural rubber and it remains today one of the important commercial elastomers. This is *cis*-1, 4-polyisoprene (8) and has the structure

$$\mathop{+\!\!\!\!\!\!\!\left[CH_2-\underset{\displaystyle CH_3}{\overset{|}{C}}=CH-CH_2\right]\!\!\!\!\!\!\!+}$$

(1-E-X)

This material has a T_g of $-70°C$ and a crystalline melting temperature of 30°C. However, it crystallizes very slowly and is usually in the amorphous form at room temperature. Natural rubber is commercially harvested from the Hevea tree in plantations concentrated in Malaysia, Liberia, and Indonesia. There is also commercially manufactured synthetic *cis*-1, 4-polyisoprene, which is widely used in a range of applications. The major application of natural rubber is in tires.

The most widely used elastomers are those based on polybutadiene

$$+\!\!\left[CH_2-CH=CH-CH_2\right]\!\!+ \quad +\!\!\left[CH_2-\underset{\displaystyle \underset{\displaystyle CH_2}{\overset{\|}{CH}}}{\overset{|}{CH}}\right]\!\!+$$

(1-E-XI)

either by itself or as a copolymer with (primarily) styrene or other monomers. Butadiene–styrene copolymers (SBR) are the most widely used synthetic commercial elastomers (93). The glass transition temperature depends upon styrene content. For 23.5% styrene, which is the most important copolymer, T_g is about $-55°C$. The material does not crystallize. Note that SBR is primarily used in tire applications. Acrylonitrile copolymers of butadiene are also of great importance because of their oil resistance and are used in mechanical goods.

Another important commercial elastomer is polychloroprene (22)

$$+\!\!\left[CH_2-\underset{\displaystyle CH_2}{\overset{\displaystyle Cl}{\overset{|}{C}}}=CH-CH_2\right]\!\!+$$

(1-E-XII)

which is used as an oil-resistant elastomer.

A final class of commercial elastomers worthy of discussion is ethylene–propylene copolymers and terpolymers.

$$+\!\!\left[CH_2-\underset{\displaystyle CH_3}{\overset{|}{CH}}\right]\!\!+ \quad +\!\!\left[CH_2-CH_2\right]\!\!+$$

(1-E-XIII)

These often contain a third monomer which possesses an additional double bond for vulcanization.

F. CHANGES IN POLYMER STRUCTURE DURING MELT FLOW

1. General Remarks

Chemical changes occur during the flow of polymer melts. The character of these changes varies. Some are, it would appear, due to purely thermal effects. Others are caused by chemical reactions with impurities brought on by elevated temperatures. Another class of chemical changes are associated with degradation or breakage of polymer chains through applied stresses. Often the detailed reactions can be complex and involve all three of these aspects, with the relative proportion varying with temperature. It is worthwhile to consider the situation for particular polymers.

2. Condensation Polymers

Polymers of this type especially polyesters, are highly subject to hydrolysis by the presence of small amounts of moisture, that is,

$$
\begin{array}{c}
\quad\quad\quad\quad O \quad\quad O \\
\quad\quad\quad\quad \| \quad\quad\ \| \\
-\!\!\!\left[O\!-\!X\!-\!O\!-\!C\!-\!X'\!-\!C\right]\!\!\!- + H_2O \longrightarrow \\
\quad\quad\quad\quad\quad O \quad\quad O \\
\quad\quad\quad\quad\quad \| \quad\quad\ \| \\
-\!O\!-\!X\!-\!OH + HO\!-\!C\!-\!X'\!-\!C\!-
\end{array}
\tag{1-F-I}
$$

The forward rate of this reaction increases rapidly with higher temperature and is very significant at the processing temperatures used for example in PET (275–290°C). Extreme care must always be taken in drying condensation polymers, especially polyesters. By reaction (1-F-1) there is a tendency for polymers to form an equilibrium molecular weight determined by the moisture content, that is, the molar concentration of H_2O through

$$
\frac{[H_2O]\left[\overset{\overset{\displaystyle O}{\displaystyle \|}}{-OC-}\right]}{[COOH][OH]} = K(T)
\tag{1-F-II}
$$

where the brackets represent concentrations. Polyamides exhibit tendencies to gel at high temperatures, as well as to hydrolyze, and one must be careful to avoid either problem (52). This is discussed by Pezzin and Gechele (69) for nylon-6. The gelling effects in nylon-66 are more serious, perhaps because of the necessity of using higher operating temperatures.

3. Vinyl Polymers

Commercial polyolefins, notably polypropylene, also exhibit chemical instability during melt flow. The presence of oxygen is necessary and greatly accelerates the rate of breakdown of α-polyolefins such as polypropylene and polybutene-1. The particular instability is associated with the removal of the tertiary hydrogen, for example,

$$-CH_2-CH-CH_2-CH \longrightarrow -CH_2-\dot{C}-CH_2-CH \longrightarrow$$
$$\underset{CH_3}{|} \qquad \underset{CH_3}{|} \qquad \underset{CH_3}{|} \qquad \underset{CH_3}{|}$$

$$-CH_2-C=CH + \cdot CH-$$
$$\underset{CH_3}{|} \qquad \underset{CH_3}{|} \qquad \text{(1-F-III)}$$

This reaction is promoted by the presence of peroxides. This behavior has been considered in some detail by Kowalski and his coworkers (*54, 80*) in the patent literature (see also Yamane and White, ref. *98*). The random degradation of polypropylene not only lowers molecular weight, but tends to produce products with M_w/M_n of 2, substantially narrowing the distribution, which originally possesses M_w/M_n of 10. Temperatures of about 260°C and higher are required. It is believed that processes of this type are used commercially to produce narrower molecular-weight distribution grades of polypropylenes.

Poly(vinyl chloride) can degrade and release HCl at elevated temperatures (*61*)

$$-CH_2-CH \longrightarrow -CH=CH- + HCl$$
$$\underset{Cl}{|} \qquad\qquad\qquad \text{(1-F-IV)}$$

if not properly stabilized. The HCl in small quantities will corrode machinery and in large quantities can injure workers' health.

4. Mechanochemical Degradation

Another type of flow-induced chemical change in polymers is direct breakup by applied stresses. This effect was first discovered and commercially exploited by Thomas Hancock (*42*) from about 1820 with natural rubber. It was not, however, until the present century that the efforts of Busse (*15, 16*) and Watson and his coworkers (*5, 71, 91*) explained the mechanism of the behavior and the key role played by oxygen. This is briefly summarized as:

$$R - R' \xrightarrow{\text{stress}} R\cdot + R'\cdot$$

$$R\cdot + O_2 \longrightarrow RO_2^\cdot \text{ (relatively stable)} \qquad \text{(1-F-V)}$$

$$R\cdot + A \longrightarrow RA\cdot \text{ or } RO_2\cdot + A \longrightarrow RO_2A$$

Oxygen acts to stabilize the free radicals and prevent them from recombining. The reaction may be hastened by the addition of efficient free-radical acceptors A (5).

While the process is most effective with cis-1,4-polyisoprene, it is known to occur with a wide range of polymers (25, 36). In masticated blends, one may form graft copolymers from interaction between radical segments and polymer chains of other species (3). Polymers dissolved in monomers can, upon mastication, break down to form free radicals which may polymerize the monomer to form a block copolymer (4).

G. MULTIPHASE SYSTEMS AND COMPOUNDS

Many polymer systems used industrially consist of more than a single polymeric component and phase. We mean here more than the obvious minor amounts of stabilizers, antioxidants, and so on, usually added. Many industrial polymer systems consist of physical blends of different polymers or physical mixtures of two different components with a thick interference of a graft copolymer. This is the case, for example, of rubber-modified (high-impact) polystyrene and ABS resins (2, 74). In still other cases, polymer systems may contain large quantities of solid particulates which are added as reinforcing agents to improve product properties (e.g., carbon black, silica or glass, aramid or cellulose fibers) or as fillers (clay) (59, 87). Many polymeric products have complex recipes involving several polymer components, different solid particulate additives, plus organic oils. In rubber compounds, one usually has cross-linking agents (sulfur plus accelerators) present as well (59).

REFFERENCES

1. R. Adams, *Rubber Chem. Tech.*, **37**(1) xxvii (1964).
2. J. L. Amos, *Polym. Eng. Sci.*, **14**, 1 (1974).
3. D. J. Angier and W. F. Watson, *J. Polym. Sci.*, **18**, 129 (1955); *Trans. I.R.I.*, **33**, 22 (1957).
4. D. J. Angier and W. F. Watson, *J. Polym. Sci.*, **20**, 235 (1956); **25**, 1 (1957); D. J. Angier, W. F. Watson and E. D. Farlie, *Trans. I.R.I.*, **34**, 8 (1958); D. J. Angier, W. F. Watson, and R. J. Ceresa, *J. Polym. Sci.*, **34**, 699 (1959); *Polymer*, **1**, 72, 397, 477, 488 (1960).
5. G. Ayrey, C. G. Moore, and W. F. Watson, *J. Polym. Sci.*, **19**, 1 (1956).
6. L. H. Baekeland, *Ind. Eng. Chem.*, **1**, 150 (1909); **2**, 932 (1910).
7. V. G. Bankar, J. E. Spruiell, and J. L. White, *J. Appl. Polym. Sci.*, **21**, 2348 (1977).
8. L. Bateman, Ed., *The Chemistry and Physics of Rubberlike Substances*, McLaren, London, 1963.
9. C. Beadle, *J. Franklin Inst.*, **B8**, 101 (1894); **143**, 1 (1897).
10. E. K. Bolton, *Ind. Eng. Chem.*, **34**, 53 (1942).

11. R. F. Boyer, *Rubber Chem. Tech.*, **34**, 53 (1942).

12. H. T. Brown and G. H. Morris, *J. Chem. Soc.*, **55**, 465 (1889).

13. C. W. Bunn, *Trans. Faraday Soc.*, **35**, 482 (1939).

14. C. W. Bunn, *Proc. Roy Soc.*, **A180**, 40 (1942).

15. C. W. Bunn and E. V. Garner, *Proc. Roy Soc.*, **A189**, 39 (1947).

16. W. F. Busse, *Ind. Eng. Chem.*, **24**, 140 (1932); W. F. Busse and E. N. Cunningham, *Proc. Rubber Technol. Conf.*, 288 (1938).

17. M. J. R. Cantow, Ed., *Polymer Fractionation*, Academic, New York, 1964.

18. W. H. Carothers, *J. Am. Chem. Soc.*, **51**, 2548 (1929).

19. W. H. Carothers, U.S. Patent 2,130,523 (1938); U.S. Patent 2,130,947 (1938); U.S. Patent 2,149,273 (1938).

20. W. H. Carothers and G. S. Arvin, *J. Am. Chem. Soc.*, **51**, 2560 (1929); W. H. Carothers and J. W. Hill, *J. Am. Chem. Soc.*, **54**, 1559 (1932).

21. W. H. Carothers and G. J. Berchet, *J. Am. Chem. Soc.*, **52**, 5289 (1930); W. H. Carothers and J. W. Hill, *J. Am. Chem. Soc.*, **54**, 2556 (1932).

22. W. H. Carothers, I. Williams, A. M. Collins, and J. E. Kirby, *J. Am. Chem. Soc.*, **53**, 4203 (1931).

23. M. Carter, *Essential Fiber Chemistry*, Dekker, New York, 1971.

24. W. A. Caspari, *J. Chem. Soc.*, **105**, 2139 (1914).

25. R. J. Ceresa and W. F. Watson, *J. Appl. Polym. Sci.*, **1**, 101 (1959).

26. C. F. Chandler, *Ind. Eng. Chem.*, **8**, 178 (1916).

27. H. de Chardonnet, U.S. Patent 394,559 (1888).

28. A. M. Collins, *Rubber Chem. Tech.*, **46**(2), G48 (1973).

29. C. F. Cross, E. J. Bevan, and C. Beadle, *J. Chem. Soc.*, 837 (1892); *J. Soc. Chem. Ind.*, 498 (1893); English Patent 8700 (1892).

30. R. P. Daubeny, C. W. Bunn, and C. J. Brown, *Proc. Roy. Soc.* **A226**, 531 (1955).

31. M. Faraday, *Q. J. Sci. Arts*, **21**, 19 (1826).

32. P. J. Flory, *Principles of Polymer Chemistry*, Cornell University Press, Ithaca, NY, 1953.

33. P. H. Geil, *Polymer Single Crystals*, Wiley, New York, 1963.

34. A. N. Gent, *J. Polym. Sci.*, **A3**, 3787 (1965); **A4**, 447 (1966).

35. J. H. Gladstone and W. Hibbert, *Phil. Mag.*, **5**(28), 38 (1889).

36. P. Goodman and A. B. Bestul, *J. Polym. Sci.*, **18**, 235 (1955).

37. C. Goodyear, U.S. Patent 3,633 (1844).

38. C. Goodyear, *Gum Elastic*, privately printed, New Haven, CN, 1855.

39. K. Gottlob, *India Rubber J.*, Aug. 16, p. 17; Aug. 23, p. 16; Aug. 30, p. 21; Sept. 6, p. 15 (1919).

40. C. Hancock, English Patent 11,032 (1846); English Patent 11,575 (1847).

41. T. Hancock, English Patent 9952 (1843).

42. T. Hancock, *Personal Narrative of the Origin and Progress of the Caoutchouc or India Rubber Manufacture in England*, Longman, Brown, London, 1857.

43. T. Hashimoto, M. Shibayama, M. Fujimura, and H. Kawai, *Mem. Fac. Eng. Kyoto Univ.*, **43**, 184 (1981).

44. N. Hayward, U.S. Patent 1090 (1839).

45. W. Haynes, *Cellulose: The Chemical That Grows*, Doubleday, New York, 1953.

46. D. R. Holmes, C. W. Bunn, and D. J. Smith, *J. Polym. Sci.*, **17**, 159 (1955).

47. J. W. Hyatt, *Ind. Eng. Chem.*, **6**, 158 (1914).

48. J. W. Hyatt and I. S. Hyatt, U.S. Patent 91,341 (1869).

49. H. Inagaki, H. Matsuda, and F. Kamiyama, *Macromolecules*, **1**, 520 (1968); *Macromol. Chem.*, **125**, 286 (1969).

50. M. Kaufman, *The First Century of Plastics: Celluloid and Its Sequel*, Plastics Institute, London, 1963.

51. A. Keller, *Phil. Mag.*, **2**(8), 1171 (1957).

52. M. I. Kohan, Ed., *Nylon Plastics*, Wiley, New York, 1973.

53. T. Kotaka and J. L. White, *Macromolecules*, **7**, 105 (1974).

54. R. C. Kowalski, U.S. Patent 3,563,972 (1971).

55. G. Kraus and K. W. Rollman, *J. Appl. Polym. Sci.*, **8**, 2585 (1964).

56. W. Minoshima, J. L. White, and J. E. Spruiell, *Polym. Eng. Sci.*, **20**, 1166 (1980).

57. H. Morawetz, *Macromolecules in Solution*, 2nd ed., Wiley-Interscience, New York, 1975.

58. H. A. Morris, *J. Am. Chem. Soc.*, **60**, 2371 (1938).

59. M. Morton, Ed., *Rubber Technology*, 2nd ed., van Nostrand Reinhold, New York, 1973.

60. H. P. Nadella, H. M. Henson, J. E. Spruiell, and J. L. White, *J. Appl. Polym. Sci.*, **21**, 3003 (1977).

61. L. I. Nass, Ed., *Encyclopedia of PVC*, Dekker, New York, 1976.

62. G. Natta, *Science*, **147**, 261 (1965).

63. G. Natta, I. W. Bassi, and P. Corradini, *Atti Accad. Naxl Lince Reed Classe Sci. Fis Mat. Nat.*, **31**, 17 (1961).

64. G. Natta and P. Corradini, *J. Polym. Sci.*, **20**, 251 (1956).

65. G. Natta and P. Corradini, *Nuovo Cimento (Suppl)*, **15**, 40 (1960).

66. G. Natta and F. Danusso, *J. Polym. Sci.*, **34**, 3 (1956).

67. A. Parkes, English Patent 11,147 (1846).

68. A. Parkes, English Patent 2675 (1864).

69. G. Pezzin and G. B. Gechele, *J. Appl. Polym. Sci.*, **8**, 2195 (1964).

70. S. S. Pickles, *J. Chem. Soc.*, **97**, 1085 (1910).

71. M. Pike and W. F. Watson, *J. Polym. Sci.*, **9**, 229 (1952).

72. R. S. Porter and J. F. Johnson, *J. Appl. Phys.*, **35**, 3149 (1964).

73. J. M. Robertson, *Organic Crystals and Molecules*, Cornell University Press, Ithaca, New York, 1953.

74. S. L. Rosen, *Ann. N.Y. Acad. Sci.*, **35**, 480 (1973).

75. C. Sadron and B. Gallot, *Makromol. Chem.*, **164**, 301 (1973).

76. C. Schonbein, *Phil. Mag.*, **31**, 7 (1847); U.S. Patent, 4,874 (1846).

77. J. M. Schultz, *Polymer Materials Science*, Prentice Hall, Englewood Cliffs NJ, 1974.

78. J. Shimizu, K. Toriumi, and K. Tamai, *Seni Gakkaishi*, **33**, 208 (1977).

79. J. E. Spruiell and J. L. White, *Polym. Eng. Sci.*, **15**, 660 (1975).

80. J. C. Staton, J. P. Keller, R. C. Kowalski, and J. W. Harrison, U.S. Patent 3,551,943 (1971).

81. H. Staudinger, *Chem. Ber.*, **53**, 1073 (1920).

82. H. Staudinger, *Chem. Ber.*, **59**, 2019 (1926).

83. H. Staudinger, *Die hochmolekularen organischen Verbindugen*, Springer, Berlin, 1932.

84. H. Staudinger, *From Organic Chemistry to Macromolecules*, (translated from the German), Wiley, New York, 1970.

85. H. Staudinger and M. Luthy, *Helv. Chim. Acta.*, **8**, 41 (1925); **8**, 67 (1925).

86. C. Tanford, *Physical Chemistry of Macromolecules*, Wiley, New York, 1961.

87. W. V. Titow and B. J. Lanham, *Reinforced Thermoplastics*, Applied Science Publishers, London, 1975.

88. E. Tornquist in *Polymer Chemistry of Synthetic Elastomers*, Part 1, J. Kennedy and E. Tornquist, Eds., Interscience, New York, 1968.

89. L. R. G. Treloar, *Trans. Faraday Soc.*, **43**, 284 (1947).

90. E. Vanzo, *J. Polym. Sci. A-1*, **4**, 1727 (1966).

91. W. F. Watson, *Trans. I.R.I.*, **29**, 32 (1953).

92. R. Weil, *Ind. Eng. Chem.*, **18**, 1174 (1926).

93. G. S. Whitby, Ed., *Synthetic Rubber*, Wiley, New York, 1954.

94. G. S. Whitby and M. Katz, *Ind. Eng. Chem.*, **25**, 1240 (1933).

95. J. L. White, Rubber Ind., August, 148 (1974).

96. J. L. White, *Polym. Eng. Rev.*, **4**, 345 (1984).

97. E. C. Worden, *Nitrocellulose Industry*, van Nostrand Reinhold, New York, 1911.

98. H. Yamane and J. L. White, *Polym. Eng. Rev.*, **2**, 167 (1982).

II

PRINCIPLES OF CONTINUUM MECHANICS

A. INTRODUCTION

In order to specify and measure the characteristics of polymer fluids when they deform or flow, it is necessary to have a thorough understanding of the mechanics of deforming media. The interpretation of flow fields, measured forces, and torques in rheometers requires a knowledge of kinematics and mechanics to specify rheological properties. The derivation of predictive theories for the stresses developed in polymer systems in response to flow necessitates a formalism not only consistent with, but embedded in kinematics and mechanics. It is the purpose of this chapter to develop this background. We presume the student is familiar with the principles of vector analysis, including both differential and integral vector calculus. We begin with a historical critique of the development of continuum mechanics and then proceed in individual sections to discuss conservation of mass and continuity, kinematics, macromechanics, the stress tensor, differential force balances, differential torque balances, mechanics of membranes, the stress and deformation rate invariants, and energy balances.

B. HISTORICAL PERSPECTIVE

The principles of mechanics date back to the time of Newton and his *Principia* (*27*). Newton developed the dynamics of isolated and interacting point particles and made the first studies of the resistance observed by moving objects in material media. It is only with the Bernoullis, and notably Leonhard Euler in the early and mid-eighteenth century, that mechanics was applied to macroscopic bodies and continuous media (*32*). Euler expanded Newton's particle dynamics to a mechanics of macroscopic bodies and developed a generalized set of equations of motions which relate applied forces to the rate of change of linear momentum and applied torques to the rate of change of angular momentum (*14*) (Section 2-E). He developed a mechanics of macroscopic rigid bodies to represent the relationship between torques and spin. Euler also considered the behavior of fluids (*11, 12*), but generally neglected the effect of viscosity. In his analysis, he developed the differential equation representing continuity in a deforming fluid (*11*) (Section 2-C) and was the first to consider rates of deformation in a continuous medium (*13*) (Section 2-D).

In 1827, Augustin-Louis Cauchy developed the modern tools and formulation of the mechanics of deformation of continua. In successive papers, Cauchy, to some extent following the lead of Euler (*11–14*), devised the stress tensor (*2*), (Section 2-F), the strain tensor (*3*), and the differential force balance or equation of motion relating the stress components (*4*) (Section 2-G). In 1828–1829, Cauchy (*5*) developed the foundations of the theory of elasticity in a series of papers. By 1830, the mechanics of continuous media had achieved a stage of development wherein theories of the mechanics of particular classes of materials could be developed. Later authors, notably Stokes (*29*), applied these concepts to linear viscous fluids.

The study of energy conservation proceeded more slowly. Contemporary with Cauchy, Fourier (*17*) developed a differential energy equation for heat flow and used it to compute temperature distributions in solids. With the studies of Joule (*19*), the mechanical equivalence of heat was established. In succeeding years, differential balances, including work as well on convective flux and heat conduction terms, were developed by various investigators.

C. CONSERVATION OF MASS

Consider a deforming material medium. We define an imaginary closed surface which encapsulates a portion of this material. If we presume that mass is conserved, the rate of change of mass within the surface S is equal to the negative efflux from the surface. This may be expressed

$$\frac{\partial}{\partial t}\int dm = \frac{\partial}{\partial t}\int \rho\, dV = -\oint \rho\mathbf{v}\cdot\mathbf{n}\, da \qquad (2\text{-C-}1)$$

Here $\mathbf{n}$ is a unit normal vector on the surface (Figure 2-C-1); dm a differential element of mass and dV of volume; da an element of surface, ρ is density, and $\mathbf{v}$ is the velocity vector. The vector quantities $\mathbf{v}$ and $\mathbf{n}$ have directions as well as magnitudes. The quantity $\mathbf{v}\cdot\mathbf{n}$ is a scalar product representing velocity normal to the surface.

If matter may be considered continuous so that quantities such as density and velocity are differentiable, we may obtain great simplifications. This allows the divergence theorem* to be applied to Eq. (2-C-1). This results in

$$\int \frac{\partial \rho}{\partial t}\, dV = -\int \nabla\cdot\rho\mathbf{v}\, dV \qquad (2\text{-C-}2a)$$

where $\partial/\partial t$ has been taken within the integral. $\nabla\cdot\rho\mathbf{v}$ represents the *divergence* of the vector $\rho\mathbf{v}$ where ∇ is the operator

$$\nabla = \mathbf{e}_1 \frac{\partial}{\partial x_1} + \mathbf{e}_2 \frac{\partial}{\partial x_2} + \mathbf{e}_3 \frac{\partial}{\partial x_3} \qquad (2\text{-C-}2b)$$

and the $\mathbf{e}_j$ are unit vectors in direction $j(=1,2,3)$. We may equivalently write

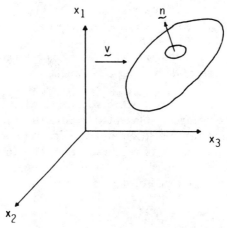

FIGURE 2-C-1. Flow of mass through a macroscopic volume. Wavy underscore indicates vectors.

*The divergence theorem, which states $\oint \mathbf{A}\cdot\mathbf{n}\, da = \int \nabla\cdot\mathbf{A}\, dV$, where $\mathbf{A}$ is a continuous, differentiable vector or tensor, is the basic theorem of the integral calculus of vectors and tensors.

Eq. (2-C-2) as

$$\int \frac{\partial \rho}{\partial t} \, dV + \int \nabla \cdot \rho \mathbf{v} \, dV = 0 \qquad (2\text{-}C\text{-}3)$$

If this integral is zero for arbitrary values of the terms in the kernel, the kernel itself may be set equal to zero, that is,

$$\frac{\partial \rho}{\partial t} + \nabla \cdot \rho \mathbf{v} = 0 \qquad (2\text{-}C\text{-}4)$$

Equation (2-C-4) is known as the *continuity equation* and is from Euler (12). In cartesian coordinates it has the form

$$\frac{\partial \rho}{\partial t} + \frac{\partial}{\partial x_1} \rho v_1 + \frac{\partial}{\partial x_2} \rho v_2 + \frac{\partial}{\partial x_3} \rho v_3 = 0 \qquad (2\text{-}C\text{-}5)$$

In Table 2-C-1, we express the continuity equation in cylindrical and spherical coordinates.

Equation (2-C-4) may be rewritten in the form

$$\frac{1}{\rho} \frac{D\rho}{Dt} = - \nabla \cdot \mathbf{v} \qquad (2\text{-}C\text{-}6)$$

where D/Dt is the substantive derivative, which is defined as

$$\frac{D}{Dt} = \frac{\partial}{\partial t} + (\mathbf{v} \cdot \nabla) = \frac{\partial}{\partial t} + v_1 \frac{\partial}{\partial x_1} + v_2 \frac{\partial}{\partial x_2} + v_3 \frac{\partial}{\partial x_3} \qquad (2\text{-}C\text{-}7)$$

This quantity represents a rate of change, not at a fixed spatial point as with $\partial/\partial t$, but with a material point or translating with a particle.

TABLE 2-C-1   Continuity Equation in Different Coordinate Systems

Cylindrical Coordinates (r, θ, z)

$$\frac{\partial \rho}{\partial t} + \frac{1}{r} \frac{\partial}{\partial r}(r v_r) + \frac{1}{r} \frac{\partial}{\partial \theta}(v_\theta) + \frac{\partial}{\partial z}(v_z) = 0$$

Spherical Coordinates (r, θ, ϕ)

$$\frac{\partial \rho}{\partial t} + \frac{1}{r} \frac{\partial}{\partial r}(r^2 v_r) + \frac{1}{r \sin \theta} \frac{\partial}{\partial \theta}(v_\theta \sin \theta) + \frac{1}{r \sin \theta} \frac{\partial}{\partial \phi}(v_\phi) = 0$$

For an incompressible fluid, the density is constant and its derivatives are zero, including

$$\frac{D\rho}{Dt} = 0 \qquad\qquad (2\text{-C-}8)$$

From Eqs. (2-C-4) and (2-C-6), this leads to

$$\nabla \cdot \mathbf{v} = \frac{\partial v_1}{\partial x_1} + \frac{\partial v_2}{\partial x_2} + \frac{\partial v_3}{\partial x_3} = 0 \qquad\qquad (2\text{-C-}9)$$

Equation (2-C-9) represents the continuity equation for an incompressible medium.

D. RATE OF DEFORMATION

Consider a differential material line segment $d\mathbf{r}$ which deforms with the medium. The rate of change of a differential line element is an important representation of rates of change in a deforming medium. We may write

$$\frac{d}{dt}(dx_i) = dv_i = \sum_j \frac{\partial v_i}{\partial x_j} dx_j \qquad\qquad (2\text{-D-}1)$$

$$\frac{d}{dt}(ds)^2 = 2 \sum_i \sum_j d_{ij} dx_i dx_j \qquad\qquad (2\text{-D-}2)$$

where

$$\mathbf{d} = \tfrac{1}{2}[\nabla\mathbf{v} + (\nabla\mathbf{v})^{\mathrm{T}}], \qquad d_{ij} = \frac{1}{2}\left[\frac{\partial v_i}{\partial x_j} + \frac{\partial v_j}{\partial x_i}\right] \qquad (2\text{-D-}3)$$

The quantity $\mathbf{d}$ is known as the rate of deformation tensor and is associated with Euler (13) and Stokes (29). It is a second-order quantity, associated with two directions, rather than a first-order quantity associated with a single direction, as is the velocity vector. Hence it has two subscripts with independent physical meaning. Equation (2-D-3) is presented in cylindrical and spherical coordinates in Table 2-D-1.

We may express the velocity gradient tensor $\nabla\mathbf{v}$ as the sum of a symmetric tensor $\mathbf{d}$ (where $d_{ij} = d_{ji}$) and an antisymmetric tensor $\boldsymbol{\omega}$ that is, as

$$\nabla\mathbf{v} = \mathbf{d} + \boldsymbol{\omega}, \qquad \frac{\partial v_i}{\partial x_j} = d_{ij} + \omega_{ij} \qquad\qquad (2\text{-D-}4)$$

Hence $\boldsymbol{\omega}$ is defined by

$$\boldsymbol{\omega} = \tfrac{1}{2}[(\nabla\mathbf{v}) - (\nabla\mathbf{v})^{\mathrm{T}}], \qquad \omega_{ij} = \frac{1}{2}\left[\frac{\partial v_i}{\partial x_i} - \frac{\partial v_j}{\partial x_i}\right] \qquad (2\text{-D-}5)$$

TABLE 2-D-1 Rate of Deformation Tensor in Different Coordinate Systems

Cylindrical Coordinates

$$d_{rr} = \frac{\partial v_r}{\partial r}; \qquad d_{\theta\theta} = \frac{1}{r}\left[\frac{\partial v_\theta}{\partial \theta} + v_r\right]; \qquad d_{zz} = \frac{\partial v_z}{\partial z}$$

$$d_{r\theta} = \frac{1}{2}\left[\frac{\partial v_r}{\partial \theta} + r\frac{\partial}{\partial r}\left(\frac{v_\theta}{r}\right)\right]; \qquad d_{rz} = \frac{1}{2}\left[\frac{\partial v_r}{\partial z} + \frac{\partial v_z}{\partial r}\right]$$

$$d_{\theta z} = \frac{1}{2}\left[\frac{\partial v_\theta}{\partial z} + \frac{1}{r}\frac{\partial v_z}{\partial \theta}\right]$$

Spherical Coordinates

$$d_{rr} = \frac{\partial v_r}{\partial r}; \qquad d_{\theta\theta} = \frac{1}{r}\left[\frac{\partial v_\theta}{\partial \theta} + v_r\right]$$

$$d_{\phi\phi} = \frac{1}{r\sin\theta}\frac{\partial v_\phi}{\partial \phi} + \frac{v_r}{r} + \frac{v_\phi\cot\theta}{r}$$

$$d_{r\theta} = \frac{1}{2}\left[\frac{1}{r}\frac{\partial v_r}{\partial \theta} + r\frac{\partial}{\partial r}\left(\frac{v_\theta}{r}\right)\right]$$

$$d_{r\phi} = \frac{1}{2}\left[\frac{1}{r\sin\theta}\frac{\partial v_r}{\partial \phi} + r\frac{\partial}{\partial r}\left(\frac{v_\phi}{r}\right)\right]$$

$$d_{\theta\phi} = \frac{1}{2}\left[\frac{\sin\theta}{r}\frac{\partial}{\partial \theta}\left(\frac{v_\phi}{\sin\theta}\right) + \frac{1}{r\sin\theta}\frac{\partial v_\theta}{\partial \phi}\right]$$

known as the *vorticity tensor*. As noted by Stokes (29), $\boldsymbol{\omega}$ represents a rate of rotation of a line element analogous to **d**, representing the rate of deformation of the element.

E. EULER'S LAWS OF MOTION

Most treatises on mechanics (18, 31, 35) are based upon Newton's dynamics of point particles and his relationship between the rate of change of the product of mass and linear velocity and the resulting force, that is,

$$\mathbf{F} = \frac{d}{dt}(m\mathbf{v}) \qquad (2\text{-E-}1)$$

In addition to the problem of varying rectilinear velocity and linear momentum

$m\mathbf{v}$ causing forces, one must deal with the problem of changing spin or angular velocity $\omega(\mathbf{r} \times \mathbf{v})$ (where $\mathbf{r}$ is the vector from an axis to the point particle) inducing turning moments or torques $\mathbf{M}$ equivalent to $\mathbf{r} \times \mathbf{F}$. In Newtonian particle dynamics, this is treated by forming a cross product of $\mathbf{r}$ with Eq. (2-E-1) to yield

$$\mathbf{M} = \mathbf{r} \times \mathbf{F} = \mathbf{r} \times \frac{d}{dt}(m\mathbf{v}) = \frac{d}{dt}(\mathbf{r} \times m\mathbf{v}) \qquad (2\text{-E-}2)$$

where we have noted that $\mathbf{v} \times \mathbf{v}$ is identically zero. Equation (2-E-2) relates torque to rate of change of angular momentum $\mathbf{r} \times m\mathbf{v}$.

We now turn to macroscopic bodies. One method of procedure usually found in the mechanics literature (18, 31, 35) is to consider the body to be a system of particles to which Eq. (2-E-1) applies, and to add together these expressions for each of the particles of the total mass. This leads to

$$\mathbf{F} = \sum \mathbf{F}_i = \frac{d}{dt}\sum m_i \mathbf{v}_i \qquad (2\text{-E-}3a)$$

If we also presume that the interparticle forces are directed along the lines connecting these particles, it follows that

$$\mathbf{M} = \sum \mathbf{r}_i \times \mathbf{F}_i = \frac{d}{dt}\sum \mathbf{r}_i \times m_i \mathbf{v}_i \qquad (2\text{-E-}3b)$$

The sums of Eqs. (3-E-3a) and (3-E-3b) may readily be re-expressed as integrals with the m_i becoming differential mass element dm.

There are some problems with the basic hypothesis given above. It must first be presumed that the ultimate particles of matter always obey Newtonian mechanics. This contradicts much of twentieth century physics, wherein it is shown that small particles are governed by quantum mechanical considerations (21). Furthermore, the presumption about all forces lying along lines of centers might rule out certain types of electromagnetic interactions, such as those suggested by Ampere's law (26, 33).

Euler (14) proposed an alternative approach in which he hypothesized two separate laws of motion

$$\mathbf{F} = \frac{d}{dt}\int \mathbf{v}\, dm$$
$$\qquad (2\text{-E-}4)$$

$$\mathbf{M} = \frac{d}{dt}\int \mathbf{r} \times \mathbf{v}\, dm \qquad (2\text{-E-}5)$$

He then represented Newtonian particle dynamics as an asymptote of

Eqs. (2-E-4) and (2-E-5). This approach has come to be accepted in continuum mechanics.

F. STRESS TENSOR

The force **F** acting on a body may be expressed as a sum of contact forces acting on the surface of the body and gravitational forces which act directly on the elements of mass. It follows that

$$\mathbf{F} = \oint \mathbf{t}\, da + \int \mathbf{f}\, dm \qquad (2\text{-F-}1)$$

where **t** is the stress vector and **f** is the body force per unit mass. The latter is usually associated with gravitation. The stress tensor $\boldsymbol{\sigma}$ of Cauchy relates the unit normal **n** on the surface S surrounding volume V to the stress vector **t**, acting on the same element (Fig. 2-F-1). This is through the relation (2)

$$\mathbf{t} = \sum_i t_i \mathbf{e}_i = \boldsymbol{\sigma} \cdot \mathbf{n} = \sum_i \left(\sum_j \sigma_{ij} n_j \right) \mathbf{e}_i \qquad (2\text{-F-}2)$$

The stress tensor $\boldsymbol{\sigma}$ has nine components σ_{ij} and may be expressed [after Cauchy $(2,4)$ as an array]

$$\boldsymbol{\sigma} = \begin{bmatrix} \sigma_{11} & \sigma_{12} & \sigma_{13} \\ \sigma_{21} & \sigma_{22} & \sigma_{23} \\ \sigma_{31} & \sigma_{32} & \sigma_{33} \end{bmatrix} \qquad (2\text{-F-}3)$$

Here the first subscript indicates the direction of the components t_i and the second subscript the direction of the normal to the surface of contact. The

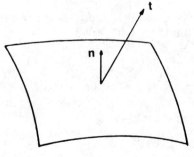

FIGURE 2-F-1. Stress vector, unit normal, and stress tensor.

stresses σ_{ii}, that is, σ_{11}, σ_{22}, and σ_{33} when t_i and n_i are parallel, are called normal stresses. The stresses σ_{ij} ($i \neq j$), that is, $\sigma_{12}, \sigma_{21}, \sigma_{13}, \sigma_{31}, \sigma_{23}$, and σ_{32}, where t_i is parallel to the surface and perpendicular to unit normal n_j, are called shear stresses.

The convention used here follows Cauchy and other nineteenth century researchers in taking stress to be positive. There is an alternative convention which takes stress to be negative. Bird, Stewart, and Lightfoot (1) are the main protagonists of this view. It derives from James Clerk Maxwell's theory (25) that the viscous stresses in flowing gases are due to negative momentum fluxes. However, Maxwell himself never used this convention (26), and it does not seem to be appropriate for polymer melts, where the molecular mechanisms of stress are different. The negative stress notation has never been used with solids. Nothing makes more sense than a consistency of sign notation for all materials. Obviously that sign notation should be that of Cauchy, which has been consistently used with solids and in most cases for fluids.

Generally, the stresses acting on an element are considered as the sum of a uniform hydrostatic pressure p which acts to compress the element and extra stresses which act to distort it. The hydrostatic pressure contributes to the normal stresses, but not to the shear stresses. If we denote the extra stress tensor $\mathbf{P}$ with components P_{ij}, it may be seen to be related to $\boldsymbol{\sigma}$ through

$$\boldsymbol{\sigma} = -p\mathbf{I} + \mathbf{P} \qquad (2\text{-}F\text{-}4)$$

where $\mathbf{I}$ is the unit matrix defined as

$$\mathbf{I} = \begin{vmatrix} 1 & 0 & 0 \\ 0 & 1 & 0 \\ 0 & 0 & 1 \end{vmatrix} \qquad (2\text{-}F\text{-}5)$$

We may define a *deviatoric stress* tensor $\mathbf{T}$ by the relationship

$$\boldsymbol{\sigma} = -p_1\mathbf{I} + \mathbf{T} \qquad (2\text{-}F\text{-}6)$$

where p_1 is referred to as the isotropic pressure. Here the components of $\mathbf{T}$ are defined by:

$$T_{11} + T_{22} + T_{33} = 0 \qquad \text{or} \qquad \text{tr}\,\mathbf{T} = 0 \qquad (2\text{-}F\text{-}7)$$

where we express the sum of the diagonal components as the trace (tr) of the tensor $\mathbf{T}$. This necessitates that p_1 have the form

$$p_1 = -\tfrac{1}{3}(\sigma_{11} + \sigma_{22} + \sigma_{33}) = -\tfrac{1}{3}\text{tr}\,\boldsymbol{\sigma} \qquad (2\text{-}F\text{-}8)$$

The deviatoric stress tensor is distinct and different from the extra stress. In general, the extra stress tensor does not satisfy Eq. (2-F-7). However, the difference in stress components will be the same, for example,

$$\sigma_{ii} - \sigma_{jj} = P_{ii} - P_{jj} = T_{ii} - T_{jj} \qquad (2\text{-}F\text{-}9)$$

G. CAUCHY'S LAW OF MOTION

The force applied to a macroscopic volume may be expressed using Eq. (2-F-1) as the sum of contact and body forces. If we introduce the stress tensor of Eq. (2-F-2) to represent the contact forces we obtain

$$\mathbf{F} = \oint \boldsymbol{\sigma} \cdot \mathbf{n}\, da + \int \rho \mathbf{f}\, dV \tag{2-G-1}$$

The divergence theorem may be applied to the surface integral in this equation. We may express the force $\mathbf{F}$ now as

$$\mathbf{F} = \int [\nabla \cdot \boldsymbol{\sigma} + \rho \mathbf{f}]\, dV \tag{2-G-2}$$

If the net force acting on the body, $\mathbf{F}$, is zero, it follows that the kernel of the integral of Eq. (2-G-2) is zero, that is,

$$0 = \nabla \cdot \boldsymbol{\sigma} + \rho \mathbf{f} \tag{2-G-3a}$$

or

$$0 = \sum_i \mathbf{e}_i \left(\sum_m \frac{\partial \sigma_{im}}{\partial x_m} + \rho f_i \right) \tag{2-G-3b}$$

This result was given by Cauchy (4) in 1827.

If the net force acting on the body is unbalanced, we must consider the rate of change of momentum terms. If the macroscopic volume of material points is roughly stationary, the time derivative d/dt of Eq. (2-E-4) may be taken inside of the integral as $\partial/\partial t$, that is,

$$\mathbf{F} = \int \frac{\partial \mathbf{v}}{\partial t}\, dm = \int \rho \frac{\partial \mathbf{v}}{\partial t}\, dV \tag{2-G-4}$$

Using Eq. (2-G-4) and the same arguments leading to Eq. (2-G-3), we may rewrite the expression given above as

$$\rho \frac{\partial \mathbf{v}}{\partial t} = \nabla \cdot \boldsymbol{\sigma} + \rho \mathbf{f} \tag{2-G-5}$$

This equation is suitable for small deformations, such as in the case of solids, and was stated by Cauchy (5) in 1828.

We now turn to the more complex situation where the volume of material points translates through space. If we consider a fixed spatial volume through which the system of material points moves, not only the change in momentum within the volume must be considered, but the flux through the surface. Mass conservation was represented by Eq. (2-C-1), which is equivalent to

$$\frac{d}{dt}\int \rho \, dV = \int \frac{\partial \rho}{\partial t} \, dV + \oint \rho \mathbf{v} \cdot \mathbf{n} \, da \tag{2-G-6}$$

Analogously, for momentum ρv, we may suppose:

$$\frac{d}{dt}\int \rho \mathbf{v} \, dV = \frac{\partial}{\partial t}\int \rho \mathbf{v} \, dV + \oint \rho \mathbf{v}\mathbf{v} \cdot \mathbf{n} \, da \tag{2-G-7}$$

where we consider momentum fluxes through the surface. This result is a special case of what is more generally called the Reynolds Transport Theorem (*33*). Applying the divergence theorem to the surface integral gives

$$\frac{d}{dt}\int \rho \mathbf{v} \, dV = \int \left[\frac{\partial \rho v}{\partial t} + \nabla \cdot \rho \mathbf{v}\mathbf{v} \right] dV \tag{2-G-8}$$

We may simplify the kernel of the integral of Eq. (2-G-8) using the continuity equation, Eq. (2-C-4). This leads to

$$\frac{d}{dt}\int \rho \mathbf{v} \, dV = \int \rho \left[\frac{\partial v}{\partial t} + (\mathbf{v}\cdot\nabla)\mathbf{v} \right] dV$$

$$= \int \rho \frac{Dv}{Dt} \, dV \tag{2-G-9}$$

where we have used the sustantive derivative of Eq. (2-C-7).

Using these results, we have in place of Eq. (2-G-5)

$$\rho \frac{D\mathbf{v}}{Dt} = \rho \left[\frac{\partial \mathbf{v}}{\partial t} + (\mathbf{v}\cdot\nabla)\mathbf{v} \right] = \nabla \cdot \boldsymbol{\sigma} + \rho \mathbf{f} \tag{2-G-10}$$

Equation (2-G-10) is generally called Cauchy's Law of Motion (though the term $\mathbf{v}\cdot\nabla\mathbf{v}$ does not appear in his papers). It has the cartesian components for the i direction:

$$\rho \left[\frac{\partial v_i}{\partial t} + \sum_m v_m \frac{\partial v_i}{\partial x_m} \right] = \sum_m \frac{\partial \sigma_{im}}{\partial x_m} + \rho f_i \tag{2-G-11}$$

or

$$\rho \left[\frac{\partial v_i}{\partial t} + v_1 \frac{\partial v_i}{\partial x_1} + v_2 \frac{\partial v_i}{\partial x_2} + v_3 \frac{\partial v_i}{\partial x_3} \right] = \frac{\partial \sigma_{i1}}{\partial x_1} + \frac{\partial \sigma_{i2}}{\partial x_2} + \frac{\partial \sigma_{i3}}{\partial x_3} + \rho f_i$$

$$\tag{2-G-12}$$

This represents 3 equations for $i = 1, 2$, and 3.

The components of Cauchy's Law of Motion in cylindrical and spherical coordinates are summarized in Table 2-G-1.

H. DIFFERENTIAL TORQUE BALANCE

We may treat Euler's Second Law relating the torque to angular momentum in a similar manner to that applied in the previous section. The most obvious approach would be to treat all torques as arising from the stress vector and replace linear momentum in Eqs. (2-G-7) and (2-G-8) with $\mathbf{r} \times \rho\mathbf{v}$. We may then write from Eq. (2-E-5):

$$\mathbf{M} = \oint \mathbf{r} \times \mathbf{t}\, da + \int \mathbf{r} \times \mathbf{f}\, dm \tag{2-H-1}$$

$$= \int [\nabla \cdot (\mathbf{r} \times \boldsymbol{\sigma}) + \mathbf{r} \times \rho\mathbf{f}]\, dV \tag{2-H-2}$$

where we have introduced the stress tensor and applied the divergence theorem. The rate of change of angular momentum of Eq. (2-E-5) is

$$\frac{d}{dt} \int \mathbf{r} \times \mathbf{v}\, dm = \int \left[\frac{\partial}{\partial t}(\mathbf{r} \times \rho\mathbf{v}) + \nabla \cdot \mathbf{r} \times \rho\mathbf{v}\mathbf{v} \right] dV \tag{2-H-3}$$

If we equate Eqs. (2-H-2) and (2-H-3), we obtain

$$\frac{\partial}{\partial t}(\mathbf{r} \times \rho\mathbf{v}) + \nabla \cdot \mathbf{r} \times \rho\mathbf{v}\mathbf{v} = \nabla \cdot \mathbf{r} \times \boldsymbol{\sigma} + \rho\mathbf{r} \times \mathbf{f} \tag{2-H-4}$$

After some rearrangement, we obtain

$$\mathbf{r} \times \mathbf{v}\left[\frac{\partial \rho}{\partial t} + \nabla \cdot \rho\mathbf{v} \right] + \rho \frac{D\mathbf{r}}{Dt} \times \mathbf{v} + \rho\mathbf{r} \times \frac{D\mathbf{v}}{Dt} = \nabla \cdot \mathbf{r} \times \boldsymbol{\sigma} + \rho\mathbf{r} \times \mathbf{f} \tag{2-H-5}$$

The first term on the left-hand side of the equation goes to zero because of the continuity equation, Eq. (2-C-4), and the second term is $\rho\mathbf{v} \times \mathbf{v}$, which is identically zero. If we introduce Cauchy's Law of Motion, Eq. (2-G-10), for $\rho\, D\mathbf{v}/Dt$, this now simplifies to

$$\mathbf{r} \times \nabla \cdot \boldsymbol{\sigma} = \nabla \cdot \mathbf{r} \times \boldsymbol{\sigma}; \qquad \varepsilon_{ijk} x_j \frac{\partial \sigma_{km}}{\partial x_m} = \frac{\partial}{\partial x_m} \varepsilon_{ijk} x_j \sigma_{km} \tag{2-H-6}$$

where we have introduced cartesian tensor notation with summation notation

TABLE 2-G-1 Cauchy's Law of Motion in Different Coordinate Systems

Cylindrical Coordinates (r, θ, z)

r Component

$$\rho\left(\frac{\partial v_r}{\partial t} + v_r\frac{\partial v_r}{\partial r} + \frac{v_\theta}{r}\frac{\partial v_r}{\partial \theta} - \frac{v_\theta^2}{r} + v_z\frac{\partial v_r}{\partial z}\right)$$

$$= \frac{1}{r}\frac{\partial}{\partial r}(r\sigma_{rr}) + \frac{1}{r}\frac{\partial \sigma_{r\theta}}{\partial \theta} - \frac{\sigma_{\phi\phi}}{r} + \frac{\partial \sigma_{rz}}{\partial z} + \rho f_r$$

θ Component

$$\rho\left(\frac{\partial v_\theta}{\partial t} + v_r\frac{\partial v_\theta}{\partial r} + \frac{v_\theta}{r}\frac{\partial v_\theta}{\partial \theta} + \frac{v_r v_\theta}{r} + v_z\frac{\partial v_\theta}{\partial z}\right) = \frac{1}{r^2}\frac{\partial}{\partial r}(r^2\sigma_{\theta r}) + \frac{1}{r}\frac{\partial \sigma_{\theta\theta}}{\partial \theta} + \frac{\partial \sigma_{\theta z}}{\partial z} + \rho f_\theta$$

z Component

$$\rho\left(\frac{\partial v_z}{\partial t} + v_r\frac{\partial v_z}{\partial r} + \frac{v_\theta}{r}\frac{\partial v_z}{\partial \theta} + v_z\frac{\partial v_z}{\partial z}\right) = \frac{1}{r}\frac{\partial}{\partial r}(r\sigma_{zr}) + \frac{1}{r}\frac{\partial \sigma_{z\theta}}{\partial \theta} + \frac{\partial \sigma_{zz}}{\partial z} + \rho f_z$$

Spherical Coordinates (r, θ, ϕ)

r Component

$$\rho\left(\frac{\partial v_r}{\partial t} + v_r\frac{\partial v_r}{\partial r} + \frac{v_\theta}{r}\frac{\partial v_r}{\partial \theta} + \frac{v_\phi}{r\sin\theta}\frac{\partial v_r}{\partial \phi} - \frac{v_\theta^2 + v_\phi^2}{r}\right)$$

$$= \frac{1}{r^2}\frac{\partial}{\partial r}(r^2\sigma_{rr}) + \frac{1}{r\sin\theta}\frac{\partial}{\partial \theta}(\sigma_{r\theta}\sin\theta) + \frac{1}{r\sin\theta}\frac{\partial \sigma_{\theta\phi}}{\partial \phi} - \frac{\sigma_{\theta\theta} + \sigma_{\phi\phi}}{r} + \rho f_r$$

θ Component

$$\rho\left(\frac{\partial v_\theta}{\partial t} + v_r\frac{\partial v_\theta}{\partial r} + \frac{v_\theta}{r}\frac{\partial v_\theta}{\partial \theta} + \frac{v_\phi}{r\sin\theta}\frac{\partial v_\phi}{\partial \phi} + \frac{v_r v_\theta}{r} - \frac{v_\phi^2\cot\theta}{r}\right)$$

$$= \frac{1}{r^2}\frac{\partial}{\partial r}(r^2\sigma_{\theta r}) + \frac{1}{r\sin\theta}\frac{\partial}{\partial \theta}(\sigma_{\theta\theta}\sin\theta) + \frac{1}{r\sin\theta}\frac{\partial \sigma_{\theta\phi}}{\partial \phi}$$

$$+ \frac{\sigma_{r\theta}}{r} - \frac{\cot\theta}{r}\sigma_{\phi\phi} + \rho f_\theta$$

φ Component

$$\rho\left(\frac{\partial v_\phi}{\partial t} + v_r\frac{\partial v_\phi}{\partial r} + \frac{v_\theta}{r}\frac{\partial v_\phi}{\partial \theta} + \frac{v_\phi}{r\sin\theta}\frac{\partial v_\phi}{\partial \phi} + \frac{v_\phi v_r}{r} + \frac{v_\theta v_\phi}{r}\cot\theta\right)$$

$$= \frac{1}{r^2}\frac{\partial}{\partial r}(r^2\sigma_{\phi r}) + \frac{1}{r}\frac{\partial \sigma_{\phi\theta}}{\partial \theta} + \frac{1}{r\sin\theta}\frac{\partial \sigma_{\phi\phi}}{\partial \phi} + \frac{\sigma_{\phi r}}{r} + \frac{2\cot\theta}{r}\sigma_{\phi\theta} + \rho f_\phi$$

(*10, 28*). Equation (2-H-6) is equivalent to

$$0 = \varepsilon_{ijk}\sigma_{kj}$$

or

$$\sigma_{ij} = \sigma_{ji} \tag{2-H-7}$$

The stress tensor is symmetric. This result is due to Cauchy (*3*) and is sometimes known as Cauchy's Second Law. We have used this result in constructing Table 2-G-1.

It has been noted by various investigators, beginning with Maxwell (*26*) and Voigt (*34*), that additional body torques and surface torques could exist, so that one should in general write

$$\mathbf{M} = \oint [\mathbf{r} \times \mathbf{t} + \mathbf{m}] \, da + \int [\mathbf{r} \times \mathbf{f} + \mathbf{l}] \, dm \tag{2-H-8}$$

where $\mathbf{m}$ is a couple stress vector and $\mathbf{l}$ a body couple. If we introduce a couple stress tensor $\boldsymbol{\mu}$ analogous to $\boldsymbol{\sigma}$ through

$$\mathbf{m} = \boldsymbol{\mu} \cdot \mathbf{n} \tag{2-H-9}$$

We may define a third-order tensor in place of $\boldsymbol{\mu}$ (*10, 33*).

We now replace Eq. (2-H-1) with Eq. (2-H-8). This leads in place of Eq. (2-H-6) to

$$\mathbf{r} \times \nabla \cdot \boldsymbol{\sigma} = \nabla \cdot (\mathbf{r} \times \boldsymbol{\sigma}) + \nabla \cdot \boldsymbol{\mu} + \rho \mathbf{l} \tag{2-H-10}$$

indicating the stress tensor is not symmetric in this case.

Nonsymmetric stress tensors arise in ordered materials such as liquid crystals. The balance laws for materials of this type are described by Ericksen (*7, 8*) and Leslie (*22*).

I. THEORY OF THIN SHELLS AND MEMBRANES

Another aspect of Cauchy's balance law deserves our attention. This is the application to a point on a thin shell of thickness h, which possesses an arbitrary curvature and a distribution of surface forces and torques. This problem has received considerable attention in the literature (*9, 16, 23, 24*).

The most important case involves the absence of surface torques and an axially symmetric shell known as a membrane. Consider the coordinate system of Figure 2-I-1, where θ is the angle from the axis and ϕ the angle from meridian. If we set up a force balance on a differential element on a membrane of thickness

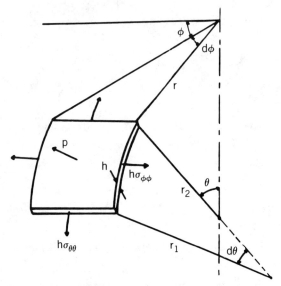

FIGURE 2-I-1. Basic coordinates and stress components of membrane.

n, or properly integrate Cauchy's equations of motion, we obtain

$$\frac{\partial}{\partial\theta}(rh\sigma_{\theta\theta}) + r_1\frac{\partial}{\partial\phi}h\sigma_{\theta\phi} - r_1\sigma_{\phi\phi}\cos\theta + p_\theta rr_1 = 0 \qquad (2\text{-I-1a})$$

$$\frac{\partial}{\partial\theta}(rh\sigma_{\phi\theta}) + r_1\frac{\partial}{\partial\phi}h\sigma_{\phi\phi} + r_1\sigma_{\theta\phi}\cos\theta + p_\phi rr_1 = 0 \qquad (2\text{-I-1b})$$

$$\frac{h\sigma_{\phi\phi}}{r_1} + \frac{h\sigma_{\theta\theta}}{r_2} = p_r \qquad (2\text{-I-1c})$$

where r_1 and r_2 are the principal ϕ and θ coordinate radii of curvature and r is the distance from the axis of rotation to a point on the surface. Also, $p_\theta, p_r,$ and p_ϕ are external load components on the surface of the membrane.

In the special case where there is symmetry, $\partial/\partial\phi = 0$ and there are no shear stresses, this reduces to

$$\frac{d}{d\theta}(rh\sigma_{\theta\theta}) - r_1\sigma_{\phi\phi}\cos\theta + p_\theta rr_1 = 0 \qquad (2\text{-I-2a})$$

$$\frac{h\sigma_{\phi\phi}}{r_1} + \frac{h\sigma_{\theta\theta}}{r_2} = p_r \qquad (2\text{-I-2b})$$

Equations (2-I-2a) and (2-I-2b) represent the stress field in a membrane.

J. STRESS AND DEFORMATION RATE ELLIPSOIDS AND INVARIANTS

The stress and rate of deformation tensors exhibit certain useful characteristics which we will develop in this section. These involve specifically the representation in terms of ellipsoids and the existence of invariant combinations of components of these tensors.

The representation of stress and deformation rates as ellipsoids dates to the origins of these concepts by Cauchy (2, 3) and Stokes (29). Consider a symmetric second-order tensor $\mathbf{a}$ and position vector $\mathbf{r}$. We may write

$$\begin{aligned}
\mathbf{a \cdot r \cdot r} &= \sum_i \sum_j a_{ij} x_i x_j \\
&= a_{11} x_1^2 + a_{22} x_2^2 + a_{33}^2 + 2a_{12} x_1 x_2 + 2a_{13} x_1 x_3 + 2a_{23} x_2 x_3 \quad \text{(2-J-1)} \\
&= K^2
\end{aligned}$$

This may be recognized as the equation of a second-order quadratic surface. If

$$a_{11} = a_{22} = a_{33} = a > 0 \qquad \text{then} \qquad a_{12} = a_{13} = a_{23} = 0 \qquad \text{(2-J-2)}$$

the surface may be seen to be a sphere of radius $K/\sqrt{a}$. If a_{11}, a_{12}, and a_{33} are all positive and a_{12}, a_{13}, and a_{23} are zero, Eq. (2-J-1) may be recognized as the equation of an ellipsoid whose principal axes correspond to the coordinate axes. More generally, Eq. (2-J-1) represents an ellipsoid whose principal axes do not correspond to the coordinate axes. The principal axes are oriented in

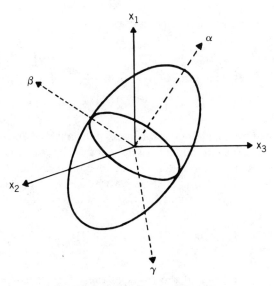

FIGURE 2-J-1. Tensor ellipsoid.

space relative to them. On the basis of these considerations, we speak of the stress and rate of deformation tensors as ellipsoids (Fig. 2-J-1).

Consider the product $\mathbf{a \cdot r}$ of tensor $\mathbf{a}$ and a position vector $\mathbf{r}$. Of interest is the special case where this vector $\mathbf{r}$ is parallel to $\mathbf{r}$, that is,

$$\mathbf{a \cdot r} = \lambda \mathbf{r} \tag{2-J-3}$$

This is equivalent to the set of algebraic equations

$$
\begin{aligned}
(a_{11} - \lambda)x_1 + a_{12}x_2 + a_{13}x_3 &= 0 \\
a_{12}x_1 + (a_{22} - \lambda)x_2 + a_{22}x_3 &= 0 \\
a_{13}x_1 + a_{23}x_2 + a_{23}x_2 + (a_{33} - \lambda)x_3 &= 0
\end{aligned}
\tag{2-J-4}
$$

It is well known that this set of equations only has a solution when the determinant of the coefficients is zero, that is,

$$\det|a_{ij} - \lambda\delta_{ij}| = \begin{bmatrix} a_{11} - \lambda & a_{12} & a_{13} \\ a_{12} & a_{22} - \lambda & a_{23} \\ a_{13} & a_{23} & a_{33} - \lambda \end{bmatrix} = 0 \tag{2-I-5}$$

This gives rise to the *characteristic equation* of the tensor (6), that is,

$$\lambda^3 - I_1\lambda^2 + I_2\lambda - I_3 = 0 \tag{2-J-6}$$

The coefficients I_1, I_2, and I_3 have the values

$$I_1 = a_{11} + a_{22} + a_{33} = \operatorname{tr}\mathbf{a} \tag{2-J-7a}$$

$$
\begin{aligned}
I_2 &= a_{11}a_{22} + a_{11}a_{33} + a_{22}a_{33} - a_{12}^2 - a_{13}^2 - a_{23}^2 \\
&= (\operatorname{tr}\mathbf{a})^2 - \operatorname{tr}\mathbf{a}^2
\end{aligned}
\tag{2-J-7b}
$$

$$I_3 = \det a_{ij} \tag{2-J-7c}$$

where $\operatorname{tr}\mathbf{a}$ defined by Eq. (2-J-7a) is the trace of the $\mathbf{a}$ tensor and $\det a_{ij}$ represents the determinant of the components of a_{ij}. Clearly, the quantities I_1, I_2, and I_3 must be independent of the initial choice of coordinate axes as they depend only on the roots of the characteristic equation. They are thus called invariants. The existence of the invariants of Eq. (2-J-7) were first pointed out by Cauchy (3). Furthermore, from Eqs. (2-J-7), if follows that

$$I_a = \operatorname{tr}\mathbf{a}; \qquad II_a = \operatorname{tr}\mathbf{a}^2; \qquad III_a = \operatorname{tr}\mathbf{a}^3 \tag{2-J-8a–c}$$

are also invariants of tensor $\mathbf{a}$.

What is the significance of the roots of the characteristic equation. If the three roots, $\lambda_\alpha, \lambda_\beta$, and λ_γ are known, we may compute three unit vectors $\mathbf{e}_\alpha, \mathbf{e}_\beta$, and $\mathbf{e}_\gamma$, representing directions of $\mathbf{r}$ that satisfy Eq. (2-J-3). We shall presume in general the values of $\lambda_\alpha, \lambda_\beta$, and λ_γ are different. Equation (2-J-3) is equivalent to the three equations

$$(\mathbf{a} \cdot \mathbf{e}_\alpha)_i = \sum_j a_{ij} l_{\alpha j} = \lambda_\alpha l_{\alpha i}$$

$$(\mathbf{a} \cdot \mathbf{e}_\beta)_i = \sum_j a_{ij} l_{\beta j} = \lambda_\beta l_{\beta i} \qquad \text{(2-J-9a–c)}$$

$$(\mathbf{a} \cdot \mathbf{e}_\gamma)_i = \sum_j a_{ij} l_{\gamma j} = \lambda_\gamma l_{\gamma i}$$

Multiplying through Eqs. (2-J-9a, b) by the directional cosines $l_{\beta i}$ and $l_{\alpha i}$ and adding yields:

$$\sum_i \sum_j a_{ij} l_{\alpha j} l_{\beta i} = \sum_i \lambda_\alpha l_{\alpha i} l_{\beta i}; \qquad \sum_i \sum_j a_{ij} l_{\beta j} l_{\alpha i} = \sum_i \lambda_\beta l_{\beta i} l_{\alpha i} \qquad \text{(2-J-10a, b)}$$

Clearly,

$$\sum_i \sum_j a_{ij} l_{\alpha j} l_{\beta i} = \lambda_\alpha \sum_i l_{\alpha i} l_{\beta i} = \lambda_\beta \sum_i l_{\beta i} l_{\alpha i} \qquad \text{(2-J-11)}$$

if λ_α and λ_β are different, we must have

$$\sum_i l_{\alpha i} l_{\beta i} = 0 \qquad \text{(2-J-12)}$$

This will only be true if the vectors $\mathbf{e}_\alpha$ and $\mathbf{e}_\beta$ are orthogonal. A similar argument may be made with regard to $\mathbf{e}_\alpha$ and $\mathbf{e}_\gamma$ or $\mathbf{e}_\beta$ and $\mathbf{e}_\gamma$.

Note that it also follows from the arguments leading to Eq. (2-J-11) that

$$\sum_i \sum_j a_{ij} l_{\alpha i} l_{\alpha j} = \lambda_\alpha \sum_i l_{\alpha i} l_{\alpha i} = \lambda_\alpha$$

$$\sum_i \sum_j a_{ij} l_{\beta i} l_{\beta j} = \lambda_\beta \qquad \text{(2-J-13a–c)}$$

$$\sum_i \sum_j a_{ij} l_{\gamma i} l_{\gamma j} = \lambda_\gamma$$

The quantities $\lambda_\alpha, \lambda_\beta$, and λ_γ may thus be seen to represent diagonal components of a_{ij} in the coordinate system defined by $\mathbf{e}_\alpha, \mathbf{e}_\beta$, and $\mathbf{e}_\gamma$, that is, $a_{\alpha\alpha}, a_{\beta\beta}$, and $a_{\gamma\gamma}$. The off-diagonal components are found to be zero.

This implies that the characteristic equation, Eq. (2-J-6), is the diagonal form of the more general matrix or tensor equation

$$\mathbf{a}^3 - I_1 \mathbf{a}^2 + I_2 \mathbf{a} - I_3 \mathbf{I} = 0 \qquad \text{(2-J-14)}$$

This result is known as the Cayley–Hamilton Theorem (6), which is usually stated "A matrix satisfies its own characteristic equation."

K. ENERGY EQUATION

We may use concepts and procedures equivalent to those developed in the preceding sections of this chapter to develop an expression for the balance of energy at a point in a deforming continuous medium. In essence, balances of this type date from Fourier (17), who obtained all of the thermal based terms but did not realize the interconversion of work and heat. Written in terms of rates of change and flow of energy, the conservation of energy may be expressed

$$\frac{dE}{dt} = Q + P \tag{2-K-1}$$

where E is the total energy content of the system, Q is the heat flow into the system, and P is the work per unit time or power applied to the system. E is the sum of internal energy (ε per unit mass), kinetic energy, and potential energy (ϕ per unit mass) and may be written

$$E = \int \left[\varepsilon + \frac{v^2}{2} + \phi \right] dm \tag{2-K-2}$$

and P may be expressed

$$Q = -\oint \mathbf{q} \cdot \mathbf{n} \, da \tag{2-K-3}$$

$$P = \oint \mathbf{t} \cdot \mathbf{v} \, da \tag{2-K-4}$$

where we neglect surface torques not arising from $\mathbf{t}$. These are included in the treatments of Truesdell and Toupin (33) and Eringen (10).

Applying arguments similar to those leading to Eq. (2-G-10), where (1) the stress tensor is introduced into Eq. (2-K-4), (2) the divergence theorem is applied to surface integrals including Eqs. (2-K-3) and (2-K-4) and those arising from dE/dt, and (3) the continuity equation is used, leads to

$$\rho \frac{D\varepsilon}{Dt} + \rho \frac{D}{Dt} \frac{v^2}{2} + \rho \frac{D\phi}{Dt} = -\nabla \cdot \mathbf{q} + \nabla \cdot (\boldsymbol{\sigma} \cdot \mathbf{v}) \tag{2-K-5}$$

If ϕ is a time-independent potential, it follows that

$$\frac{D\phi}{Dt} = \frac{\partial \phi}{\partial t} + \mathbf{v} \cdot \nabla \phi = -\mathbf{v} \cdot \mathbf{f} \tag{2-K-6}$$

It also may be noted that

$$\frac{D}{Dt} \frac{v^2}{2} = \mathbf{v} \cdot \frac{D\mathbf{v}}{Dt} \tag{2-K-7}$$

Equations (2-K-6) and (2-K-7) suggest introducing the scalar product of **v** and Cauchy's Law of Motion, Eq. (2-G-10). This leads to the simplification of Eq. (2-K-5) to

$$\rho \frac{D\varepsilon}{Dt} = -\nabla \cdot \mathbf{q} + \boldsymbol{\sigma} : \nabla \mathbf{v} \tag{2-K-8a}$$

where

$$\boldsymbol{\sigma} : \nabla \mathbf{v} = \nabla \cdot \boldsymbol{\sigma} \cdot \mathbf{v} - \mathbf{v} \cdot \nabla \cdot \boldsymbol{\sigma} = \sum_i \sum_j \sigma_{ij} \frac{\partial v_i}{\partial x_j} \tag{2-K-8b}$$

This is the generally quoted form of the energy equation.

The internal energy may be expressed as a function of temperature T and density ρ for a homogeneous medium

$$\varepsilon = \varepsilon(T, \rho)$$

$$d\varepsilon = c_v dT + \frac{\partial \varepsilon}{\partial \rho} d\rho \tag{2-K-9}$$

where c_v is the heat capacity at constant density. This leads to

$$\rho c_v \frac{DT}{dt} = -\nabla \cdot \mathbf{q} + \boldsymbol{\sigma} : \nabla \mathbf{v} - \rho \frac{\partial \varepsilon}{\partial \rho} \frac{D\rho}{Dt} \tag{2-K-10}$$

If we apply Eq. (2-C-4), we obtain

$$\rho c_v \frac{DT}{dt} = -\nabla \cdot \mathbf{q} + \boldsymbol{\sigma} : \nabla \mathbf{v} + \rho^2 \frac{\partial \varepsilon}{\partial \rho} \nabla \cdot \mathbf{v} \tag{2-K-11}$$

The general formalism of the energy equation and equivalent formulations are discussed by Toor (*31*) and by Bird, Stewart, and Lightfoot (*1*), which gives a most excellent treatment.

Through classical thermodynamics it may be shown from Eq. (2-J-9) that (*1,31*)

$$\rho^2 \left[\frac{\partial \varepsilon}{\partial \rho} \right] T = -\left[T \left[\frac{\partial p}{\partial T} \right] c_v - p \right] \tag{2-K-12}$$

Furthermore,

$$p\nabla \mathbf{v} + \boldsymbol{\sigma} : \nabla \mathbf{v} = \mathbf{p} : \nabla \mathbf{v} \tag{2-K-13}$$

where **P** is the extra stress. Equation (2-K-11) becomes

$$\rho c_v \frac{DT}{Dt} = -\nabla \cdot \mathbf{q} + \mathbf{P} : \nabla \mathbf{v} - T \left[\frac{\partial p}{\partial T} \right] c_v \nabla \cdot \mathbf{v} \tag{2-K-14}$$

TABLE 2-K-1   Energy Equations in Different Coordinate Systems

Cylindrical Coordinates

$$\rho c_v \left(\frac{\partial T}{\partial t} + v_r \frac{\partial T}{\partial r} + \frac{v_\theta}{r} \frac{\partial T}{\partial \theta} + v_z \frac{\partial T}{\partial z} \right) - \left(\frac{1}{r} \frac{\partial}{\partial r}(r q_r) + \frac{1}{r} \frac{\partial q_\theta}{\partial \theta} + \frac{\partial q_z}{\partial z} \right)$$

$$- T \left(\frac{\partial p}{\partial T} \right)_v \left(\frac{1}{r} \frac{\partial}{\partial r}(r v_r) + \frac{1}{r} \frac{\partial v_\theta}{\partial \theta} + \frac{\partial v_z}{\partial z} \right)$$

$$+ \left\{ \left(P_{rr} \frac{\partial v_r}{\partial r} + P_{\theta\theta} \frac{1}{r} \left(\frac{\partial v_\theta}{\partial \theta} + v_r \right) + P_{zz} \frac{\partial v_z}{\partial z} \right) \right.$$

$$+ P_{r\theta} \left[r \frac{\partial}{\partial r} \left(\frac{v_\theta}{r} \right) + \frac{1}{r} \frac{\partial v_r}{\partial \theta} \right] + P_{rz} \left(\frac{\partial v_z}{\partial r} + \frac{\partial v_r}{\partial z} \right)$$

$$\left. + P_{\theta z} \left(\frac{1}{r} \frac{\partial v_z}{\partial \theta} + \frac{\partial v_\theta}{\partial z} \right) \right\}$$

Spherical Coordinates

$$\rho c_v \left(\frac{\partial T}{\partial t} + v_r \frac{\partial T}{\partial r} + \frac{v_\theta}{r} \frac{\partial T}{\partial \theta} + \frac{v_\phi}{r \sin\theta} \frac{\partial T}{\partial \phi} \right)$$

$$= - \left(\frac{1}{r^2} \frac{\partial}{\partial r}(r^2 q_r) + \frac{1}{r \sin\theta} \frac{\partial}{\partial \theta}(q_\theta \sin\theta) + \frac{1}{r \sin\theta} \frac{\partial q_\phi}{\partial \phi} \right)$$

$$- T \left(\frac{\partial p}{\partial T} \right)_v \left(\frac{1}{r^2} \frac{\partial}{\partial r}(r^2 v_r) + \frac{1}{r \sin\theta} \frac{\partial}{\partial \theta}(v_\theta \sin\theta) \right.$$

$$+ \frac{1}{r \sin\theta} \frac{\partial v_\phi}{\partial \phi} \right) + \left\{ P_{rr} \frac{\partial v_r}{\partial r} + P_{\theta\theta} \left(\frac{1}{r} \frac{\partial v_\theta}{\partial \theta} + \frac{v_r}{r} \right) \right.$$

$$+ P_{\phi\phi} \left(\frac{1}{r \sin\theta} \frac{\partial v_\phi}{\partial \phi} + \frac{v_r}{r} + \frac{v_\theta \cot\theta}{r} \right)$$

$$+ P_{r\theta} \left(\frac{\partial v_\theta}{\partial r} + \frac{1}{r} \frac{\partial v_r}{\partial \theta} - \frac{v_\theta}{r} \right) + P_{r\phi} \left(\frac{\partial v_\phi}{\partial r} + \frac{1}{r \sin\theta} \frac{\partial v_\phi}{\partial \phi} - \frac{v_\phi}{r} \right)$$

$$\left. P_{\theta\phi} \left(\frac{1}{r} \frac{\partial v_\phi}{\partial \theta} + \frac{1}{r \sin\theta} \frac{\partial v_\theta}{\partial \phi} - \frac{\cot\theta}{r} v_\phi \right) \right\}$$

We may express Eq. (2-K-14) in cartesian coordinates as

$$
\rho c_v \left[\frac{\partial T}{\partial t} + v_1 \frac{\partial T}{\partial x_1} + v_2 \frac{\partial T}{\partial x_2} + v_3 \frac{\partial T}{\partial x_3} \right]
$$

$$
= - \left[\frac{\partial q_1}{\partial x_1} + \frac{\partial q_2}{\partial x_2} + \frac{\partial q_3}{\partial x_3} \right] + P_{12} \left[\frac{\partial v_1}{\partial x_2} + \frac{\partial v_2}{\partial x_1} \right] + P_{13} \left[\frac{\partial v_1}{\partial x_3} + \frac{\partial v_3}{\partial x_1} \right]
$$

$$
+ P_{23} \left[\frac{\partial v_2}{\partial x_3} + \frac{\partial v_3}{\partial x_2} \right] + P_{11} \frac{\partial v_1}{\partial x_1} + P_{22} \frac{\partial v_2}{\partial x_2} + P_{33} \frac{\partial v_3}{\partial x_3}
$$

$$
- T \left[\frac{\partial p}{\partial T} \right]_v \left[\frac{\partial v_1}{\partial x_1} + \frac{\partial v_2}{\partial x_2} + \frac{\partial v_3}{\partial x_3} \right] \tag{2-K-15}
$$

We have presumed symmetry of the stress tensor. We summarize Eq. (2-K-14) in cylindrical and spherical components in Table (2-K-1).

One usually presumes incompressibility, which eliminates the last term in Eq. (2-K-15).

REFERENCES

1. R. B. Bird, W. E. Stewart, and E. N. Lightfoot, *Transport Phenomena*, Wiley, New York, 1960.

2. A. L. Cauchy, *Ex. de Math*, **2**, 42 (1827); Oeuvres, **7**, (2) 60.

3. A. L. Cauchy, *Ex. de Math*, **2**, 60 (1827); *Oeuvres*, **7** (2), 82.

4. A. L. Cauchy, *Ex. de Math*, **2**, 108 (1827); *Oeuvres*, **7** (2), 141.

5. A. L. Cauchy, *Ex. de Math*, **3**, 160 (1828); *Oeuvres*, **8** (2), 195; *Ex. de Math*, **4**, 293 (1829); *Oeuvres*, **9** (2), 243.

6. A. Cayley, *Phil. Trans. Roy. Soc.*, **148**, 17 (1858).

7. J. L. Ericksen, *Trans. Soc. Rheol.*, **5**, 23 (1961).

8. J. L. Ericksen in *Advances in Liquid Crystals*, Vol. 2, Academic, New York, 1976.

9. J. L. Ericksen and C. Truesdell, *Arch. Rat. Mech. Anal.*, **1**, 295 (1958).

10. A. C. Eringen, *Non-Linear Theories of Continuous Media*, McGraw-Hill, New York, 1962.

11. L. Euler, *Mem. Acad. Sci. Berlin*, **11**, 274 (1755); *Opera Omnia*, **12**(2), 54.

12. L. Euler, *Mem. Acad. Sci. Berlin*, **11**, 316 (1755); *Opera Omnia*, **12**(2), 92.

13. L. Euler, *Novi. Comm. Acad. Sci. Petrop.*, **14**, 270 (1764); *Opera Omnia*, **13**(2), 73.

14. L. Euler, *Novi. Comm. Acad. Sci. Petrop.*, **20**, 189 (1775).

15. J. Finger, *Sitzungber. Akad. Wiss. Wien.*, **103**(11a), 1073 (1894).

16. W. Flugge, *Stresses in Shells*, Springer, Berlin, 1960.

17. J. Fourier, *Analytical Theory of Heat*, translated by A. Freeman, Dover, New York, 1822.

18. H. Goldstein, *Classical Mechanics*, Addison-Wesley, Cambridge, MA, 1950.

19. J. P. Joule, *Phil. Mag.*, **23**(3), 263 (1843); **27**(3), 205 (1845).

20. H. Lamb, *Proc. London Math. Soc.*, **21**, 119 (1890).

21. L. Landau and E. M. Lifschitz, *Quantum Mechanics*, Pergamon, London, 1958.

22. F. M. Leslie in *Advances in Liquid Crystals*, Vol. 4, Academic, New York, 1979.

23. A. E. Love, *Phil. Trans. Roy. Soc.*, **A179**, 491 (1888).

24. A. E. Love, *A Treatise on the Mathematical Theory of Elasticity*, 4th ed., Cambridge University Press, 1927.

25. J. C. Maxwell, *Phil. Trans. Roy. Soc.*, **157**, 49 (1867).

26. J. C. Maxwell, *A Treatise on Electricity and Magnetism*, Constable, London, 1873.

27. I. Newton, *Mathematical Principles of Natural Philosophy and His System of the World*, 1686–1725, 3rd ed., University of California Press, Berkeley, 1956.

28. W. Prager, *Introduction to the Mechanics of Continua*, Ginn, Boston, MA, 1961.

29. G. G. Stokes, *Trans. Camb. Phil. Soc.*, **8**, 287 (1845).

30. J. L. Synge, *Classical Dynamics*, in *Handbuch der Physik*, Vol. III/1, Springer, Berlin, 1960.

31. H.L. Toor, *Ind. Eng. Chem.*, **48**, 922 (1956).

32. C. Truesdell, *Arch. Hist. Exact Sci.*, **1**, 3 (1960).

33. C. Truesdell and R. A. Toupin, "The Classical Field Theories," in *Handbuch der Physik*, Vol. III/1, Springer, Berlin, 1960.

34. W. Voigt, *Abh. ges. Wiss. Gottingen*, **34**, 100 (1887).

35. E. Whittaker, *A Treatise on the Analytical Dynamics of Particles and Rigid Bodies*, 4th ed., Cambridge University Press, 1937.

III

FABRICATION AND FLOW IN POLYMER PROCESSING

A. INTRODUCTION

The subject of rheology is studied by polymer engineers in order to understand the response of polymer melts, solutions, and compounds in processing operations. In this chapter we turn our attention to the methods used to fabricate engineering parts from polymers, including plastics, elastomers, and fibers. We begin by critically discussing the historical development of polymer processing. The presentation then continues on to consideration of the range of polymer processing technologies. The chapter concludes by discussing the characteristics of kinematics within the various processing operations. These indicate the types of flows which the polymer engineer needs to characterize in order to understand processing behavior.

B. ORIGINS OF POLYMER PROCESSING TECHNOLOGY

1. Early Rubber Industry

The polymer industry in a modern sense begins with the commercial develpment of rubber in Europe. The first pioneer of the rubber industry was the English entrepreneur Thomas Hancock (*38–46*), who began manufacturing rubber products in London, England about 1820. One finds in Hancock's patents and autobiography (*43*) the first descriptions of an "internal mixer" for mastication degradation and mixing of elastomers (*42, 46*). The apparatus designed by Hancock which he called a "pickle" is shown in Figure 3-B-1. It was originally

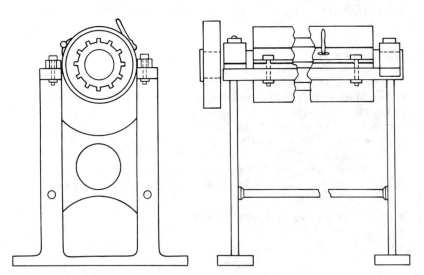

FIGURE 3-B-1. Thomas Hancock's masticator and internal mixer, pickle from English Patent 7344 (1837).

kept secret at the time to give Hancock an advantage over his competitors (*46*), but was later described in a patent of 1837 (*42*). The machine consisted of a horizontal shaft with spikes which rotated in a cylindrical box. The rubber would be torn apart by the spikes.

The second major innovation in polymer processing was the coating of rubber solutions on fabrics. This was developed in connection with the production of the Macintosh sandwich fabric (*66, 67, 89*), which was a waterproof fabric consisting of two outer layers with an intermediate rubber layer. This process was also carried out in secret, originally by the Macintosh firm (*Charles Macintosh and Company*) (*67, 89*) and later by their licensee Hancock (*39*) during the period 1825–1830. It is described in the 1837 patent of Hancock (*42*).

The use of pairs of rotating steel rolls for sheeting rubber seems to date to the 1820s and is described by Hancock (*40, 46*). However, the concept of using a pair of horizontal steam-heated iron rolls as a mastication and mixing device is from the American entrepreneur Edwin M. Chaffee (*19, 33, 91, 92*) of the Roxbury India Rubber Company and was cited in his patent of 1836 (*19*). Chaffee also described the use of several rolls (Fig. 3-B-2) to coat solid rubber onto fabric. This was the first calender. This machine was developed as a competitive process

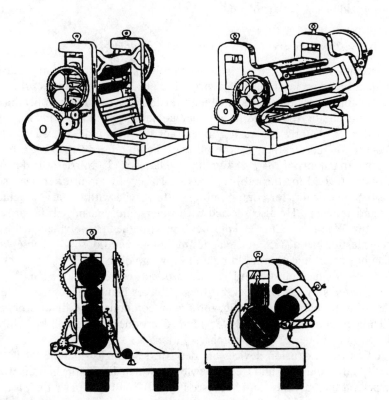

FIGURE 3-B-2. Chaffee's two-roll mill (Figs. 3 and 4) and four-roll calender (Figs. 1 and 2) from U.S. Patent 16 (1836).

to the solution-coating procedure. A fabric would be fed horizontally through the small gap between two vertically separated horizontal steel rolls rotating in opposite directions. The rubber would be metered to the fabric by the upper two rolls and a third roll. The roll mill was used as a device for preparing the compounds for the calender. Chaffee built a 30-ton calender of this type which was called the "Monster" (*33, 91, 92*).

One of Chaffee's motivations in developing the calender was to eliminate the disastrous adhesiveness of rubber products during summer months. This was associated with the use of solvents such as turpentine, which could not completely be removed. However, when calendered articles proved to be adhesive, it became apparent that this was not the case. This led, around 1839, to the invention of curing with sulfur at elevated temperatures. This development can be followed in the work of Nathaniel Hayward and Charles Goodyear (*32, 33*) in the United States and Thomas Hancock (*44, 46*) in England, all of whom claimed independent invention. Posterity has generally given credit to Goodyear. The name vulcanization, which describes this process, was coined by Hancock's colleague, William Brockedon (*46*). The development of vulcanization rapidly led to the addition of sulfur to compounds and the introduction of ovens and/or curing presses in rubber factories throughout the world.

2. Early Plastics Industry: Gutta Percha and Celluloid

The preceding section described the origin in rubber processing and of processing techniques based on multiple-roll devices. The beginnings of plastics processing and extrusion processes is associated with the introduction of gutta percha into England during the 1840s and its commercial development as an insulation for electrical wire. Among the early pioneers of the new industry was Thomas Hancock's younger brother, Charles, one of the founders of the Gutta Percha Company. In patents of 1846–1847, Charles Hancock (*35–37*) described process technology devised for the rubber industry, largely by his brother. He used a pickle-type masticator for compounding gutta percha with additives including sulfur and softeners. He also sheeted with rollers and vulcanized the products with sulfur. We find in Charles Hancock's patents the first polymer blends and rubber–plastics blends (*35*) and plastics molding operations (*36*). These would include both ram injection molding and blow molding.

The first foamed plastic and rubber products were developed in 1846 in separate patents by the Hancock brothers. Charles Hancock (*36*) foamed gutta percha using ammonium carbonate and similar compounds. William Brockedon and Thomas Hancock (*9*) produced foamed products using sulphur chloride dissolved in a rubber or gutta percha solution.

The first ram extrusion devices are described in English patents of 1845 by Richard A. Brooman (*10*) and Henry Bewley (*5*), which discuss the manufacture of gutta percha thread, tubes, and hose (see Fig. 3-B-3). Brooman's patent is a five-hole die which produces five simultaneous continuous filaments which are extruded into a bath and taken up on a roll. Bewley extrudes tubes and hose.

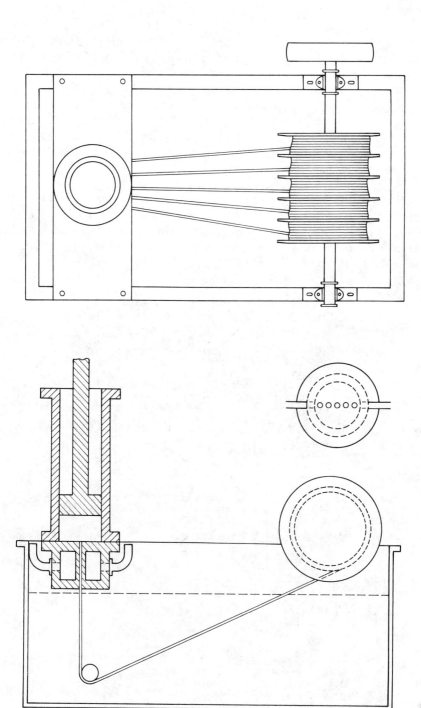

FIGURE 3-B-3. Brooman's ram extruder for producing thread from a multihole die from English Patent 10,582 (1845).

Charles Hancock, who was a partner of Bewley in the Gutta Percha Company, is said to have developed wire coating intended for electrical insulation using Bewley's extrusion methods (58). Methods of coating wires are described in patents by Barlow and Forster (3) and by Siemens (80) in 1848–1850.

The development of continuous extrusion of plastics using screw extruders begins with gutta percha and natural rubber and dates to the 1870s. The concept of screw pumping would seem to date to Archimedes, and the earlier use of screw pumps in the soap industry is described in the patent literature. The first patent for screw extrusion of thermoplastics is that of Matthew Gray (34) of London in 1879 (Fig. 3-H-4). It was intended for wire coating using gutta percha. Interestingly, the reason for the invention cited by Gray is the elimination of voids in extruded products. The extruder was fed from a two-roll mill or calendering device. There seems to have been independent developments of the screw extruder for extrusion of thermoplastics in Germany and the United States at about the same time (43, 58), but Gray's patent is the first clear statement.

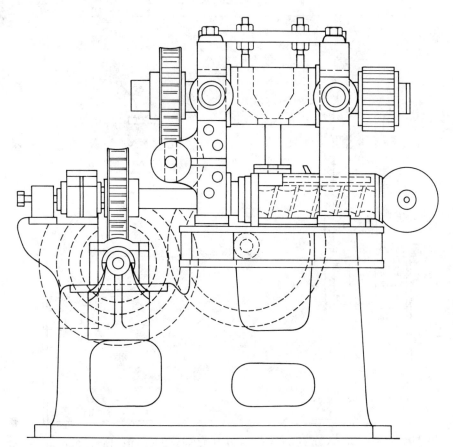

FIGURE 3-B-4. Matthew Gray's screw extruder from English Patent 5056 (1879).

An 1887 patent by Willoughby S. Smith (*81*) describes the use of a gear pump for gutta percha for the purpose of insulating electrical wires. Smith incorporated Gray's heated roll-feed system.

The next stage in the development of polymer processing methods came with the commercial development of cellulose nitrate as a plastic. The first efforts in this direction during the 1860s by Alexander Parkes (*73, 74*) and Daniel Spill (*82*) in England met with only limited success (*54*). Cellulose nitrate could not be melted and they used a range of volatile solvents that would evaporate from their products. These would then have high levels of residual stresses and shrink and crack. Parkes and Spill had rubber backgrounds and apparently used rubber processing machinery. In the United States, John Wesley Hyatt and his brother Isaiah Smith Hyatt (*49–55*) found that compounds or solutions of cellulose nitrate in nonvolatile camphor produced more desirable products. This was called *Celluloid*. The Celluloid Manufacturing Company was formed in the 1870s in Newark, New Jersey to exploit this product which proved to be a great success.

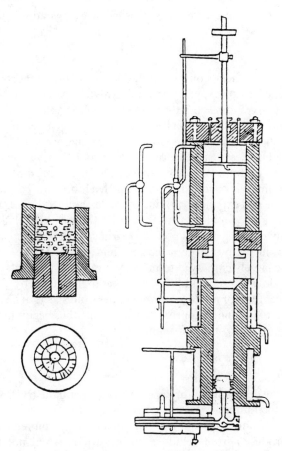

FIGURE 3-B-5. Stuffing machine of the Hyatt brothers from U.S. Patent 133,229 (1872).

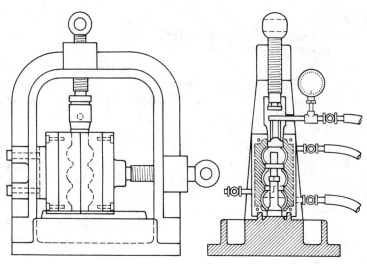

FIGURE 3-B-6. William B. Carpenter's blow-molding apparatus from U.S. Patent 237,168 (1881).

The Hyatts and their colleagues developed many important industrial processing operations to exploit Celluloid. An 1872 patent (Fig. 3-B-5) by the Hyatt brothers (49) contains both the reinvention of the ram extruder and the ram injection molding machine. They called this a *stuffing* machine. Shortly thereafter, a compression molding process was patented by the brothers (50). Numerous patents appeared in these years to form molded products of various types (51, 94).

In an 1878 patent, John Wesley Hyatt (53) described extruding Celluloid from the stuffing machine over a mandrel coated with a lubricant. This mandrel could be programmable and expand to produce complex hollow shapes. This led to the development of blow molding (Fig. 3-B-6) in 1881 by the Hyatts' colleague William B. Carpenter (17). Here a preform extruded tube is placed in a mold. This is expanded to fill the model by a heated fluid pumped into the tube. These inventions were largely employed to produce a range of products, including the components of dolls (16, 65) and liners for pipes (64). Patents by Thurber (88) at the turn of the century for making hollow shapes from sheets of Celluloid using a plunger and mold would appear to represent the beginning of modern thermoforming (matched-mold forming).

3. Early Fiber Industry

The 1880s saw the development of the man-made fiber industry. Brooman's 1845 patent (10) for the formation of gutta percha thread makes clear procedures for producing fibers from the melt. The man-made fibers commercialized in this period were produced from cellulose nitrate, which could not be melted. A method of producing fibers by extruding acetic-acid solutions of cellulose nitrate into a water of alcohol coagulation bath was described by Joseph Wilson Swan (*84*) in 1883. Swan's patent describes the later carbonization of the fibers with heat and thus represents the beginning of the carbon fiber industry. Swan's application was filaments for incadescent lights. Shortly thereafter, in France, the Count de Chardonnet (*20*) described a process for forming fibers from ether alcohol solutions by extruding the solutions into a water coagulation bath. Chardonnet produced much finer fibers than Swan. He formed a company and commercialized them as artificial silk. Later Chardonnet (*21*) described a dry-spinning process in which the filaments were extruded into the air where the solvent was evaporated. Also during the 1890s, Stearn collaborating with Cross, Bevan, and Beadle (*26*) invented a reactive spinning process in which cellulose was dissolved in a mixture of sodium hydroxide and carbon disulfide to form cellulose xanthate, which is extruded into an acid coagulating bath that regenerates the cellulose. This material became known as rayon. It was first commercialized by the English firm Courtaulds.

4. Early Tire Industry

A major change took place in the rubber industry at the turn of the century with the development of the pneumatic tire for the bicycle and later the automobile (*25*). Pneumatic devices based on leather had been long known, and as early as 1835 Thomas Hancock (*41*) had patented pneumatic air cushions from rubber. In 1845, Robert W. Thomson (*85*) patented pneumatic tires made of a rubber–canvas–leather composite for use on horse-drawn carriages. This was far ahead of its time, found little application, and was forgotten. The pneumatic tire was reinvented 40 years later by John Boyd Dunlop (*25*), leading to the founding of the Pneumatic Tyre Company, later the Dunlop Rubber Company. The Dunlop tires were made entirely of rubber and fabric. This led to a great increase in the sale of rubber, the subsequent growth of the rubber industry, and important roles for high-quality mixing, calendering, and extrusion.

Large internal mixers containing two counter-rotating rotors of special design were introduced into the rubber industry by the Werner and Pfleiderer Company of Stuttgart, Germany (*48*). In the second decade of the present century, a major new design was developed by Fernley H. Banbury (*2, 60*), whose machine was manufactured by the Birmingham Iron Foundry and its successor companies Farrel-Birmingham and Farrel. These internal mixers were made necessary by increased production needs, the poisonous accelerators used to hasten vulcanization, and the increasing use of large quantities of finely divided particles, especially in the tire industry. To some extent, these represent a new generation of Hancock's pickle mixers (*39, 43*), suitable for a new period.

5. Modern Plastics and Fiber Industry

By 1920, the essential features of the processing of rubber, plastics, and fibers had been established. The past 60 years has seen additional major innovations. These are associated with the development of new synthetic thermoplastics which were largely introduced from 1920 to 1960.

Screw extrusion devices using two co- or counterrotating screws were developed and applied as continuous mixing devices (47, 48, 90).

Intermeshing counterrotating twin-screw extruders were developed by Leistritz working with the IG Farbenindustrie (Hoechst) in the 1930s (11, 47, 48, 59, 90). Intermeshing corotating twin-screw extruders were introduced by LMP in Italy in the same period (24). Welding Engineers developed and commercialized tangential counterrotating twin-screw machines in the 1940s (30, 47, 48, 90). The 1960s saw the development of the tangential counterrotating Farrel continuous mixer (1, 47, 48, 90). In the post-war period, Bayer working with Werner and Pfleiderer developed the modular intermeshing corotating twin-screw extruder containing kneading discs (7, 28, 47, 48, 68, 71, 90).

The modern screw-injection molding machine for thermoplastics was invented by Hans Beck (4) of the IG Farbenindustrie (BASF) in 1943 though there was significant earlier work.

Numerous innovations in forming technology have been developed. E. I. DuPont de Nemours, notably through the efforts of Carothers, developed a commercial process for continuously forming highly oriented synthetic fibers from the melt (8, 14, 15). In the same period, films were produced by Ernst Studt (83) of the Norddeutsche Seekabelwerke by extruding melt from an annular die and expanding the melt with air.

The early process technologies mixed, pumped, and shaped products. New process technologies developed since the mid-1960s go beyond this to produce products with desirable mechanical properties by inducing certain types of molecular orientation. The manufacture of melt-spun and drawn fibers is a forerunner of such technologies. Stretch-blow molding of bottles is a key commercial development of this type (63, 95). Injection molding (23) and extrusion (18, 27) processes with rotating mold components have been invented which lead to biaxially oriented products. Most spectacular has been the invention of liquid crystalline process technology by Kwolek, Morgan, and their coworkers (6, 22, 70) of E. I. DuPont de Nemours, which has been applied to the manufacture of high modulus and tensile strength fibers.

C. MODERN DAY POLYMER PRODUCT FABRICATION

We now turn to present-day methods of polymer product manufacture. It must first be noted that the procedures used to fabricate parts from polymers are highly varied, both in type and in the number and character of the steps. Many manufacturing procedures involve only a single operation, while others involve numerous interacting steps. In the remainder of this chapter we will describe the more important examples of manufacture of polymeric products. The predecessor of many fabrication technologies is a compounding operation. We will begin by discussing compounding technology and then consider various fabrication procedures.

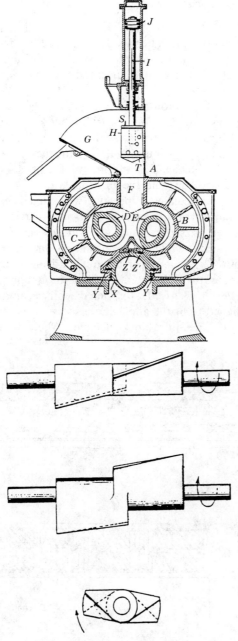

FIGURE 3-C-1. Fernley H. Banbury's internal mixer from U.S. Patents 1,200,070 (1916) and 1,905,955 (1933).

1. Compounding Operations

A range of procedures is used to blend polymers and add particulate fillers or reinforcing agents to polymeric materials. These procedures generally are divided into two parts: (1) batch mixers including two roll mills and more commonly internal mixers, such as that of Banbury, and (2) single-screw of twin-screw extrusion devices.

An internal mixer with two rotors containing blades, as shown in Figure 3-C-1, is used commercially to compound carbon black into rubber and blend elastomers with each other as well as with plastics. The rotors, which have screwlike configurations, distribute material in the mixer by inducing a circulating motion (69). Particulates and immiscible phases are distributed by passing between the flight tips and the chamber wall.

Compounding of particulates, especially with dimensions larger than a micron, and other additives into plastics is usually carried out in extruders, increasingly twin-screw extruders. Such apparatus is shown in Figure 3-C-2. Twin-screw extruders are divided into corotating and counterrotating screw devices. Compounding is usually carried out with an intermeshing modular corotating screw apparatus and partially intermeshing/tangential counterrotating machines (1, 7, 28, 30, 47, 61, 71, 90). Such machines contain modular screws which may be designed to fit the application, screw elements below a hopper, which feed pellets to kneading block elements where they are melted and dispersed.

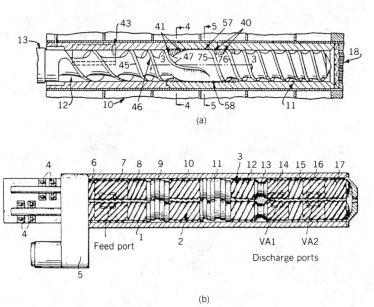

FIGURE 3-C-2. Single- and twin-screw extruders: (*a*) single-screw extruder; with mixing section, from U.S. Patent 2,496,625 (1950); (*b*) Erdmenger's modular intermeshing twin-screw extruder with kneading disc and screw elements from U.S. Patent 3,536,680 (1970).

2. Simple Machine Fabrication Operations Based upon Screw Extrusion Devices

a. General Remark

We now come to fabrication processes where screw extruders of the type shown in Figure 3-C-1a are used to pump and feed polymer melt to a shaping die and often post die secondary shaping.

Many such manufacturing operations for polymer products involve only a single processing machine. These manufacturing operations use plastics. Often, a small entrepreneur will operate a business based on the production of several similar identical machines of this type. Large manufacturers may use greater numbers of not dissimilar machines.

In these manufacturing operations, the screw extruder serves as both a melting device and a pump, and often as a mixer for additives. The basic operation of the screw pump is supplied by the friction of the barrel, and the material is pumped much as a nut is moved by a wrench along the length of the screw. In some cases, special static mixing devices are added following the extruder screw to aid in dispersing these added ingredients. Bags of pelletized plastics are introduced to the hopper and the melts produced are pumped into the shaping apparatus.

b. Extrusion through Dies

Here we have the simplest of the manufacturing operations of this type. The melt is pumped by the screw extruder and acted upon by a shaping device known as a die at its end. Typical fabrication machinery and methods based upon single-screw extruders coupled with dies are shown in Figure 3-C-2a. A simple example of a shaping die is the coathanger die of Figure 3-C-3a in which the melt emerging from the extruder is formed into a sheet. The die is designed to ensure a uniform sheet thickness laterally along the width of the die. The melt first flows into a manifold and then gradually distributes itself along the width of the die. Similar designs may be used to extrude complex-shaped profiles. Another example is the wire coating die of Figure 3-C-3b. A wire is continuously moved through a die. Melt is pumped through from the screw, which attaches itself and emerges with the wire.

Several screw extruders may be combined as in Figure 3-C-4 to produce a multilayer sheet. This is called coextrusion (22, 78).

Special-purpose dies with movable parts exist. Dies with rotating members to produce helical orientations in extruded tubes have also been reported (18, 27) and are usually used for fiber-filled melts. It is also possible to achieve this using specially designed dies with nonmoving parts (31).

c. Extrusion with Post-Die Shaping

i. Tubular Film   Post-die devices that draw down the dimensions of emerging simple shapes are rather broadly used on plastics. Take up rolls are often used to

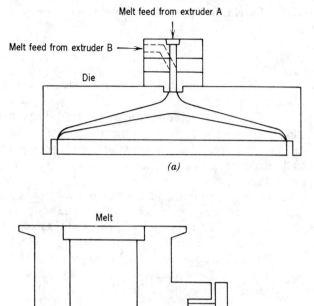

Melt feed from extruder A

Melt feed from extruder B

Die

(a)

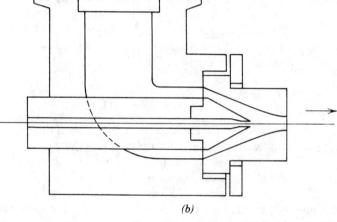

Melt

(b)

FIGURE 3-C-3. Coathanger and wire coating dies: (*a*) coathanger die; (*b*) Schematic cross section of a "cross-head" wire coating die: *A*, die body, cross-head; *B*, guider tip; *C*, die; *D*, die retaining ring; *E*, die retaining bolt; *F*, wire; *G*, core tube. (Reprinted with permission from P. N. Richardson, *Introduction to Extrusion*, Society of Plastics Engineers, Greenwich, CN., 1974).

produce thin film from melt emerging from a slit die. More commonly, melt is extruded through an annular die to form a tube into which air is blown. The air pressure expands the tube to several times its diameter and acts together with the draw down rolls to decrease the film thickness. This is shown in Figure 3-C-5. It is, of course, possible to produce multilayer film using these methods by combining melts emerging from two or more extruders (77).

ii. Blow Molding   Hollow objects such as bottles are produced in a process which dates at least to the efforts of Carpenter (*17*). In modern machines, pellets are melted and pumped by a screw device into an annular die, from which they

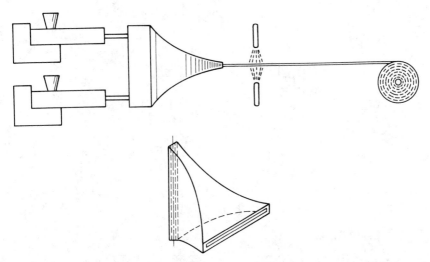

FIGURE 3-C-4. Coextrusion of sheet from two extruders from Chisolm and Schrenk, U.S. Patent 3,557,265 (1971).

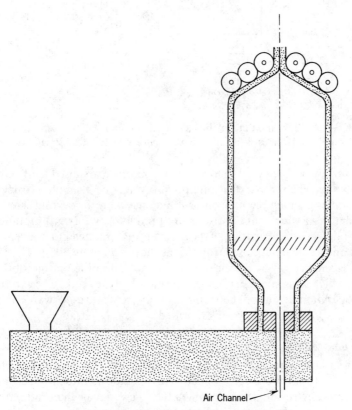

Air Channel

FIGURE 3-C-5. Tubular film extrusion.

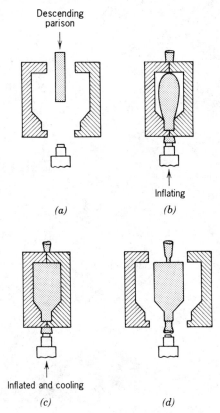

FIGURE 3-C-6. Blow molding of bottles from thermoplastics. (Reprinted with permission from W. A. Holmes–Walker, *Polymer Conversion*, Halstead Press, London, 1975).

emerge vertically downward to form a parison, which is subsequently molded to form a bottle. The screw-extrusion apparatus may be a simple extruder which shuts off after forming the parison or it may have a reciprocating screw and an accumulator, in which case parisons are produced by forward motions of the screw. Following the formation of the parison, air is injected into the center of the annulus and a cold mold closes around it as shown in Figure 3-C-6. The top of the parison is snipped off by the closing of the mold, and the bottom of the mold, through which air is injected, becomes the mouth of the bottle. The extruder then begins operation again (or the reciprocating screw moves forward) and a new parison is produced. The cycle is repeated.

d. Injection Molding

Another variant on shaping apparatus is a mold into which melt is injected from the extruder to form a desired shade. This method is presently accomplished using a screw-extrusion device in which the screw may be intermittently moved along its axis (reciprocating screw) to inject a polymer melt into a mold (see

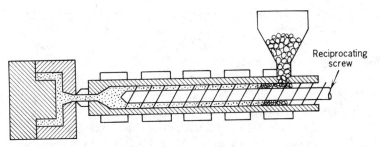

FIGURE 3-C-7. Screw-fed injection molding machine (reprinted with permission from W. A. Holmes–Walker, *Polymer Conversion*, Halsted, London, 1975).

Figure 3-C-7). In this processing method, pellets are added to the hopper and are melted and pumped in the manner of a screw extruder. Melt is accumulated in front of the screw. At regular intervals, the screw moves along its axis and pushes melt through a "runner" and "gate" into a cold mold where it gradually solidifies. When the melt solidifies at the gate, the screw retracts and continues to rotate and pump melt into the accumulation region. After the melt solidifies in the mold, the mold opens and ejects the part. The cycle now repeats itself. Through injection molding, intricate shapes may be produced in a rapid and inexpensive manner.

Injection-molding processes exist with molds containing rotating members (*33*). The purpose is to develop biaxial molecular orientation in molded parts by subjecting the melt to shearing in two orthogonal directions.

If gas is dissolved in the melt and allowed to come out of the solution in the mold, very large parts may be produced (*86*). Skin-core parts involving different polymers may be produced using two screw-injection molding machines which sequentially (and sometimes simultaneously) inject two melts into a mold. This is known as sandwich molding (*72, 96*).

3. Two-Machine Polymer Product Manufacture

a. General Remarks

There are many cases in which polymer products are produced using two sequentially operating processing machines. The first of these is an extrusion device which makes a preform. In the second step, the preform, which solidifies, is heated, sometimes above its softening temperature, and is further deformed to produce the final product. There is some arbitrariness in our classification, as we have included tubular-film extrusion and traditional blow molding in the previous section on single-machine manufacture rather than here. There are two major reasons for carrying out such two-stage manufacturing operations. One is to obtain a desired set of mechanical properties not achievable in processing operations involving single-machine manufacturing operations; that is we can shape filaments, film, and bottles in such a manner but often can not develop desired levels of molecular orientation needed for the required properties. The second reason is the formation of intricate shapes which are difficult to produce through injection molding.

b. Melt Spinning/Drawing of Fibers

Synthetic fibers require a two-step manufacturing process, as shown in Figure 3-C-8. In the first step, melt is extruded using a screw extruder and/or a gear pump through a multihole die called a spinneret. The emerging extrudates descend vertically to a rotating roll where they are taken up at a much higher velocity than they were extruded from the die. This results in both the production

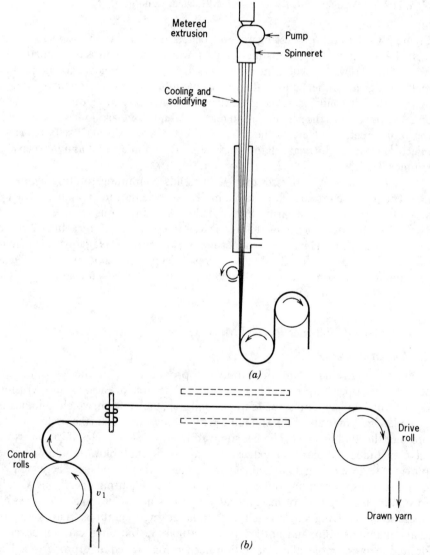

FIGURE 3-C-8. Melt spinning and drawing of fibers: (a) melt spinning; (b) drawing.

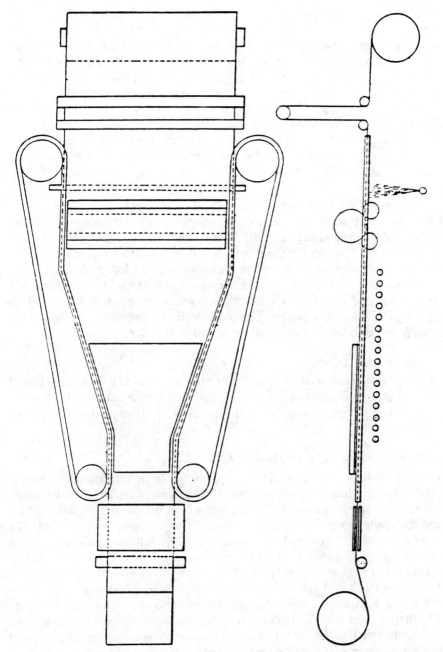

FIGURE 3-C-9. Tentering frame for biaxially oriented film from U.S. Patent 2,328,827 (1942).

of smaller diameters and enhanced anisotropy. This process is called melt spinning.

In the second processing stage, filaments are run from slow rolls over a hot plate or through a hot tube, which softens but does not melt the filament, and onto a fast roll. This further draws down the diameter and enhances the mechanical properties and anisotropy. This process is known as drawing.

The use of high speed melt spinning with very large drawdown ratios can alleviate the need for the second drawing step.

In some drawing operations, a rotating spindle twists the filament over a hot plate or in a hot tube and gives the filament helical in addition to transverse anistropy. This is called draw-texturing.

c. Two-Stage Film Formation

Films are often manufactured in two-stage manufacturing operations. Pellets are melted and pumped through a slit die, forming a film that vertically descends onto a rotating roll. The film then moves from a slow roll into a region of elevated temperature from where it moves onto a faster roll. The film is clamped in the transverse direction and is sequentially (in most cases) or simultaneously stretched in and transverse the machine direction, as shown in Figure 3-C-9. This type of film manufacturing process produces products that have high moduli and equal properties in all directions in the plane of the film.

d. Stretch-Blow Molding

In this process, preforms are extruded or injection molded before being placed in the stretch-blow-molding press. The inflation is carried out in such a manner as to biaxially stretch and orient the material as it forms a bottle (13,95) (Figure 3-C-10).

e. Thermoforming and Cold Forming (87)

In thermoforming, our concern is with producing large intricate shapes from a plastic sheet produced by a screw extruder and a sheeting die. The extruded sheets are sliced to specified lengths after emerging from the die and are carried by a conveyor belt and perhaps partially manually to a position over a mold. The sheet is heated by thermal radiation slightly above its softening point and pressed into the mold (Figure 3-C-11). This may be done by impressing a vacuum (vacuum forming) on the mold which draws the sheet into it. Alternatively, a plunger matched in shape to the mold (matched-mold forming) is used to push the heated sheet into the mold. Gas pressure may also be used (pressure forming). The shaped sheet is allowed to cool and is then manually moved from the mold.

In cold forming, the sheet is deformed below its softening temperature and pushed into a mold with a plunger matched in shape to the mold.

4. Multistage Processing Including Compounding and Curing Rubber

In the rubber industry, one generally requires several independent processing steps and machines. One begins with elastomers (generally solid bales) and

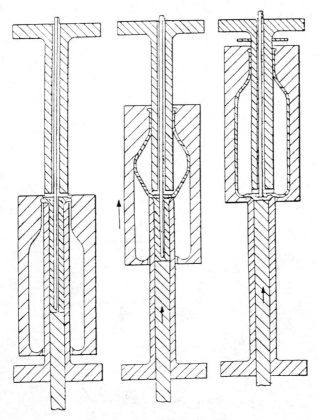

FIGURE 3-C-10. Stretch blow molding from Wyeth and Roseveare, U.S. Patent 3,849,530 (1974).

compounding ingredients (e.g., carbon black, oils, curatives) and produces a shaped, cured (cross linked) product. Compounding is usually carried out in an internal mixer, such as that of Banbury, combined with roll mills or an extruder. The product is then shaped, usually with a screw extruder or with the rolls of a calender and sometimes a screw-injection molding machine. Calenders are generally used as a step in the multistage processing of compounds (e.g., rubber and PVC). An extruded or rubber-calendered product is subsequently vulcanized in a press or by dielectric heating (Fig. 3-C-12). Injection-molded products are generally cured during the mold filling step.

The most complex fabricated rubber product is probably the pneumatic tire. This involves several components including a tread, carcass (rubber-coated fabric), sidewall, and linear, each of which has a different composition including distinct polymers and compounding ingredients and proportions. These are separately compounded in internal mixers. The resulting treads and sidewalls are coextruded, and the carcass rubber is calendered onto textile fabrics. The various

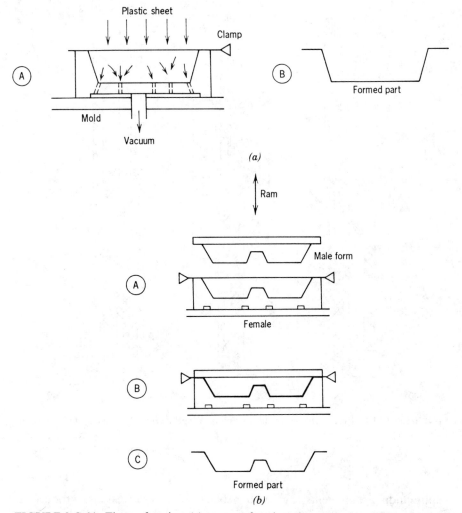

FIGURE 3-C-11. Thermoforming: (*a*) vacuum forming; (*b*) matched mold forming.

components are built together on a drum and the annular product is placed on a deformable membrane in a mold. When the mold closes, the membrane is inflated by steam pressure and fills the mold and forms a tire. The heated rubber composite cross-links and forms a pneumatic tire. Following the curing process, the mold cools, the membrane deflates, and the finished tire is removed.

Multistage processing including compounding and shaping steps is often found in thermoplastic fabrication. Compounding of reinforcing fillers such as fibers is often carried out in twin-screw extruders. The compounds so produced are subsequently fabricated using extrusion through dies, injection molding, or similar processing methods.

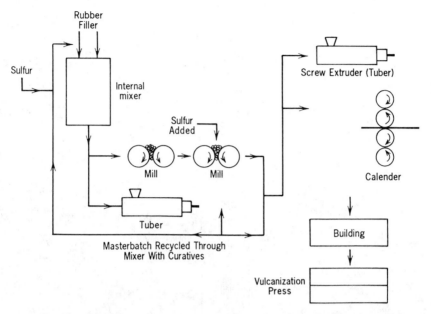

FIGURE 3-C-12. Fabrication processes for conversion of rubber into tires.

5. Solution Processing

In many cases, it is not possible to fabricate polymers in the molten state. Processing technology exists largely in the fiber area to produce products from polymer solutions. This is a multistep procedure involving dissolution, filament formation, coagulation, and solid-state drawing. Filament coagulation is either carried out by extruding the filament through a bath of liquid that is miscible with the solvent but not the polymer. The solvent is extracted, leaving a coagulated filament. Alternatively, the solvent may be evaporated by air streams as the filament leaves the spinneret. These processes are known as wet and dry spinning and are from Chardonnet (*20, 21*) (see Fig. 3-C-13). Processes also exist where there is an air gap between the spinneret and the bath (*6*). The purpose of this air gap is apparently to allow the development of high levels of molecular orientation. Films, including biaxially oriented films, may also be prepared in this manner (*29*).

D. KINEMATICS IN POLYMER PROCESSING OPERATIONS

Generally, flow in polymer fabrication machinery may be subdivided into two major classifications. These are: (1) internal flows where melt flows under pressure between the steel walls of an extrusion die or the walls of a mold; (2) external flows where the melt is situated in the air and is deformed by applied forces, as in melt spinning of a fiber, tubular film extrusion, or blow molding.

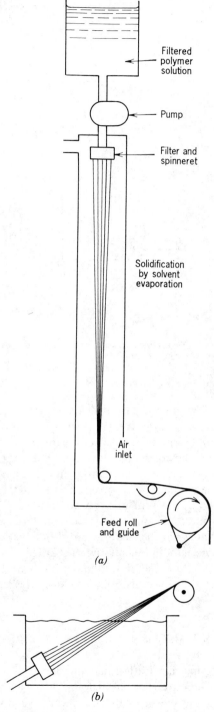

Filtered polymer solution

Pump

Filter and spinneret

Solidification by solvent evaporation

Air inlet

Feed roll and guide

(a)

(b)

FIGURE 3-C-13. Solution spinning processes: (*a*) dry spinning of fibers; (*b*) wet spinning of fibers.

There are major differences in the characteristics of these two flows. Polymer melts like Newtonian liquids generally adhere to solid-steel walls. The melts flow through dies and molds under pressure, suggesting velocity fields of the form

$$\mathbf{v} = v_1[x_2, x_3]\mathbf{e}_1 + 0\mathbf{e}_2 + 0\mathbf{e}_3 \qquad (3\text{-E-}1)$$

The deformation rate tensor is

$$\mathbf{d} = \tfrac{1}{2}\begin{bmatrix} 0 & \dot{\gamma}_2 & \dot{\gamma}_3 \\ \dot{\gamma}_2 & 0 & 0 \\ \dot{\gamma}_3 & 0 & 0 \end{bmatrix} \qquad (3\text{-E-}2)$$

Here subscript 1 denotes the direction of flow and 2 and 3 represent the normal directions. $\dot{\gamma}_2$ and $\dot{\gamma}_3$ represent rates of shear in the 2 and 3 directions perpendicular to flow known as *shearing*.

External flows[†] involve no walls and the material deforms in the direction of the applied stresses. The kinematics are thus of the form

$$\mathbf{v} = v_1(x_1)\mathbf{e}_1 + v_2(x_2)\mathbf{e}_2 + v_3(x_3)\mathbf{e}_3 \qquad (3\text{-E-}3)$$

$$\mathbf{d} = \begin{bmatrix} \partial v_1/\partial x_1 & 0 & 0 \\ 0 & \partial v_2/\partial x_2 & 0 \\ 0 & 0 & \partial v_1/\partial x_3 \end{bmatrix} \qquad (3\text{-E-}4)$$

These are extensional or elongational flows. This expression is restricted by the continuity equation, Eq. (2-C-9)

$$\frac{\partial v_1}{\partial x_1} + \frac{\partial v_2}{\partial x_2} + \frac{\partial v_3}{\partial x_3} = 0 \qquad (3\text{-E-}5)$$

If the forces are only applied in a single direction, as in the melt-spinning of fibers, elongation will only occur in a single direction and the responses will by symmetric about it

$$\frac{\partial v_2}{\partial x_2} = \frac{\partial v_3}{\partial x_3} \qquad (3\text{-E-}6)$$

From Eq. (3-E-6) we have

$$\frac{\partial v_2}{\partial x_2} = \frac{\partial v_3}{\partial x_3} = -\tfrac{1}{2}\frac{\partial v_1}{\partial x_1} \qquad (3\text{-E-}7)$$

This is called uniaxial extension or elongational flow.

For operations such as tubular film and blow molding, we cannot make this assumption. More complex kinematics needed to be considered. Consider a sheet

---

[†]The reader should note that our use of this term is different from that commonly used in Newtonian fluid mechanics, where it means flow around submerged objects. Here it means flow in regions beyond the die where the melt does not interact with metal walls.

with a coordinte system, in which directions 1 and 2 are in the plane of the film and 3 is normal to it. For planar extension

$$\frac{\partial v_2}{\partial x_2} = 0 \qquad (3\text{-E-}8)$$

By continuity

$$\frac{\partial v_3}{\partial x_3} = -\frac{\partial v_1}{\partial x_1} \qquad (3\text{-E-}9)$$

In equal biaxial extension we have

$$\frac{\partial v_2}{\partial x_2} = \frac{\partial v_1}{\partial x_1} \qquad (3\text{-E-}10)$$

and from continuity

$$\frac{\partial v_3}{\partial x_3} = -2\frac{\partial v_1}{\partial x_1} \qquad (3\text{-E-}11)$$

There are other differences between shear and elongational flows beyond the kinematics described thus far. (1) Internal flows in dies and to a lesser extent in molds and screw extruders are roughly isothermal in character. On the other hand, external flows involve rapid cooling of the melt at rates often of hundreds of degrees per second. (2) Internal flows frequently exist with roughly constant kinematics over a reasonable duration, while the kinematics of external flow generally rapidly change with time.

A result of these distinctions is that careful rheological measurements carried out under controlled conditions are of much more direct usefulness in analyses of shearing as opposed to elongational flows.

REFERENCES

1. E. H. Ahlefield, A. J. Baldwin, P. Hold, W. S. Rapetzky, and H. P. Schurer, U.S. Patent 3,154,808 (1964).
2. F. H. Banbury, U.S. Patent 1,200,070 (1916).
3. W. H. Barlow and T. Foster, English Patent 12,136 (1848).
4. H. Beck, German Patent 858,310 (1952).
5. H. Bewley, English Patent 10,825 (1845).
6. H. Blades, U.S. Patent 3,767,756 (1972).
7. H. Boden, H. Ocker, and W. Worz, U.S. Patent 3,305,844 (1967).
8. E. K. Bolton, Ind. Eng. Chem., 34, 53 (1942).
9. W. Brockedon and T. Hancock, English Patent 11,455 (1846).
10. R. A. Brooman, English Patent 10,582 (1845).
11. F. Burghauser, U.S. Patent 2,115,006 (1938).
12. K. Burkhardt, H. Herrmann, and S. Jakopin, Plast. Compd., Nov./Dec., 73 (1978).

13. M. Cakmak, J. L. White, and J. E. Spruiell, *J. Appl. Polym. Sci.*, **30**, 3679 (1985).

14. W. H. Carothers, U.S. Patent 2,130,948 (1938); U.S. Patent 2,137,235 (1938).

15. W. H. Carothers and J. W. Hill, *J. Am. Chem. Soc.*, **54**, 1579 (1932).

16. W. B. Carpenter, U.S. Patent 235,933 (1880).

17. W. B. Carpenter, U.S. Patent 237,168 (1880).

18. L. R. Cessna, U.S. Patent 3,651,187 (1972).

19. E. M. Chaffee, U.S. Patent 16 (1836).

20. H. de Chardonnet, U.S. Patent 394,559 (1888).

21. H. de Chardonnet, U.S. Patent 531,158 (1894).

22. D. Chisolm and W. J. Schrenk, U.S. Patent 3,557,265 (1971); W. J. Schrenk, *Plast. Eng.*, **30**(3), 65 (1974).

23. K. J. Cleereman, *SPE J.*, **23**(10), 43 (1967); U.S. Patent 3,307,726 (1967); U.S. Patent 3,907,952 (1975).

24. R. Colombo Italian Patent 370,578 (1937); U.S. Patent 2,563,396 (1951).

25. A. duCros, *Wheels of Fortune*, Chapman and Hall, London, 1938.

26. C. F. Cross, E. J. Bevan, and C. Beadle, English Patent 8700 (1892).

27. H. J. Donald, U.S. Patent 3,278,501 (1966).

28. R. Erdmenger U.S. Patent 3,122,356 (1964); U.S. Patent 2,670,188 (1954); U.S. Patent 2,814,472 (1957).

29. J. E. Flood, J. L. White, and J. F. Fellers, *J. Appl. Polym. Sci.*, **27**, 2965 (1982).

30. L. J. Fuller, U.S. Patent 2,441,222 (1948); U.S. Patent 2,615,199 (1952).

31. L. A. Goettler and A. J. Lambright, U.S. Patent 4,056,591 (1977); U.S. Patent 4,057,610 (1977); L. A. Goettler, R. I. Leib, and A. J. Lambright, *Rubber Chem. Technol.*, **52**, 838 (1979).

32. C. Goodyear, U.S. Patent 2622 (1844).

33. C. Goodyear, *Gum Elastic*, privately printed, New Haven, CN, 1855.

34. M. Gray, English Patent 5056 (1879).

35. C. Hancock, English Patent 11,032 (1846).

36. C. Hancock, English Patent 11,208 (1846).

37. C. Hancock, English Patent 11,575 (1847).

38. T. Hancock, English Patent 4451 (1820).

39. T. Hancock, English Patent 4768 (1823).

40. T. Hancock, English Patent 5045 (1824).

41. T. Hancock, English Patent 6849 (1835).

42. T. Hancock, English Patent 7344 (1837).

43. T. Hancock, English Patent 7549 (1838).

44. T. Hancock, English Patent 9952 (1843).

45. T. Hancock, English Patent 11,135 (1846).

46. T. Hancock, *Personal Narrative of the Origin and Progress of Caoutchouc or India Rubber Manufacture in England*, Longmans, London, 1857.

47. H. Herrmann, *Schneckenmaschinen in Verfahrenstechnik*, Springer, Berlin, 1972.

48. H. Herrmann in *Kunststoffe ein Werkstoff macht karrier*, W. Glenz, Ed., Hanser, Munich, 1985.

49. I. S. Hyatt and J. W. Hyatt, U.S. Patent 133,229 (1872).

50. J. W. Hyatt, U.S. Patent 138,245 (1873); I. S. Hyatt and J. W. Hyatt, U.S. Patent 152,232 (1874).

51. J. W. Hyatt, U.S. Patent 199,909 (1878).

52. J. W. Hyatt, U.S. Patent 202,441 (1878); J. W. Hyatt and C. Burroughs, U.S. Patent 204,229 (1878).

53. J. W. Hyatt, U.S. Patent 204,228 (1878).

54. J. W. Hyatt, Ind. Eng. Chem., **6**, 158 (1914).

55. J. W. Hyatt and I. S. Hyatt, U. S. Patent 91,341 (1869).

56. F. Jones in *History of the Rubber Industry*, P. Schidrowitz and T. R. Dawson, Eds., Heffer, Cambridge, 1952, Chapter 1.

57. M. Kaufman, *The First Century of Plastics Institute*, London, 1963.

58. M. Kaufman, *Plastics and Polymers*, **37**, 43 (1969).

59. S. Kiesskalt, H. Tampke, K. Winnacker, and E. Weingartner, German Patent 652,990 (1937).

60. D. H. Killhefer, *Banbury, the Master Mixer*, Palmerton, New York, 1962.

61. H. Koch, U.S. Patent 3,608,868 (1971).

62. S. L. Kwolek, U.S. Patent 3,671,542 (1972).

63. S. Lee, U.S. Patent 4,144,298 (1979).

64. M. C. Lefferts, U.S. Patent 281,529 (1883).

65. M. C. Lefferts and W. B. Carpenter, U.S. Patent 237, 559 (1881).

66. C. Macintosh, English Patent 4804 (1823).

67. G. Macintosh, *Biographical Memoir of the Late Charles Macintosh, FRS* Blackie, Glasgow, 1847.

68. W. Meskat and J. Pawlowski, German Patent 949,162 (1956).

69. K. Min and J. L. White, *Rubber Chem. Technol.*, **58**, 1024 (1985).

70. P. W. Morgan, *Macromolecules*, **10**, 1381 (1977).

71. H. W. G. Ocker, U.S. Patent 3,525,124 (1970).

72. D. F. Oxley and D. J. H. Sandiford, *Plast. Polym.*, **39**, 288 (1971); D. F. Oxley, U.S. Patent 3,751,534 (1977).

73. A. Parkes, English Patent 2675 (1864).

74. A. Parkes, English Patent 1313 (1865).

75. F. F. Pease, U.S. Patent 2,048,286 (1936).

76. B. D. Porrit and H. Rogers, *India Rubber J.*, March 8, 382 (1924).

77. G. E. Raley, U.S. Patent 3,223,761 (1965).

78. W. J. Schrenk and T. Alfrey, *SPE J.*, **29**(6), 78 (1973); **29**(7), 43 (1973).

79. H. A. Schurer, *Rubber J.*, **130**, 132 (1956); **133**, 272 (1957); **133**, 306 (1957).

80. E. W. Siemens, English Patent 17,768 (1887).

81. W. S. Smith, English Patent 17,768 (1887).

82. D. Spill, U.S. Patent 91,377 (1869); U.S. Patent 97,454 (1869).

83. E. Studt, German Patent 655,013 (1938).

84. J. W. Swan, English Patent 5978 (1883).

85. R. W. Thomson, English Patent 10,990 (1845).

86. J. L. Throne, *J. Cell. Plast.*, **10**, 208 (1972); **14**, 161 (1976).

87. J. L. Throne, *Thermoforming*, Hanser, Munich, 1986.

88. C. Thurber, U.S. Patent 669, 330; U.S. Patent 669,331 (1901).

89. J. L. White, *Rubber Ind.*, August, 148 (1974); *Polym. Eng. Rev.*, **4**, 345 (1984).

90. J. L. White, W. Szydlowski, K. Min, and M. Kim, *Adv. Polym. Technol.*, **7**, 295 (1987).

91. R. F. Wolf, *India Rubber World*, August 1, 39 (1936).

92. R. F. Wolf, *India Rubber Man*, Caxton, Caldwell, Idaho, 1939.

93. W. Woodruff, *The Rise of the British Rubber Industry During the Nineteenth Century*, Liverpool University Press, Liverpool, UK, 1958.

94. E. C. Worden, *Nitrocellulose Industry*, van Nostrand Reinhold, New York, 1911.

95. N. C. Wyeth and R. N. Roseveare, U.S. Patent 3,489,530 (1974).

96. S. S. Young, J. L. White, E. S. Clark, and Y. Oyanagi, *Polym. Eng. Sci.*, **20**, 789 (1980).

97. H. G. Zimmerman, U.S. Patent 3,170,566 (1965).

EXPERIMENTAL POLYMER RHEOLOGY

IV

RHEOLOGICAL CHARACTERIZATION OF POLYMER FLUIDS

A. INTRODUCTION

In this chapter we turn our attention to experimental measurements of rheological properties in polymer fluids including elastomers, molten plastics, polymer solutions, and their suspensions with small particles. We seek to do this with minimal presumptions about the detailed rheological behavior of these materials beyond the existence of the property to be measured. This chapter is not concerned with the fluid mechanics of materials of known rheological behavior and is purposely placed prior to that on rheological constitutive equations.

Most rheological measurements are carried out in shear flow. This is almost always the case in low-viscosity fluids and largely the situation for polymer melts. Shear flow measurements replicate the environment and response of polymer systems in extruder screw channels, dies, and injection molds. We will discuss both rheological measurements in shear flow and the possibility of wall slippage. However, the importance of elongational flow in polymer processing has led to extensive studies of tensile stretching modes. We survey experiments on small strains, including sinusoidal oscillation measurements, which now have a long history as a class of rheological investigations. The first studies of the mechanics of materials subjected to small static or dynamic deformations were for solids. However, as many polymer melts and solutions appear to have properties between solids and fluids, these techniques were applied to them.

B. SHEAR FLOW

1. Kinematics and Rheological Functions

By shear flows, we mean fluid motions kinematically equivalent to (Fig. 4-B-1)

$$\mathbf{v} = v_1(x_2)\mathbf{e}_1 + 0\mathbf{e}_2 + 0\mathbf{e}_3 \tag{4-B-1}$$

The rate of deformation tensor has the form

$$\mathbf{d} = \frac{1}{2} \begin{bmatrix} 0 & \dot{\gamma} & 0 \\ \dot{\gamma} & 0 & 0 \\ 0 & 0 & 0 \end{bmatrix} \tag{4-B-2}$$

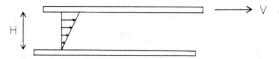

FIGURE 4-B-1. Schematic of shear flow field.

where $\dot{\gamma}$ is the rate of shear, dv_1/dx_2. Our great concern for this class of motions is that *internal* flows of the type described in Section 3-E are herein included. The flows we will consider here are equivalent or similar to those occurring in dies and molds.

Shear flow between parallel plates and pressure flow in a slit are included in this description. There are other important classes of shear flows, including torsional flow between discs and a cone and a plate, flow between coaxial cylinders, and pressure flow in a capillary. In torsional flow, we must consider, in cylindrical coordinates, the velocity field v_θ varying linearly with axial distance, that is, it is $\dot{\gamma}x_2$, with zero radial and axial velocities. In cone–plate flow, we have, in spherical coordinates, $v_\phi = v_\phi(\theta)$ with zero radial (v_r) and angular (v_θ) components. For flow between coaxial cylinders (*Couette*), we have in cylindrical coordinates a circumferential velocity $v_\theta = r\omega(r)$, where (ωr) is an angular velocity with zero axial (v_1) and radial (v_r) velocities. Axial flow in the coaxial cylinder geometry (*Pochettino*) due to an axially moving cylinder is another shearing motion. Here we have axial velocity $v_1(r)$ and zero radial (v_r) and angular (v_θ) velocities. In pressure-induced flow in a tube we have the velocity field $v_1(r)$, where the radial (v_r) and angular (v_θ) velocity components are zero. Pressure-induced laminar flow in tube, slit, or annulus is referred to as *Poiseuille* flow.

In response to a shearing flow, the corresponding shear stresses are developed. Normal stresses arise as well. The stress field in response to shear flow has been noted by Markovitz (*132*) and Coleman and Noll (*46*) (see also Coleman et al., ref. *45*), largely from symmetry and initial isotropy arguments, to be of the form

$$\boldsymbol{\sigma} = \begin{bmatrix} \sigma_{11} & \sigma_{12} & 0 \\ \sigma_{12} & \sigma_{22} & 0 \\ 0 & 0 & \sigma_{33} \end{bmatrix} \tag{4-B-3}$$

The sign of the shear stress varies with the sign of the shear rate, that is,

$$\sigma_{12}(\dot{\gamma}) = -\sigma_{12}(-\dot{\gamma}) \tag{4-B-4}$$

This suggests the formulation between σ_{12} and $\dot{\gamma}$ as

$$\sigma_{12} = \eta(\dot{\gamma}^2)\dot{\gamma} \tag{4-B-5}$$

where η is the shear viscosity

We may define three independent combinations of normal stresses. As the normal stresses are the sum of a pressure and an extra stress,

$$\sigma_{ii} = -p + P_{ii} \tag{4-B-6}$$

our three functions are logically a measure of this pressure p and two differences between normal stresses. These may be taken as

$$p = -\tfrac{1}{3}(\sigma_{11} + \sigma_{22} + \sigma_{33}) \tag{4-B-7}$$

$$N_1 = \sigma_{11} - \sigma_{22} = P_{11} - P_{22} \tag{4-B-8}$$

$$N_2 = \sigma_{22} - \sigma_{33} = P_{22} - P_{33} \tag{4-B-9}$$

N_1 is known as the *principal normal stress* difference and N_2 as the *second normal stress difference*. N_1 and N_2 will be uniquely defined by the rheological behavior of the material, while p is largely determined by the externally applied hydrostatic pressure and the balance of forces during flow.

The normal stresses are independent of the direction of shear. Thus

$$N_i(\dot{\gamma}) = N_i(-\dot{\gamma}) \tag{4-B-10}$$

and

$$N_1 = N_1(\dot{\gamma}^2) \tag{4-B-11}$$

$$N_2 = N_2(\dot{\gamma}^2) \tag{4-B-12}$$

One may define from this normal stress coefficients Ψ_1 and Ψ_2 through:

$$N_1 = \Psi_1(\dot{\gamma}^2)\dot{\gamma}^2 \tag{4-B-13}$$

$$N_2 = \Psi_2(\dot{\gamma}^2)\dot{\gamma}^2 \tag{4-B-14}$$

The purpose of shear-flow rheological measurements is to determine σ_{12} and N_1 and N_2 as a function of shear rate $\dot{\gamma}$, and through them the material functions η, Ψ_1 and Ψ_2. We also seek the time-dependent or transient behavior of these functions.

In addition to the rheological functions described above, it was also important to determine the boundary conditions along metal surfaces. We shall describe the attempt to characterize slip in our discussion of several instruments.

2. Historical Perspective[†]

The first rheological measurements were carried out on Newtonian fluids in a shear-flow mode. The Newtonian fluid has only one rheological characteristic,

---

[†]Aspects of this development are also described by Markovitz (*135*).

a constant viscosity coefficient. While the concept of viscosity dates to Newton (*165*) in the seventeenth century, rational attempts to measure this coefficient follow the formulation and acceptance of the Navier–Stokes equations and the development of solutions in various geometries. This occurred with the work of Stokes (*211–213*) in 1845–1851, who also established the no-slip boundary condition at solid walls. However, some earlier measurements could be interpreted in terms of viscosity. Inspired by the careful early experiments on capillary flow by Poiseuille (*180*) (who did not use Newtonian fluid mechanics to interpret his data), various investigators from about 1860, such as Hagenbach (*89*), Reynolds (*190*), and Thorpe and Rodgers (*222*), used cylindrical tubes to measure viscosity. New instruments also were developed. In 1866, Clerk Maxwell (*140*) determined viscosity using an oscillating disc. In 1888, Couette (*48*) devised a rotating concentric cylinder viscometer for viscosity measurement. This instrument was used shortly thereafter by Schwedoff (*202*). Pochettino (*179*) described the use of the relative axial motion of coaxial cylinders as a measurement of viscosity.

The first experimental studies indicating that the flow behavior of certain suspensions and colloidal solutions may be more complex than Newtonian appears to be found in the work of Theodor Schwedoff (*202*) in 1890 on gelatin solutions. Schwedoff found the torque in a Couette instrument not to be proportional to the angular velocity. He interpreted this as arising from the material exhibiting a yield value, that is, in steady flow .

$$\sigma_{12} = Y + \eta_p \dot{\gamma} \qquad (4\text{-}B\text{-}15)$$

From 1900, various investigators studying gelation systems, blood, and other colloids found deviations from Newtonian flow in their experiments. Beginning in 1910, Hess (*97*) reported extensive studies using capillary flow which showed deviations from the predictions of Newtonian fluid behavior. From 1913, Emil Hatschek and his coworkers (*94, 95, 104*) described Couette viscometer measurements showing deviations similar to that found by Schwedoff (*202*). Hatschek described this apparently anomalous behavior as being due to shear-rate-dependent viscosity functions. Eugene C. Bingham and his coworkers (*19, 20*) reported capillary flow studies of suspensions of clay and paints from 1916 on, which exhibited nonlinear pressure-drop–flow-rate behavior. They interpreted this behavior as resulting from yield values. Unlike Schwedoff, the researches of Hatschek and especially Bingham were widely read. Bingham (*20*) published a monograph entitled *Fluidity and Plasticity* in 1922 which outlined his views and summarized experimental studies of viscosity measurement.

A rational viscometry of rheologically complex fluids began to develop in the 1920s. The Schwedoff–Bingham yield value concept was used by Buckingham (*34*) to analyze flow in a capillary instrument and by Reiner and Riwlin (*188*) in a Couette coaxial cylinder instrument. Direct comparisons of theory and experiment were not possible, but rheological parameters could be determined. Power-law relationships between shear stress and shear rate, that is,

$$\sigma_{12} = K\dot{\gamma}^n \qquad (4\text{-}B\text{-}16)$$

were used by Porter and Rao (*184*) and by Farrow et al. (*72*) to analyze flow in these instruments.

Farrow, Lowe, and Neale (72) reported a study in 1928 of the flow of a starch paste in both concentric cylinder and capillary instruments, analyzed using power-law models. They showed that the viscosity–shear rate behavior obtained from both instruments was consistent. This was a most significant conclusion and forms an experimental basis for the concept of viscometric shear-flow measurements on complex fluids.

Weissenberg[†] was the first to realize the importance of determining the form of the viscosity function, $\eta(\dot{\gamma})$, without presuming particular models for analyzing flow in instruments. In 1929 his research team reported a procedure for capillary measurements which he had developed (*68, 185*). A similar analysis was given 2 years later by Mooney (*152*) for the Couette viscometer and in 1935 by Burgers (*35*) for a torsional flow instrument.

The rheometry developed in the 1920s continued to use Stokes' no-slip boundary condition. This was seriously questioned and examined again about 1930, notably by Schofield and Scott Blair (*200*) and Mooney (*152*). The former authors note that phase separation in a suspension can give the appearance of slippage. Mooney (*152–160*) would return again and again to the problem of slippage in elastomers throughout his career.

The first serious measurements of the rheological behavior of bulk polymers and compounds appeared in the 1930s. The investigations of Dillon and his coworkers (*61*) and especially Mooney (*153, 154, 158, 160*) in this period are to be cited. Mooney's 1936 study on natural rubber (*154*) using a coaxial cylinder viscometer would appear to be the first careful, properly determined viscosity shear rate curve for a bulk polymer. In the same period, Mooney developed a disc viscometer for characterizing the flow of rubber (*154–156, 158*).

In the 1940s observations of rod climbing in the coaxial cylinder flow of soap–gasoline colloidal suspensions (*79, 246–248*) led to a new era in shear-flow rheology. Weissenberg (*246–248*) recognized in this behavior (see Fig. 4-B-2) the existence of normal stress effects.[‡] Working with Russel (*194, 195*) and Freeman (*76*), Weissenberg devised instruments to measure these effects, eventually settling on total force measurements in a cone–plate geometry (*76, 95, 248*). This instrument, which could measure two different stress combinations in shear flow, was called a rheogoniometer. The use of goniometer was undoubtedly suggested by Weissenberg's earlier career in x-ray diffraction (*124, 245*).

Shear-flow rheology in the 1950s was dominated by normal stress measure-

---

[†]See the footnotes of Rabinowitsch (*185*) and Eisenschitz (*66*).
[‡]Actually, similar effects had been observed by earlier investigators, such as in Mooney's (*154, 249*) studies of Couette flow of rubber, but were not recognized as being due to normal stresses.

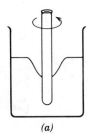 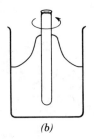

(a)   (b)

FIGURE 4-B-2. Weissenberg effect: (a) Newtonian fluid; (b) fluid exhibiting normal stresses.

ments. Early measurements were reported on Weissenberg's cone–plate total thrust instrument by Pollett and Cross (*181–183*). Later studies on polymer solutions were made by Brodnyan, Gaskins, and Philippoff (*31, 32*). Jobling and Roberts (*106*), and others. Philippoff and his coworkers (*29, 30, 80, 177*) played a very important role in the 1950s in emphasizing the importance of normal stresses and the ability to measure them using different experimental techniques. Radial variations of the normal pressure distribution were initiated by Green-smith and Rivlin (*88*) using torsional flow, and were used by Markovitz and Williamson (*136*) in both cone–plate flow and torsional flow parallel-disc studies. Kotaka, Kurata, and Tamura (*113*) described total thrust measurements in torsional flow. A new generation of normal stress measurements on polymer melts in cone–plate geometries began with the work of King (*111*) in the late 1960s. Since that time, numerous papers have appeared, largely using total thrust measurements in the cone–plate and torsional flow geometries.

In a series of papers appearing in 1968, it was shown by Broadbent, Kaye, Lodge, and Vale (*29, 109*) that the Greensmith–Rivlin–Markovitz procedure (*88, 132, 133, 136*) of measurement of normal stresses, using capillary pressure rise to determine normal pressure variations, was subject to errors associated with secondary flows developed by the presence of the holes. This has led to Weissenberg's total thrust measurements becoming the preferred experimental procedure for normal stress measurements. Miller and Christiansen (*151*) have shown that flush-mounted transducers may be used in place of capillary pressure rise to give correct measurements.

There have been a long series of efforts, beginning with Philippoff and Gaskins (*80, 177*) in 1958–1959, to develop methods to determine normal stresses using extrusion devices. The papers of Metzner et al. (*144, 203*), Sakiadis (*198*), Han (*90–93*) and Higashitani and Lodge (*98*) may be cited. However, none of these methods has gained broad acceptance.

3. Drag-Induced Rectilinear Shear Flow

a. Sandwich Viscometer

The simplest viscometers used for polymer melts are the parallel plate and sandwich instruments described by Zakharenko et al. (*260*), Middleman (*147*),

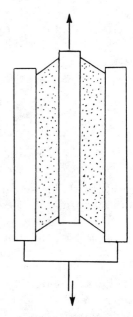

FIGURE 4-B-3. Parallel plate sandwich viscometer.

and later authors (*53, 78, 85, 223*) (Fig. 4-B-3). In this instrument a very viscous fluid, generally an elastomer, is sheared between two parallel steel plates or in a sandwich between three plates.

$$\dot{\gamma} = \frac{V}{H} \tag{4-B-17}$$

where V is the velocity of the moving member and H the interplate distance.

If the moving member is programmed at a constant velocity, a constant shear rate is obtained. The applied force F may be determined as a function of time. This yields, after normalization with area A, the shear stress as a function of time, that is,

$$\sigma_{12} = \frac{F}{A} \quad \text{(parallel plate viscometer)}$$

$$= \frac{F}{2A} \quad \text{(sandwich viscometer)} \tag{4-B-18}$$

Transient start-up shear stresses are determined which, if the material proves stable between the plates, give at long times the steady-state stress. If this is achieved, the viscosity may be determined as a function of shear rate. In many cases, rather large times and apparatus shear strains are required. In general, high machine aspect ratios L/H are needed.

An advantage of this type of instrument is the ability to achieve very low shear rates. This is, of course, also a limitation. It is most useful for materials such as elastomers, which generally exhibit slippage in nonpressurized rotational instruments such as those based on the cone–plate geometry. Generally, to avoid slippage of elastomeric test specimens, the contact surfaces are best knurled and the plates pushed together, pressurizing the system. The pressurization forces the material into the interstices between the knurl. This prevents slippage.

As the experiment proceeds and the steel plates slide past each other, part of the surface of the polymer material becomes uncovered. This probably tends to slip (see the discussion of Toki and White, ref. *223*). We should write for a sandwich rheometer,

$$\sigma_{12} = \frac{F}{A_0 - WVt} \tag{4-B-19}$$

where W is the width of the instrument. Clearly, large instruments where L (equal to A_0/W) is much greater than Vt are desirable.

Very recently Dealy and his coworkers (*53*) have described an instrument, of this type based on a shear stress transducer, which may overcome many of the traditional problems of this instrument. Data on polyethylene melts are presented, including results at shear rates of $50 \sec^{-1}$.

b. Pochettino Viscometer

A viscometer similar in principle, but more infrequently seen, consists of a solid cylinder which is dropped or pushed in a coaxial manner through a viscous fluid in a cylindrical container (Fig. 4-B-4). Such instruments were devised by Pochettino (*179*). A useful instrument of this type for low-viscosity fluids is the falling cylinder viscometer devised by Lohrenz et al. (*127*). Instruments intended for polymer melts are described by Fox and Flory (*74*) and Myers and Faucher (*163*). The shear stress in this instrument at the surface of the inner cylinder is

$$\sigma_{12}(R) = \frac{F}{2\pi RL} \tag{4-B-20}$$

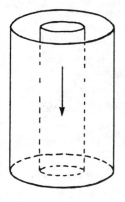

FIGURE 4-B-4. Pochettino viscometer.

where R is the radius and L is the length of the inner cylinder. The shear rate is to a first approximation given by

$$\dot{\gamma} = \frac{V}{h} \tag{4-B-21}$$

where V is the velocity of the moving member and h the clearance. This expression is only adequate for $h \rightarrow 0$. The Pochettino instrument is useful only at low shear rates and Newtonian fluid approximations have usually been applied to estimate $\dot{\gamma}$. This approximation leads to

$$\dot{\gamma} = \left[\frac{dv_1}{dx_2} \right]_{r=R} = \frac{V}{R} \log \left[\frac{R+h}{R} \right] \tag{4-B-22}$$

Ashare et al. (7) have integrated the equations of motion in this geometry for particular non-Newtonian viscosity models and also introduced a differentiation technique to evaluate the cylinder wall shear rate independent of the rheological model. We do not describe any detailed analyses here, as the instrument has not as yet received sufficient attention to justify it. Indeed, although the work of Ashare et al. represents a good start, much remains to be done in analyzing this instrument. An experimental study comparing data obtained on this instrument with known viscosity results for different non-Newtonian fluids is required.

As stated above, this instrument is limited to very low shear rates. There are also inherent problems owing to errors in coaxiality of the two cylinders. This could lead to angular pressure variations and localized slip in materials such as elastomers.

4. Rotational Rheometers

a. Cone–Plate and Biconical Rheometers

i. General Remark   A series of important laboratory and industrial rheological instruments attempt to measure flow properties in a cone and plate or biconical geometry (Fig. 4-B-5). The earliest analysis of flow in this geometry was by Mooney and Ewart (160), who propsed using a biconical region to give a known end effect in a coaxial cylinder instrument. The earliest true instruments for measuring viscosity were described by Piper and Scott (178) and Weissenberg (76, 248). Weissenberg also developed a method for normal stress measurement. It was first applied to polymer melts by Pollett and Cross (181–183) and later by King (111). Bruker and Lodge (33) have described a "rheodilatometer"

[†]An interesting general treatment of rotational shearing flows as related to both viscometry and to practical machines is given by Leonov (123).

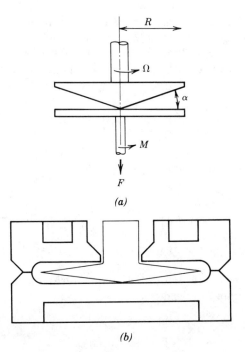

FIGURE 4-B-5. (*a*) Cone–plate viscometer; (*b*) Biconical viscometer.

intended for polymer melts which is based on this geometry. This measures stress-induced volume changes as well as shear stress and normal stress differences. Generally, this is a low-shear-rate instrument being effectively used for polymer melts at shear rates less than $5\,\mathrm{sec}^{-1}$ (because of flow instabilities) and for lower-viscosity fluids at shear rates less than $500\,\mathrm{sec}^{-1}$.

The generally presumed flow pattern in a cone–plate instrument may be expressed in spherical coordinates (*75, 126, 146, 240*) as

$$v_r = v_\theta = 0, \qquad v_\phi = r\omega(\theta)\sin\theta \qquad (4\text{-B-}23)$$

In essence, we have in this equation presumed the angular velocity of the fluid to be independent of spherical coordinate radius, but to vary with the angle that measures distances from the surface of the cone to the plate. This motion is one of laminar shearing with the shear rate from the rate of deformation tensor being

$$d_{\phi\theta} = d_{12} = \tfrac{1}{2}\dot{\gamma} = \tfrac{1}{2}(\sin\theta)\frac{d\omega}{d\theta} \qquad (4\text{-B-}24)$$

For such small cone angles, the apparent shear rate $\dot{\gamma}$ may be written

$$\dot{\gamma} = \sin\theta\frac{d\omega}{d\theta} \sim \sin 90°\frac{\Omega}{\alpha} \sim \frac{\Omega}{\alpha} \qquad (4\text{-B-}25)$$

where we have presumed the fluid to adhere to both surfaces. This constant shear rate leads to constant shear stress throughout the gap.

We may equate the ϕ direction with 1, the θ direction with 2, and the r direction with 3. The equations of motion in spherical coordinates are then as follows:

$$0 = \frac{\partial \sigma_{12}}{\partial \theta} + 2 \cot \theta (\sigma_{12}) \qquad \text{(4-B-26a)}$$

$$- \rho r^2 \omega^2 \sin^2 \theta = - \frac{\partial p}{\partial r} + \frac{2 P_{33} - P_{11} - P_{22}}{r} \qquad \text{(4-B-26b)}$$

$$- \rho r^2 \omega^2 \sin \theta \cos \theta = - \frac{1}{r} \frac{\partial p}{\partial \theta} + \frac{\partial P_{22}}{\partial \theta} + \frac{\cot \theta}{r} (P_{22} - P_{11}) \qquad \text{(4-B-26c)}$$

where we have neglected body forces.

ii. Evaluation of Apparent Viscosity   The equation for the shearing stress may be immediately integrated to yield

$$\sigma_{12} = \frac{C}{\sin^2 \theta} \qquad \text{(4-B-27)}$$

where the constant C is independent of r and ϕ as well as θ. The percentage difference, P, in shear stress between the cone ($\theta = 90° - \alpha$) and plate ($\theta = 90°$) surface is

$$\frac{P}{100} = \frac{\sigma_{12}(90° - \alpha) - \sigma_{12}(90°)}{\sigma_{12}(90°)} = \frac{1}{\sin^2(90 - \alpha)} - 1 = \tan^2 \alpha \qquad \text{(4-B-28)}$$

From this we may construct Table 4-B-1 (3, 75). It may be seen that up to angles as large as 4°, the shear stress only varies by 0.5%.

The torque in this instrument is related to the shear stress by

$$M = \int_0^R 2\pi r \cdot r \sigma_{12} \, dr = \frac{2\pi R^3}{3} (\sigma_{12}) \qquad \text{(4-B-29a)}$$

and

$$\eta = \frac{3M}{2\pi R^3} \frac{\alpha}{\Omega} \qquad \text{(4-B-29b)}$$

Equations (4-B-25) and (4-B-29) provide the basis for obtaining the non-Newtonian viscosity η in this geometry.

The instrument as described above is not pressurized. Turner and Moore (234) have described a pressurized biconical rheometer intended for elastomers. One major purpose was to investigate slippage.

TABLE 4-B-1 Shear Stress Variations in Gap of
Cone–Plate Rheometer and Errors in Presuming Simple
Shear Flow

α	P^a
0.5	0.0076
1	0.0306
2	0.1218
4	0.4886
7	1.5
10	3.1

[a] $\dfrac{P}{100} = \dfrac{\sigma_{12}(90° - \alpha) - \sigma_{12}(90°)}{\sigma_{12}(90°)}.$

iii. Evaluation of Normal Stresses The major feature of the cone–plate geometry is the additional measurement of the upward force on the plate in this
geometry.

$$F = \int_0^R 2\pi r \sigma_{\theta\theta}\, dr = \int_0^R 2\pi r (p - P_{22})\, dr \qquad (4\text{-B-}30)$$

If inertial forces are neglected in Eq. (4-B-26c), it may be integrated to yield

$$p(r) = p(R) + [2P_{33} - P_{11} - P_{22}]\ln \frac{r}{R} \qquad (4\text{-B-}31)$$

where we have used the uniformity of shear rate throughout the gap to presume
the normal stresses independent of radius and neglected centrifugal forces.
Substitution of this expression into Eq. (4-B-30) yields

$$\frac{F}{\pi R^2} = p(R) + (2P_{33} - P_{11} - P_{22})2\int_0^1 x \ln x\, dx - P_{22}$$

$$\frac{2F}{\pi R^2} = P_{11} - P_{22} + p(r) - P_{33}$$

$$= P_{11} - P_{22} - \sigma_{33} \qquad (4\text{-B-}32)$$

If surface tension effects may be neglected, we have, because σ_{33} is atmospheric
pressure, simply

$$N_1 = P_{11} - P_{22} = \frac{2F}{\pi R^2}; \qquad \psi_1 = \frac{2F}{\pi R^2}\left(\frac{\alpha}{\Omega}\right)^2 \qquad (4\text{-B-}33\text{a,b})$$

Equations (4-B-32) and (4-B-33) are due to Weissenberg (76, 248).

Ginn and Metzner (84) and Slattery (204) have carefully analyzed the significance of edge effects, including surface tension and centrifugal forces. Ginn and Metzner in particular argued that Eq. (4-B-33) should be replaced by

$$N_1 = \frac{2}{\pi R^2}\left[F + 2\pi \int_0^R \left[\kappa \frac{1}{r_e} - \frac{1}{r'_e} + \frac{\rho \omega^2 \delta}{2}(R^2 - r^2) \right] r \, dr \right] \qquad (4\text{-B-}34)$$

where κ is the surface tension, r_e, r'_e = fluid–air interface radius in the sheared and unsheared state, and

$$\delta = \left(\frac{\theta - \pi/2}{\alpha} \right)^2 \qquad (4\text{-B-}35)$$

Hutton (103) and Suetsugu and White (210) (compare Lobe and White (125)) have found problems measuring normal stresses in greases and particle-filled polymer melts with yield values by the total thrust technique. The zero of N_1 is uncertain to within an amount dependent upon the yield value. This arises because normal force measurement requires a vertical relative axial movement of the plates. Such movement requires the test liquid to flow rapidly between the platens. The stresses so induced decay not to zero, but to a finite value. This induces uncertainty in N_1 as determined by F.

As noted earlier, experimentalists carried out measurements of normal stresses in the cone–plate and similar geometries using capillary tubes or pressure taps. Such instrumentation leads to systematic "hole" errors (29, 109). Miller and Christiansen (180) have overcome this problem through the use of flush-mounted transducers. They can directly measure in this way $\sigma_{\theta\theta}$ or $\sigma_{22}(r)$, which is from Eq. (4-B-31):

$$\sigma_{22}(r) = \sigma_{22}(R) + (N_1 - 2N_2)\ln\frac{r}{R} \qquad (4\text{-B-}36)$$

Various experimental problems exist with this type of instrument which deserve consideration. Bearing imperfections lead to axial movements of the rotating member and to squeeze-film lubrication effects (3).

The problem of transient buildup at normal stresses at the start-up of the flow and the experimental problems associated with measurement difficulties due to instrument compliance have been considered by Meissner (142), Sakai et al. (196), and Tanner (217).

iv. Secondary Flows   Secondary flows in the biconical geometry arise from two sources. Mooney and Ewart (160) (compare Slattery, ref. 204) pointed out that the equations of motion for laminar shear flow were incompatible with centrifugal forces. It has, in fact, been found in Newtonian fluids that by raising the cone rotation speed, the fluid motion breaks down into secondary flows (149). A second source of flow breakdown in this geometry was found by Oldroyd (169) and Ericksen (69). The point made by these authors is that the independent appearance of normal stress terms in Eqs. (4-B-26b) and (4-B-26c) cannot be simply compensated for by pressure variations and that, in fact, the following additional restriction (readily obtained by substituting Eq. 4-B-31 into Eq. 4-B-26c) holds:

$$\frac{\partial}{\partial \theta}(2P_{33} - P_{11} - P_{22})\ln r = \frac{\partial P_{22}}{\partial \theta} + \cot \theta (P_{22} - P_{11}) \qquad (4\text{-}B\text{-}37)$$

Secondary flows arising from this mechanism were observed by Giesekus (82). Walters and Waters (243) consider the magnitude of the experimental errors these flows cause and the necessary corrections for Eqs. (4-B-25) and (4-B-29). Generally, these may be neglected except for large cone angles.

The limitation of this instrument for Newtonian fluids and polymer solutions is due to centrifugal forces, which throw the fluid out of the gap at high angular velocities.

A second instability has been observed by various investigators using the instrument on polymer melts. As first found by King (111), the melt seems to ball up and emerge from the gap at low shear rates (about $1-10 \sec^{-1}$) (12, 102, 122). Similar observations have been made on polymer solutions. Hutton (102) describes this phenomenon in greases and suggests that this type of breakdown of flow is due to a fracture of the material.

v. Viscous Heating Admirable studies of viscous heating in a cone–plate viscometer have been carried out by Turian and Bird (233) and Middleman (146). These authors note that the energy equation for viscous dissipation heating in steady laminar shear flow between a cone and plate is, in spherical coordinates (compare Table 2-K-1),

$$0 = k\left[\frac{1}{r}\frac{\partial}{\partial r}r^2\frac{\partial T}{\partial r} + \frac{1}{r^2 \sin \theta}\frac{\partial}{\partial \theta}\sin \theta \frac{\partial T}{\partial \theta}\right] + \sigma_{\theta\phi}\left[\frac{1}{r}\frac{\partial v_\theta}{\partial \theta} + \frac{\cot \theta}{r}v_\phi\right] \qquad (4\text{-}B\text{-}38)$$

with boundary conditions

$$\frac{\partial T}{\partial r}(r = R, \theta) = 0 \qquad (4\text{-}B\text{-}39)$$

$$T(r, 90°) = T(r, 90° - \alpha) = T(0, \theta) = T_0$$

The significance of these boundary conditions is that there is not heat loss across the air–liquid interface and cone and plate assembly is maintained at temperature T_0. Turian and Bird (233) devised a variational principle equivalent to Eq. (4-B-38) for small cone angles and obtained an approximate solution for the temperature field in terms of the dissipation function for arbitrary temperature-independent fluid properties. They show that the maximum temperature rise

$T_{max} - T_0$ is

$$T_{max} - T_0 = \frac{3N\Omega\alpha}{16\pi\kappa R} \tag{4-B-40}$$

Turian and Bird also take into consideration the temperature dependence of viscosity and obtain two different approximate simultaneous solutions of the momentum and energy equations for Newtonian and non-Newtonian fluids (239). Turian and Bird have compared their mathematical solutions to data obtained on National Bureau of Standards calibration oils with the Ferranti–Shirley viscometer. They note that viscous heating can lead to errors in viscosity of up to 15%.

However, it is difficult to apply theoretical studies of viscous heating because of the unknown magnitudes of heat-transfer boundary conditions on solid surfaces.

b. Coaxial Cylinder (Couette) Rheometer

i. General Remarks   The coaxial viscometer is one of the most important of rheological characterization instruments (Fig. 4-B-6). It is, of course, one of the oldest viscometers, having been developed and widely used in the late nineteenth century. The coaxial cylinder may involve a rotating inner (bob) or outer cylinder (cup). This instrument was first applied to polymer systems by Mooney (154) in 1936, who reported the first true viscosity–shear rate data obtained on such a material (natural rubber) to appear in the literature. Few investigators have used

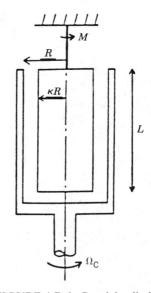

FIGURE 4-B-6. Coaxial cylinder.

this instrument since the time of Mooney. Philippoff and Gaskins (176) reported experimental studies in the mid-1950s using a coaxial cylinder viscometer with polyethylene. More recently, Cogswell (42) has used this type of instrument to study viscosity under conditions of controlled pressure. Many interesting variants on the coaxial cylinder rheometer have been discussed in the literature. Among these are Nguyen and Boger's (167) vane rheometer and methods of evaluating surface shape to determine normal stress (36, 107). These techniques are limited to low-viscosity fluids at room temperature.

Various experimental problems are reported with the use of this instrument. One of the most striking, which is reported by Mooney, is due to the Weissenberg normal stress effect. It must be remembered that the coaxial cylinder instrument possesses the classical geometry which exhibits this effect. Large normal stresses are developed in shear flow between the bob and cup and these give rise to "rod climbing." Mooney found that with natural rubber, the polymer climbed out of the apparatus during shearing (154). This necessitated pressurization at both ends.

Angular flow caused by a rotating cylinder is usually specified in cylindrical coordinates by

$$v_r = v_z = 0, \qquad v_\theta = r\omega(r) \tag{4-B-41}$$

This was first analyzed by Newton (incorrectly, refs. 165 and 187) and Stokes (211), who essentially first solved the Navier–Stokes equations for this problem. A satisfactorily general treatment for evaluating non-Newtonian viscosity was suggested by Mooney (152, 154) in 1931 and later improved by Pawlowski (174) and by Krieger and his coworkers (115–117). Although this instrument was first mentioned as a quantitative method of determining normal stresses in laminar shear flows by Padden and deWitt (173), thus far it has really only yielded results of qualitative value (45, 132, 134).

It may be shown that the fluid motion specified by Eq. (4-B-41) is a laminar shearing flow with

$$\dot{\gamma} = -r\frac{d\omega}{dr} \tag{4-B-42}$$

We may equate the θ direction with 1, the r direction with 2, and the z direction with 3. The equations of motion are

$$-\rho r\omega^2 = \frac{-\partial p}{\partial r} + \frac{\partial P_{22}}{\partial r} + \frac{P_{22} - P_{11}}{r}; \tag{4-B-43a}$$

$$0 = -\frac{\partial p}{\partial z}; \qquad 0 = \frac{\partial \sigma_{12}}{\partial r} + \frac{2\sigma_{12}}{r} \tag{4-B-43b,c}$$

where we have neglected body forces.

ii. Evaluation of Apparent Viscosity   The equation for the shearing stresses may be integrated to yield

$$\sigma_{12} = \frac{C}{r^2} \tag{4-B-44}$$

This variation of the shear stress from the outer (cup) to the inner (bob) cylinder is

$$\frac{P}{100} = \frac{\sigma_{12}(\kappa R) - \sigma_{12}(R)}{\sigma_{12}(R)} = \frac{(1/\kappa R)^2 - (1/R)^2}{(1/R^2)} = \frac{1 - \kappa^2}{\kappa^2} \tag{4-B-45}$$

This is summarized Table 4-B-2.

If the bob radius is 99% of the cup radius, the assumption of a uniform shear stress is a good approximation. However, if the radius is even 90%, the presumption is inadequate. If it is small enough that $\dot{\gamma}$ may be considered uniform, we may evaluate the shear rate by noting that

$$\dot{\gamma}(\kappa R) = -r\frac{d\omega}{dr} \sim \kappa R \frac{\Omega_B - \Omega_C}{R - \kappa R} = \frac{\Omega_B - \Omega_C}{1/\kappa - 1} \tag{4-B-46}$$

where Ω_C is the angular speed of the cup and Ω_B the angular speed of the bob. The bob shear stress is obtained from the torque M:

$$M = 2\pi\kappa RL(\kappa R\sigma_{12}) \tag{4-B-47}$$

$$\sigma_{12}(\kappa R) = \frac{M}{2\pi\kappa^2 R^2 L} \tag{4-B-48}$$

We now turn to the problem of obtaining shear stress–shear rate data for the case in which the uniform stress approximation is not valid. This problem consists of evaluating the shear rate at one of the cylinders, where the shear stress is known from applying Eq. (4-B-48) to torque data. The difference in angular

TABLE 4-B-2   Shear Stress Variations in a Coaxial Cylinder Viscometer

$\kappa\left(\dfrac{\text{Inner Radius}}{\text{Outer Radius}}\right)$	$P = \dfrac{\sigma_{12}(\kappa R) - \sigma_{12}(R)}{\sigma_{12}(R)}$
0.99	2.03
0.98	4.12
0.95	10.8
0.90	23.5

velocities between the outer and inner cylinders is

$$\Omega_C - \Omega_B = \int_{\Omega_B}^{\Omega_C} d\omega = \int_{\kappa R}^{R} \frac{d\omega}{dr} dr = -\int_{\kappa R}^{R} \dot\gamma \, d\ln r \qquad (4\text{-B-}49)$$

from Eq. (4-B-44); it follows that

$$d\ln r = -\tfrac{1}{2} d\ln \sigma_{12} \qquad (4\text{-B-}50a)$$

and

$$\Omega_C - \Omega_B = \tfrac{1}{2} \int_{\sigma_{12}(\kappa R)}^{\sigma_{12}(R)} \dot\gamma \, d\ln \sigma_{12} \qquad (4\text{-B-}50b)$$

Equation (4-B-50b) is due to Mooney (152), who also included slippage in his analysis. This equation forms the basis of modern coaxial cylinder viscometer theory.

For the case of a bob rotating in an infinite fluid, Eq. (4-B-50b) is

$$\Omega_B = \tfrac{1}{2} \int_0^{\sigma_{12}(R_B)} \dot\gamma \, d\ln \sigma_{12} \qquad (4\text{-B-}51)$$

and differentiating with respect to the logarithm of the bob shear stress gives (using the Leibnitz rule for differentiating integrals)

$$\dot\gamma(R_B) = 2 \frac{d\Omega_B}{d\ln \sigma_{12}(R_B)} = 2\left[\frac{d\ln \Omega_B}{d\ln M}\right]\Omega_B \qquad (4\text{-B-}52)$$

This may be seen to be the asymptote opposite to that of Eq. (4-B-46) with $2(d\ln \Omega_b/d\ln M)$ replacing $[1/(1/\kappa) - 1]$ as the coefficient of the bob angle velocity.

We now consider another possible coaxial cylinder viscometer in which the bob is stationary and the cup rotates. We wish to measure the shear rate at the bob where the torque is being simultaneously measured. Equation (4-B-52) is

$$\Omega_C = \tfrac{1}{2} \int_{\sigma_{12}(\kappa R)}^{\kappa^2\sigma_{12}(\kappa R)} \dot\gamma \, d\ln \sigma_{12} \qquad (4\text{-B-}53)$$

where we express shear rate in terms of local shear stress rather than position. We may differentiate Eq. (4-B-53) with respect to $\sigma_{12}(\kappa R)$ to obtain

$$\frac{d\Omega_c}{d\ln \sigma_{12}(\kappa R)} = \left(\frac{d\ln \Omega_c}{d\ln M}\right)\Omega_c = \tfrac{1}{2}\{\dot\gamma[\sigma_{12}(\kappa R)] - \dot\gamma[\kappa^2\sigma_{12}(\kappa R)]\} \qquad (4\text{-B-}54)$$

where the shear is expressed in terms of local shear stress rather than position.

This equation may be inverted by substituting $\kappa^{2s}M$ for M in Eq. (4-B-54) and then adding from n equal to zero to n equal to infinity. We then obtain

$$\dot{\gamma}[\sigma_{12}(\kappa R)] = \Omega_C\left[2\sum_{s=0}^{s=\infty}\frac{d\ln\Omega_C}{d\ln M}(\kappa^{2s}M)\right] \qquad (4\text{-}B\text{-}55)$$

where we have noted that

$$\lim_{s\to\infty}\dot{\gamma}[\kappa^{2s+2}\sigma_{12}(\kappa R)] = 0 \qquad (4\text{-}B\text{-}56)$$

Krieger and Elrod (116), noting that Eq. (4-B-55) is a slowly convergent expansion, show that it may be rewritten

$$\dot{\gamma}[\sigma_{12}(\kappa R)] = \frac{\Omega_C}{\ln(1/\kappa)}\left[1 + \frac{d\ln\Omega_C}{d\ln M}\left(\ln\left(\frac{1}{\kappa}\right)\right) + \frac{d^2\Omega_C}{d(\ln M)^2}\frac{(\ln(1/\kappa))^2}{3\Omega_C}\right.$$
$$\left. - \frac{d^4\Omega_C}{d(\ln M)^4}\frac{(\ln(1/\kappa))^4}{45\Omega_C} + \cdots\right] \qquad (4\text{-}B\text{-}57)$$

It should be pointed out that Eq. (4-B-57) represents a higher-order correction to the narrow-gap asymptote. Krieger and Elrod note that when $(d\ln\Omega_b/d\ln M)\ln(1/\kappa)$ is less than 0.2, the third- and higher-order terms may be neglected.

More recently, Krieger (115) has shown that power-law fluid representations, Eq. (4-B-16) (due to Farrow et al., ref. 72), is a good approximation and may be usefully directly applied to obtain viscosity–shear rate data. For the case of a stationary bob and a rotating cup, Krieger finds and recommends

$$\dot{\gamma}(\kappa R) = \frac{2/n}{1 - \kappa^{2/n}}\Omega_C \qquad (4\text{-}B\text{-}58)$$

Darby (52) has examined the errors in Krieger's formulation for fluids exhibiting yield values in shear flow.

An alternate procedure of obtaining non-Newtonian viscosity data from the coaxial cylinder instrument was given by Krieger and Maron (117). This method, which involves using a series of different bobs, consists of differentiating Eq. (4-B-53) with respect to the radius ratio at constant cup shear stress (torque):

$$-2\left[\frac{\partial\ln\Omega_C}{\partial\ln\kappa}\right]\Omega_C = \dot{\gamma}(\kappa R) \qquad (4\text{-}B\text{-}59)$$

Krieger and Maron (117) describe the use of an approximate two-bob method.

iii. End Effect   There is an unavoidable end-effect problem in this geometry. At the interface of the fluid with the bottom of the apparatus or at both ends if it is doubly capped, there is additional shearing due to the solid capping surfaces. This will be of unknown magnitude and vary with the rheological properties of the fluid. The simplest procedure is to calibrate the instrument with a fluid of known rheological properties and to determine an instrument constant C or C'.

$$\sigma_{12}(\kappa R) = CM = C' \frac{M}{2\pi(\kappa R)^2 L} \qquad (4\text{-B-}60)$$

The problem with this method is as implied above—the instrument constant may vary with the viscosity shear-rate behavior. This effect is usually considered small.

Another procedure is to carry out a series of experiments involving bobs of varying length and subtract out the end effect. If we write Eq. (4-B-60)

$$M = [2\pi(\kappa R)^2 \sigma_{12}(\kappa R)] L + M_{ends} \qquad (4\text{-B-}61)$$

$\sigma_{12}(\kappa R)$ may be determined from the slope of a plot of M versus L.

Mooney and Ewart (*160*) have described an alternate approach to end effects in this geometry. These authors have built an instrument in which the coaxial cylinders culminate in a biconical section. This gives an end correction region of almost constant shear stress and shear rate. A suitable choice of angle can lead to a shear rate equivalent to that in the gap and an end correction equivalent to adding a ΔL to L, which may be calculated from the analysis.

iv. Evaluation of Normal Stresses   With the exception of researchers at the Mellon Institute, few have given attention to normal stress measurement in this geometry. Padden and deWitt (*173*) first described an instrument to make measurements in 1954 and Markovitz (*132, 134*) discussed it in later publications. Only very limited quantitative data on moderately concentrated polymer solutions has been reported. The basis of normal stress measurement in this instrument is to evaluate the difference in radial gauges or capillary tubes. By measuring this difference, we have

$$\Delta P = \sigma_{22}(\kappa R) - \sigma_{22}(R) = \int_R^{\kappa R} (P_{11} - P_{22}) d\ln r - \int_{\kappa R}^R \rho r \omega^2 \, d\ln r \quad (4\text{-B-}62)$$

and an averaged value of the normal stress difference N_1 may be obtained. Equation (4-B-62) can be rewritten by means of Eq. (4-B-50a) and differentiated with respect to the shear stress or torque at the rotating cylinder to yield (neglecting inertia)

$$\frac{d\Delta P}{d\ln M} = [P_{11} - P_{22}](R) - [P_{11} - P_{22}](\kappa R) \qquad (4\text{-B-}63)$$

which is similar to Eq. (4-B-52). This expression may be inverted as done with the angular velocity–shear rate equation by Krieger and Elrod. We refer to a paper by Markovitz (134) for the details.

v. Breakdown of Flow   The breakdown of laminar shear flow in a coaxial cylinder instrument with a rotating inner cylinder is one of the classic problems of Newtonian fluid dynamics and is closely connected with the name of G. I. Taylor (219). In 1923 Taylor published a stability analysis of such flows and found that essentially at a critical Reynolds number (the form of which is now called a Taylor number N_{Ta}), the laminar shear motion is caused by the centrifugal forces to break down into secondary flows (Taylor vortices). Increasing the rotation speed of the bob eventually makes the fluid motion turbulent. An extensive review of hydrodynamic stability theory of Newtonian fluids in this geometry, together with some discussion of experiments, is given in the treatise by Chandrasekhar (37). For narrow gaps with rotating bobs, the Taylor number criterion is

$$N_{Ta} = \frac{4R^4(1-\kappa)^4\Omega_B 2\rho^2}{\kappa^3}\left[\frac{\kappa^2}{1-\kappa^2}\right] = 3390 \qquad \text{(4-B-64a)}$$

whereas for wide gaps it is

$$N_{Ta} = \frac{64}{9}\frac{R^4\kappa^4\Omega_B 4\rho^2}{\eta^2} = 33,100 \qquad \text{(4-B-64b)}$$

The cases of the outer cylinder rotating and both cylinders rotating have also been considered. Rotation of the outer cylinder in the same direction as the inner cylinder is stabilizing.

Thomas and Walters (221) and later investigators (83, 148) studied this same hydrodynamic stability problem for slightly viscoelastic fluids, that is, dilute polymer solutions. They find that both the presence of non-Newtonian viscosity and elasticity in the liquid lowers the critical Taylor number at which the instability occurs.

The breakdown of flow of a polymer melt in a Couette viscometer has been described by Cogswell (42) for a high-density polyethylene. There is a sharp drop in torque apparent at a critical shear stress. This may well correspond to the onset of slippage.

vi. Viscous Heating   Studies of viscous heating in coaxial cylinder viscometers are given by Frederickson (75) and Middleman (146). These authors note that the energy equation in steady laminar flow in this geometry is of the form

$$0 = k\left[\frac{1}{r}\frac{\partial}{\partial r}\left(r\frac{\partial T}{\partial r}\right)\right] + \sigma_{12}r\frac{d\omega}{dr} \qquad \text{(4-B-65)}$$

with the following boundary conditions

$$T(R) = T_0 \qquad \text{(4-B-66a)}$$

and

$$T(\kappa R) = T_0 \quad \text{or} \quad T_0 + \Delta T$$

or

$$\frac{\partial T}{\partial r}(\kappa R) = 0 \qquad \text{(4-B-66b)}$$

The significance of the first set of boundary conditions is that the coaxial cylinder assembly is maintained at constant temperature; the significance of the second set is that there will be some finite temperature buildup at the inner cylinder.

The maximum temperature rise (at the inner cylinder) at the steady state for an adiabatic bob is for a power-law fluid (Eq. 4-B-16):

$$T_{\text{max}} - T_0 = \frac{n^2 K^{2-n}}{4}\left[\frac{2/n}{1-\kappa^{2/n}}\right]^n \frac{R^{1-n}\kappa(R\Omega)^n}{k}\left[1 - \kappa^{-2/n} - \frac{2}{n}\kappa^{-2/n}\ln\kappa\right] \qquad \text{(4-B-67)}$$

where we neglect the influence of temperature on K and n.

c. Torsional and Disc Rheometers

i. General Remarks   It is possible to measure viscosity in torsional flow between a stationary and a rotating disc. Instruments of this type have been reported in the literature for more than a century. The first such instrument to be used on bulk polymers was Mooney's 1934 "shearing disc" viscometer (*55, 153, 155–158, 164, 220, 253*) (Fig. 4-B-7). In this instrument a serrated disc is rotated in a sample fixed in a pressurized cavity. It was developed for application to rubber and its compounds and both pressurization and serrations were meant to prevent slippage. This has become the standard quality-control instrument of the rubber industry. In a separate development, Russell (*194, 195*), Greensmith and Rivlin (*88*) and notably Kotaka, Kurata, and Tamura (*113*) used unpressured torsional flow to measure normal stresses in polymer solutions. Beginning in the late 1960s, various investigators (*23, 122, 197*) used unpressurized smooth parallel discs to determine viscosity and normal stresses in shear flow of molten plastics. Traditionally, the parallel disc instruments have been limited to low shear rates— less than $5\,\text{sec}^{-1}$ for polymer melts, because of flow instabilities, and less than $500\,\text{sec}^{-1}$ for lower-viscosity fluids. More recently, Connelly and Greener (*47*) described one instrument which can achieve $\dot{\gamma}$ of $50{,}000\,\text{sec}^{-1}$ for low-viscosity fluids. Very small gaps, $\sim 50\,\mu\text{m}$, and large disc radii, $\sim 2.5\,\text{cm}$ are used.

The velocity field in torsional flow between parallel discs (see Fig. 4-B-7) may be expressed in cylindrical coordinates as

$$v_r = v_z = 0, \qquad v_\theta = rf(z) \qquad \text{(4-B-68)}$$

This is a laminar shear flow in which

$$\dot{\gamma} = \frac{dv_\theta}{dz} = r\frac{df}{dz} \qquad \text{(4-B-69)}$$

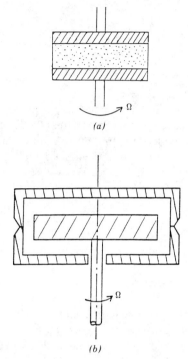

FIGURE 4-B-7. (*a*) Torsional flow rheometer; (*b*) Mooney viscometer.

We may equate the θ direction with 1, the z direction with 2, and the r direction with 3. The equations of motion are

$$- \rho r \omega^2 = - \frac{\partial p}{\partial r} + \frac{\partial P_{33}}{\partial r} + \frac{P_{33} - P_{11}}{r} \tag{4-B-70a}$$

$$0 = - \frac{\partial p}{\partial z} + \frac{\partial P_{22}}{\partial z}, \qquad 0 = \frac{\partial \sigma_{12}}{\partial z} \tag{4-B-70b,c}$$

where we have neglected body forces.

Flow in a disc viscometer is, after Mooney (*153, 156, 157*), generally divided into three regions. The first region, I, consists of the area between the upper and lower surfaces of the disc and the top and bottom of the cavity. Here the flow may be considered torsional. In the second region, II, between the sides of the disc and the walls of the cavity, the flow is approximately Couette flow. The third region, III, which is in the neighborhood of sharp corners, is more difficult to analyze.

ii. Evaluation of Apparent Viscosity The equation for the shearing stresses in the torsional flow region may be integrated to yield

$$\sigma_{12} = C(r) \tag{4-B-71}$$

Because the shear stress depends only upon radius and is independent of z, the shear rate $\dot{\gamma}$ must be

$$\dot{\gamma} = \frac{r\Omega}{h}; \quad v_\theta = \dot{\gamma}\left[1 - \frac{z}{h}\right] \tag{4-B-72}$$

Now the basic problem of torsional flow viscometry differs from that in the coaxial cylinder instrument. In the former, the bob shear stress is obtained from torque data and some method must be devised for evaluating the bob shear rate. Here the shear rates are known as a function of position and we must determine the shear stress at some particular radius. To accomplish this, we relate the shear stress $\sigma_{\theta z}$ to torque M. The torque M is

$$M = \int_0^R 2\pi r^2 \sigma_{\theta z}\, dr \tag{4-B-73}$$

Elimination of the radius r through Eq. (4-B-72) yields

$$M = 2\pi\left[\left(\frac{h}{\Omega}\right)^3 \int^{\dot{\gamma}R} \dot{\gamma}^2 \sigma_{\theta z}\, d\dot{\gamma}\right] \tag{4-B-74}$$

Differentiation of this equation with respect to $\dot{\gamma}_R$ allows us to show that shear stress at the outer radius is

$$\sigma_{12}(R) = \frac{2M}{\pi R^3}\left(\frac{N+3}{4}\right) \tag{4-B-75a}$$

where

$$N = \frac{d \ln M}{d \ln \Omega} \tag{4-B-75b}$$

In Mooney's shearing disc viscometer, the shear stress at the outer radius of the disc may be expressed by

$$M = \int_0^R 2\pi r^2 \sigma_{\theta z}\, dr + 2\pi R^2 \Delta\sigma_{\theta r}(R), \quad \sigma_{\theta z}(R) = \frac{M}{\pi R^3} F(N) \tag{4-B-76}$$

where f represents the contributions of the outer periphery. The magnitude of this contribution has been the concern of various investigators (156, 157, 253), most recently and notably Nakajima and Harrel (164), who found using a power law approximation [i.e., Eq. (4-B-16)] that

$$F = \frac{N+3}{4\left[1 + \frac{(2/N)^N(N+3)h^N}{R^{N+1}[1 - [R/R + \Delta]^{2/N}]^N}\right]} \tag{4-B-77}$$

Here again, N is to be interpreted using Eq. (4-B-75).

iii. Evaluation of Normal Stresses   We now discuss the use of a torsional shearing-flow instrument for normal stress measurements. The best procedure for accomplishing such measurements would seem to be the upward force measurement technique devised by Kotaka et al. (*113*).

From Eq. (4-B-68), the pressure distribution may be evaluated to give (where we have neglected inertia)

$$P(r) = -\sigma_{33}(R) + P_{33}(r) + \int_R^r (P_{33} - P_{11}) \, d \ln r \tag{4-B-78}$$

and the upward force F is

$$F = -\int_0^R 2\pi r \sigma_{22} \, dr = -\int_0^R 2\pi r \left[P_{22} - P_{33} + \int_R^r (P_{11} - P_{33}) \, d \ln r\right] dr \tag{4-B-79}$$

where we have noted that the stress $\sigma_{33}(R)$ is equal to the air pressure if the surface tension is neglected. Eliminating r in terms of $\dot\gamma$ and reversing the order of integration of the second integral of Eq. (4-B-79) leads to

$$\frac{F}{\pi R^2} = \frac{1}{\dot\gamma(R)^2} \int_0^{\dot\gamma(R)} \dot\gamma(P_{11} - 2P_{22} + P_{33}) \, d\dot\gamma \tag{4-B-80}$$

Differentiating with respect to $\dot\gamma_R$ and rearranging gives the final Kotaka–Kurata–Tamura equation

$$P_{11} - 2P_{22} + P_{33} = \frac{2F}{\pi(R)^2}\left[1 + \frac{1}{2}\frac{d \ln F}{d \ln \Omega}\right]$$

$$= N_1 - N_2 = (\Psi_1 - \Psi_2)\dot\gamma(R)^2 \tag{4-B-81}$$

where $\dot\gamma(R)$ is $R\Omega/h$.

Analysis of the effects of inertia and surface tension on this geometry have been given by Slattery (*204*) and Ginn and Metzner (*84*). The latter authors replace F with F', given by

$$F' = (\Psi_1 - \Psi_2)\dot\gamma(R)^2 + 2\pi \int_0^R \left[K\left[\frac{1}{r_e} - \frac{1}{r'_e}\right] + \frac{\rho\omega^2\delta(R^2 - r^2)}{2} r \, dr\right] \tag{4-B-82}$$

Here r_e and r'_e denote the radii of curvature of the circumferential fluid–air interface when the fluid is being sheared or is at rest and K is the surface tension. Total force measurements in the cone and plate and parallel disc measurements may be used to separately determine Ψ_1 and Ψ_2. Binding et al. (18) have described a "torsional balance" rheometer for measurement of normal stresses. The applied vertical load is fixed and superposes a squeeze flow on the torsional motion. They argue that this instrument allows for measurement of normal stresses in greases with yield values without the uncertainty existing in fixed-plate rheometers. White and Tokita (253) have discussed normal stresses in disc viscometers.

More recent studies by Turner and his coworkers (101, 234) and in our laboratories (151) have extended these observations by developing pressurized disc viscometers. A pressurized reservoir above the cavity contains material maintained in contact with polymer in the cavity. The effect of pressure on shear viscosity may be studied.

iv. Secondary flows   The breakdown of flow in the parallel disc geometry for polymer melts is the same as that found for the cone–plate instrument. For Newtonian fluids and polymer solutions, the limitation is fluid being thrown out of the cap by centrifugal forces. For polymer melts, greases, and other complex fluids the low shear-rate failure at the outer radius described by King (111), Hutton (102), and later investigators occurred.

5. Extrusion Rheometers

a. Capillary Instrument

i. General Remarks   The most important industrial and laboratory instrument for measuring rheological properties is the capillary tube (Fig. 4-B-8). Its use for polymer systems dates to the 1930s (60). The basic theory of viscosity measurement using this instrument is due to Weissenberg (68, 185).

The fully developed velocity field in a capillary tube is (in cylindrical coordinates)

$$v_1 = v_1(r), \qquad v_r = v_\theta = 0 \tag{4-B-83}$$

This is a laminar shear flow in which

$$\dot{\gamma} = -\frac{dv_1}{dr} \tag{4-B-84}$$

We may equate the z direction with 1, the r direction with 2, and the θ direction with 3. The equations of motion are

$$0 = -\frac{\partial p}{\partial x} + \frac{1}{r}\frac{\partial}{\partial r}(r\sigma_{12}), \qquad 0 = -\frac{\partial p}{\partial r} + \frac{\partial P_{22}}{\partial r} + \frac{P_{22} - P_{33}}{r}, \qquad 0 = -\frac{\partial p}{\partial \theta}$$

$$\tag{4-B-85a–c}$$

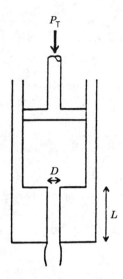

FIGURE 4-B-8. Capillary rheometer.

ii. Evaluation of Apparent Viscosity   Equation (4-B-85b) for the shearing stress may be integrated to yield

$$\sigma_{12} = \frac{r}{2}\left[\frac{\partial p}{\partial x}\right] \quad \text{and} \quad \sigma_{12}(R) = \frac{D}{4}\frac{\partial p}{\partial x} \tag{4-B-86}$$

The shearing stress thus varies linearly with radius across the tube cross section. If $\partial p/\partial x$ is known, then so is $\sigma_{12}(R)$.

The flow rate Q through the capillary may be written

$$Q = \int_0^R 2\pi r v_1 \, dr \tag{4-B-87}$$

This equation may be integrated by parts to yield

$$Q = \int_0^R \pi r^2\left[-\frac{dv_1}{dr}\right]dr + \pi R^2 v_s \tag{4-B-88}$$

where v_s is the slip velocity along the capillary wall.

By means of Eq. (4-B-86), which may be expressed

$$\frac{\sigma_{12}(r)}{\sigma_{12}(R)} = \frac{\sigma_{12}(r)}{(\sigma_{12})_w} = \frac{r}{R} \tag{4-B-89}$$

the flow rate may be rewritten as

$$\frac{8Q}{\pi D^3} = \frac{1}{(\sigma_{12})_w^3}\int_0^{(\sigma_{12})_w} (\sigma_{12})^2\left[-\frac{du}{dr}\right]d\sigma_{12} + \frac{2v_s}{D} \tag{4-B-90}$$

The wall shear rate is obtained by differentiating with respect to $(\sigma_{12})_w$ to yield, where we take v_s to be zero,

$$\left(-\frac{du}{dr}\right)_w = \dot{\gamma}_w[(\sigma_{12})_w] = \frac{3n'+1}{4n'}\left(\frac{32Q}{\pi D^3}\right); \qquad n' = \frac{d\ln(\sigma_{12})_w}{d\ln[32Q/\pi D^3]} \qquad \text{(4-B-91a, b)}$$

Equation (4-B-91) is due to Weissenberg (66, 68, 185). [See the footnotes in Eisenschitz (66) and Rabinowitsch (185)].

iii. End Corrections A major problem that arises in polymer melts is that the pressure drop usually measured is the difference between the pressure applied to force the fluid into the capillary from a reservoir and atmospheric pressure. All this pressure loss is not due to overcoming frictional drag at the capillary walls. As first pointed out by Mooney (155, 159) and later more clearly by Bagley (9), a significant part of the pressure loss is due to pushing the melt from the reservoir into the die.

One method of eliminating these end effects is to use a series of capillaries of constant diameter but varying length, and to plot total applied pressure (for a specific rate) versus capillary length–diameter ratio. From Eq. (4-B-89)

$$p_T = 4(\sigma_{12})_w\left(\frac{L}{D}\right) + \Delta p_e \qquad \text{(4-B-92)}$$

This method is due to Bagley (9). Because flow rate Q and diameter D are the same for all tubes, there will be identical tube-wall shear rates $\dot{\gamma}_w$ and shear stress in the region of fully developed flow. The slope of the data will yield the tube-wall shear stress. The intercept of this straight line with the p_T axis yields the end effect. In flowing polymer melts, kinetic energy is negligible in comparison with elastic energy and the large end effects observed in $p_T - L/D$ plots are largely due to the viscoelastic behavior of the flowing melt. This method is now widely applied (9, 12, 177) to correct extrusion pressure data in capillary rheometers.

An alternate procedure is to place flush-mounted transducers along the length of the capillary and to measure the differential pressure gradient. This procedure has been applied notably by Han and his coworkers (90–92).

For low-viscosity purely viscous fluids, considerable progress has been made in theoretically evaluating end corrections and lengths required for fully developed flow. For rapid flows, the major contribution to the corrections is usually kinetic energy. The complete problem is now a classic one in the Navier–Stokes theory of Newtonian fluids and reviews have been published by Goldstein (85). Various investigators (27, 225) have obtained solutions of non-Newtonian fluids by different procedures. They express the pressure loss in a form equivalent to

$$p_T = 4(\sigma_{12})_w\frac{L}{D} + c(n)\frac{\rho V^2}{2} \qquad \text{(4-B-93)}$$

where n is the power exponent of Eq. (4-B-16), and $c(n)$ varies from 1.3 to 0.7 as n decreases from 1.0 to 0.4.

There is also a literature for creeping (slow) flow of Newtonian and viscous non-Newtonian fluids. This is generally expressed in the form

$$p_T = 4(\sigma_{12})_w \frac{L + \delta L}{D} = 4(\sigma_{12})_w \frac{L}{D} + 4m(\sigma_{12})_w \qquad (4\text{-}B\text{-}94)$$

The parameter m is called the Couette correction. For Newtonian fluids, m is about 0.57. This has been found both experimentally (63) and theoretically (166, 193, 244). For viscous non-Newtonian fluids, Eq. (4-B-94) is again found to be valid. However, the value of m is significantly increased (26, 64, 217). This problem has not been successfully solved for viscoelastic fluid models.

Philippoff and Gaskins (177) have sought to relate Δp_e of Eq. (4-B-92) to the principal normal stress difference, N_1, through heuristic energetic arguments. White and Kondo (251) report an empirical relationship of this type for low-density polyethylene melts.

iv. Normal Stress Measurements Methods for measuring normal stresses in the capillary flow of polymer melts date to the work of Philippoff and Gaskins (80, 177), who suggested both the use of ends pressure-loss correction and extrudate swell. The studies of Metzner et al. (144, 203), Sakiadis (198), Funatsu and Mori (77), Tomita and Kobayoshi (225), Han (90–93), and Davies et al. (51) followed. Exit pressure determinations coupled with thrust measurements emerged as the better experimental approach.

The axial normal stress in a tube may be expressed

$$\sigma_{11}(r, z) = -p(r, z) + P_{11}(r, z), \qquad \sigma_{11}(r, z) = \sigma_{22}(r, z) + N_1(r) \quad (4\text{-}B\text{-}95a, b)$$

and at the exit of the tube at the capillary wall

$$\sigma_{11}(R, L) = \sigma_{22}(R, L) + N_{1w} \qquad (4\text{-}B\text{-}96)$$

where $\sigma_{22}(r, z)$ is the radial outward stress measured. More generally, using Eq. (4-B-85b)

$$\sigma_{22}(r, z) = -p(0, z) + \int_0^r N_2 d\ln r \qquad (4\text{-}B\text{-}97a)$$

$$\sigma_{11}(r, z) = -p(0, z) + N_1(r) + \int_0^r N_2 d\ln r \qquad (4\text{-}B\text{-}97b)$$

where N_2 is $P_{22} - P_{33}$

The total thrust F at the capillary exit is

$$F = \int_0^r 2\pi r \rho u^2 \, dr - \int_0^r 2\pi r \sigma_{11}(r, L) \, dr = \int_0^r 2\pi r \rho u^2 \, dr - F_\sigma \qquad (4\text{-}B\text{-}98)$$

where the first integral reflects a momentum flux. The normal stress contribution may be rewritten

$$F_\sigma = \int_0^R 2\pi r \sigma_{11}(r, L) \, dr = \pi R^2 \sigma_{22}(R, L) + \int_0^R \pi r (2N_1 + N_2) \, dr \qquad (4\text{-}B\text{-}99)$$

which may be differentiated to yield

$$[2N_1 + N_2]_w = \frac{1}{\pi R^2 (\sigma_{12})_w} \frac{d}{d(\sigma_{12})_w} [(\sigma_{12})_w^2 \{F_\sigma - \pi R^2 \sigma_{22}(R, L)\}] \qquad (4\text{-}B\text{-}100)$$

For polymer melts, F_σ may generally be neglected. It is not possible to develop an expression for N_1 or N_2 directly with the capillary instrument alone.

The major difficulty with using the arguments of this section is that they appear to be limited by the nature of the velocity profile at the die exit. It has been established in recent years that the flow rearranges near the exit of the capillary, most notably in low Reynolds number flows as occur in polymer melts. This argument is detailed by Davies, et al. (50) and by Boger and Denn (25). The magnitude of the error is not known.

v. Breakdown of Flow   The classical study of breakdown of flow in the capillary geometry is Reynolds's (189) study of turbulence for Newtonian fluids which occurs at a critical value of the Reynolds number

$$N_{Re} = \frac{D v \rho}{\eta} = 2000 \qquad (4\text{-}B\text{-}101)$$

Similar results appear to exist for polymer solutions and slurries (62), although there is evidence that the presence of the polymer dampens the turbulence, perhaps through the presence of viscoelasticity.

Unlike Newtonian fluids, slurries, and polymer solutions, high-molecular-weight polymer melts do not exhibit flow breakdown due to Reynolds turbulence, but exhibit rather a phenomenon widely known as "melt fracture," or perhaps better as "extrudate distortion." This effect, which consists of the polymer fluid exiting from the capillary and becoming rough and distorted with increase in extrusion rate, has been known in the rubber industry since the nineteenth century. However, the first detailed and quantitative investigation of this problem (which was based upon polystyrene) was presented by Spencer and Dillon (207) at the 1948 Society of Rheology Meeting. These researchers were able to show that the extrudate roughness phenomenon occurred at a critical value of the apparent wall shear stress (10^6–10^7 dyn/cm^2) that varied from melt to melt. Spencer and Dillon obtained a universal criterion available for all polystyrene melts by noting that the product of the critical wall shear stress and molecular weight was always the same. Spencer and Dillon conjecture that the extrudate roughness effect was a buckling phenomenon caused by uneven randomization of the polymer molecules in the exiting melt due to differences between the oriented outer layers and the random core. The next significant contribution was a series of papers by Tordella (227–229) in 1956–1958, showing that the Spencer–Dillon critical wall–shear stress criterion was valid for several melts in addition to polystyrene. Tordella also argued that the phenomenon does not originate in the exiting polymer but in the die entrance, and it was he who introduced the term melt fracture. This entrance-region mechanism has been supported by flow visualization studies by Tordella himself (228), Bagley and Birks (10), Ballenger and White (13), and others. Later studies by Benbow and Lamb (16) and Tordella (230) have suggested that extrudate distortion is sometimes initiated by slippage at the capillary wall and is not necessarily associated with the entry into the die. Critical if dated reviews of experimental studies on various polymer melts have been given by the author elsewhere (17, 250). The phenomenon is observed in polymer solutions as well as melts.

vi. Viscous Heating   Viscous heating is important at high shear rates.

The classic study of viscous heating in capillary tube flows was given in 1951 by Brinkman (28). The basic paper for application to non-Newtonian fluids was published by Bird (21) in 1955 and later contributions have been made by Toor (226), Gerrard and his colleagues (81), and Cox and Macosko (49). A useful critical discussion is given by Middleman (146).

The energy equation for laminar shear flow in a capillary takes the form

$$\rho c v_1 \frac{\partial T}{\partial x_1} = k\left[\frac{1}{r}\frac{\partial}{\partial r}\left(r\frac{\partial T}{\partial r} \right) + \frac{\partial^2 T}{\partial x_1^2} \right] + \sigma_{12}\frac{\partial v_1}{\partial x_2} \qquad (4\text{-B-}102)$$

which is somewhat more complicated than the expression for rotational viscometry because of the convective term. The term $\partial^2 T/\partial x_1^2$ may always be neglected because axial heat conduction is small in comparison to convection. The boundary conditions on this equation are usually taken as

$$T(0, r) = T_0, \qquad \frac{\partial T}{\partial r}(x_1, 0) = 0 \qquad (4\text{-B-}103)$$

with

$$T(x_1, R) = T_0 \qquad \text{or} \qquad \frac{\partial T}{\partial r}(x_1, R) = 0$$

Usually the velocity field in non-Newtonian fluids is determined from the power law as a first approximation. The solution of this boundary value problem and comparison with experimental data on polymer melts is given by Bird (21). Such solutions indicate that for short residence times, as occur in a capillary die, the maximum temperature rises are in the neighborhood of the die wall. Toor (*226*) modifies Bird's results to include density changes with pressure in the polymer melts as they flow along the pressure gradient.

An approximate solution more widely discussed in the German literature may be obtained by noting that for adiabatic flow the total rate of doing work on a system is ΔpQ. Thus

$$\underset{\text{heat generation}}{\rho c Q \Delta T} \quad = \quad \underset{\text{power input}}{\Delta p Q} \qquad \text{(4-B-104a, b)}$$

or

$$\Delta T = \frac{\Delta p}{\rho c}$$

This could give a cross-sectional average temperature rise. Cox and Macosko (*49*) find this still underestimates the maximum temperature rise at the surface of extrudates.

Generally, quite large temperature rises can occur at high shear rates with high-viscosity fluids such as polymer melts, as indicated by both calculation and experiment. This acts to limit the reliability of capillary measurements in the high-shear-rate range.

b. Slit Rheometer

i. General Remarks   The slit rheometer (Fig. 4-B-9) is a two-dimensional plate version of the capillary instrument just described. If a slit has thickness H and with W, we must limit ourselves to geometries for which $W > H$. The instrument has received much less attention in the literature than the capillary geometry. Measurement of viscosity in this geometry was proposed later by Mooney and Black (*159*), White and Metzner (*252*), Han (*90–93*), and Davies et al. (*51*). The instrument was developed as a practical rheological tool for polymer melts through the efforts of Han.

The fully developed velocity field is of the form

$$v_1 = v_1(x_2), \qquad v_2 = v_3 = 0 \qquad \text{(4-B-105)}$$

This is a fully developed flow for which

$$\dot{\gamma} = -\frac{dv_1}{dx_2} \qquad \text{(4-B-106)}$$

The equation of motions are

$$0 = -\frac{\partial p}{\partial x_1} + \frac{\partial \sigma_{12}}{\partial x_2}, \qquad 0 = -\frac{\partial p}{\partial x_2} + \frac{\partial P_{22}}{\partial x_2}, \qquad 0 = -\frac{\partial p}{\partial x_3} \qquad \text{(4-B-107a–c)}$$

ii. Evaluation of Viscosity   If the slit has thickness H, the wall shear stress is from Eq. (4-B-107a)

$$(\sigma_{12})_w = 2H\left(\frac{\partial p}{\partial x_1}\right) \tag{4-B-108}$$

We may derive, analogous to Eq. (4-B-91), that

$$\dot{\gamma}_w = \left(\frac{2n'' + 1}{3n''}\right)\frac{3v}{H} \tag{4-B-109a}$$

where

$$n'' = \frac{d\log(\sigma_{12})_w}{d\log 3v/H} \tag{4-B-109b}$$

and v is the mean linear velocity that is, Q/WH. This result is in essence from Mooney and Black (*159*).

iii. End Corrections   End effects for pressure may be evaluated in a manner analogous to that described for capillary flow. Either slits of varying length or flush-mounted tranducers may be used and $(\partial p/\partial x_1)$ measured directly. The latter procedure is convenient here and was used by Han (*90–92*) and others.

iv. Normal Stress Measurements   Davies et al. (*51*) and Han (*91*) have shown how "exit pressure" measurements may be used in a manner similar to the previously cited capillary studies to determine normal stresses. We refer the reader to the original papers for details. Essentially, these authors use arguments similar to those leading to Eq. (4-B-100):

$$N_{1,w} = -\sigma_{22}(H,L) - \frac{d\sigma_{22}(H,L)}{d\ln(\sigma_{12})_w} + \frac{1}{2WH}\frac{d[[(\sigma_{12})]_wF]}{d(\sigma_{12})_w} \tag{4-B-110}$$

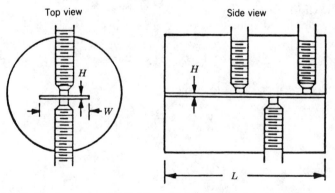

FIGURE 4-B-9. Slit rheometer (C. D. Han, *Rheology in Polymer Processing*, Academic, New York, 1976) (with permission).

where W is the width of the slit and F the thrust. For melts, F may be neglected. This gives a measure of N_{1w} directly. Han (*91, 92*) has described an instrument for carrying out such measurements.

The major problem with the use of this procedure is as noted with the capillary, the rearrangement of the velocity field near the exit of the die (*25, 52*). The magnitude of the errors involved are not known.

White and Metzner (*252*) have discussed jet expansion in this geometry and its relationship to normal stresses. This approach is limited to the high Reynolds number region.

c. Annular and Trough Rheometers

i. General Remarks   Tanner and his coworkers (*96, 110, 118, 215–217*) have given their attention to the development of rheometers which have the single purpose of measuring the second normal stress difference, N_2. These are flow in an annulus (*96, 215*) (Fig. 4-B-10) and in a trough (*110, 118, 216*) (Fig. 4-B-11). Both instruments are based on the radial imbalance of normal stresses induced in shear flow. In these geometries, the radial and angular components of Cauchy's equation of motion, including gravitation forces, is (compare Eq. 4-B-85b)

$$0 = -\frac{\partial p}{\partial r} + \frac{\partial P_{22}}{\partial r} + \frac{P_{22} - P_{33}}{r} - \rho g \cos \alpha \sin \theta \qquad \text{(4-B-111a)}$$

$$0 = -\frac{1}{r}\frac{\partial p}{\partial \theta} + \rho g \cos \alpha \sin \theta \qquad \text{(4-B-111b)}$$

where α is the angle with the horizontal.

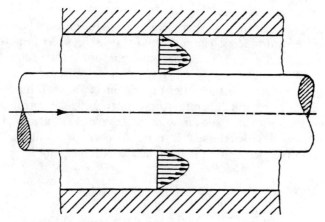

FIGURE 4-B-10. Axial flow in an annular rheometer.

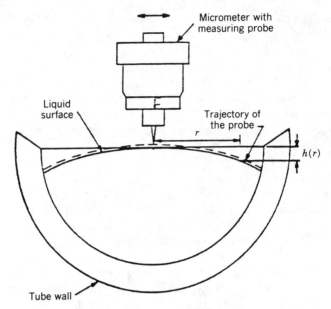

FIGURE 4-B-11. Trough rheometer.

ii. Annular Rheometer   The radial difference in σ_{22} across the annulus, that is, at κR and R, is from Eq. (4-B-111a), where

$$\sigma_{22}(\kappa R) - \sigma_{22}(R) = \int_{\kappa R}^{R} (P_{22} - P_{33}) \, d \ln r = \int_{\kappa R}^{R} N_2 \, d \ln r \quad \text{(4-B-112)}$$

This experiment has been carried out using Eq. (4-B-112) by Hayes and Tanner (96) and Tanner (216) for polymer solutions and by Ehrmann (65) for polymer melts. Usually, Eq. (4-B-112) is inverted by presuming a relationship, such as a power law between N_2 and $\dot{\gamma}$ or σ_{12}.

One must take care in this experiment to avoid the pressure hole error and use flush-mounted transducers. The early studies (96, 215) suffer from this problem, while the later study of Ehrmann uses an appropriate transducer–hole design.

iii. Trough Rheometer   In the trough experiment of Tanner (216), relatively low-viscosity fluids at room temperature flow down an inclined trough. Available measurements are limited to flow rate and the shape of the interface of the liquid. This shape is used to determine N_2. As we are concerned with the shape of the interface with the air, we use the boundary condition at sin h/r

$$\sigma_{\theta\theta}\left(r, \sin^{-1}\frac{h}{r} \right) = 0 \quad \text{(4-B-113)}$$

where h is the interface elevation over the flat surface. From Eqs. (4-B-111a) and

(4-B-111b):

$$p(r) = P_{22}(r) + \int_0^r (P_{22} - P_{33}) d \ln r - \rho g r \cos \alpha \sin \theta + C(\theta) \qquad \text{(4-B-114a)}$$

$$p(\theta) = - \rho g r \cos \alpha \sin \theta + C'(r) \qquad \text{(4-B-114b)}$$

It follows that by evaluating σ_{33} or $\sigma_{\theta\theta}$ we obtain, after noting that $r \sin \theta$ is $h(r)$ defined in Figure 4-B-11,

$$\rho g h(r) \cos \alpha = N_2 + \int_0^r N_2 \, d \ln r \qquad \text{(4-B-115)}$$

From $h(r)$ we may determine N_2.

Equation (4-B-115) may be inverted by differentiation to give N_2

$$\frac{d}{d\sigma_w} [\rho g h \cos \alpha] = \frac{dN_2}{d\sigma_w} + \frac{N_2}{\sigma_w} \qquad \text{(4-B-116)}$$

or

$$N_2|_w = \rho g h_w \cos \alpha - \frac{1}{\sigma_w} \int_0^{\sigma_w} \rho g h_w \cos \alpha \, d\sigma_w \qquad \text{(4-B-117)}$$

where the subscript w implies evaluation at the wall of the trough. Experimental studies and techniques are described by Tanner and his students (110, 118, 216) for polymer solutions to which the technique is certainly limited. However, it shows the beautiful insight of Tanner.

6. Slippage on Solid Walls

a. General Remarks

The analyses of this section have generally presumed that the melts investigated adhere to the walls or rheometers. This has classically been presumed in rheometry. This assumption is not generally valid and considerable evidence has been accumulated over the years of the occurrence of slippage in elastomers and plastics (55, 128, 151, 156, 157, 235). Mooney (152, 156–159) was early to recognize this phenomenon and its importance. He led efforts to develop experimental techniques to characterize it. We describe two of those methods in succeeding pages.

b. Capillary Rheometer

Mooney (152) argued that slippage at a capillary wall could be determined from Eq. (4-B-90) by differentiating at constant $(\sigma_{12})_w$

$$v_{1s} = \frac{\partial [4Q/\pi D^3]}{\partial (1/D)} \Bigg|_{(\sigma_{12})_w} \qquad \text{(4-B-118)}$$

where $4Q/\pi D^3$ would be plotted versus $1/D$. Slip velocities are determined from the slopes of lines of data of the same shear stress.

Equation (4-B-118) has been used by various investigators to detect the occurrence of slippage. Perhaps the most successful efforts have related to interpretation of unstable flow in high-density polyethylene (235).

As slippage undoubtedly depends upon pressure, one may well have slip at lower pressures near a die exit while the melt adheres to the wall near the die entry. Equation (4-B-113) would not be expected to be valid under such circumstances.

Another set of experiments that has been applied to characterize slippage on a steel wall is the use of markers. Ma et al. (128) have shown the occurrence of slippage of rubber compounds on steel walls. Addition of high levels of carbon black induced slippage, but this seemed not to occur in filled thermoplastics (129).

c. Rotational Rheometer

For highly viscoelastic rubbery materials, slip may occur. It is often a problem in cone–plate and parallel-plate viscometers. In a typical Mooney viscometer both the shearing disc and the surrounding cavity are serrated so as to maintain adhesion of the flowing polymer to these surfaces. Decker and Roth (55) and Mooney (156, 157) have shown that replacing the serrated disc with a polished one frequently gives rise to considerably lower torques and apparent shearing stresses. This phenomenon can only be interpreted as being due to slippage at the polished metal surface. Mooney (157) quantitatively analyzed this problem by presuming that the fluid adheres to the serrated cavity surface and slips along the disc at a velocity uniquely determined by the local shearing stress. To facilitate the analysis, Mooney presumed power-law relationships between the shear rate, the slippage, and the shear stress, that is

$$\dot{\gamma} = \frac{1}{K}(\sigma_{12})^m, \qquad v_\theta(0, r) = s[\sigma_{12}(0, r)]^p \qquad \text{(4-B-119)}$$

The velocity gradient in the torsional flow region is then

$$\dot{\gamma}_1 = \frac{r\Omega}{h} - \frac{s}{h}[\sigma_{12}(0, r)]^p \qquad \text{(4-B-120)}$$

This is a polynomial expression which, in general, is not analytically soluble, but in the case of relatively small slippage, we have

$$\dot{\gamma}_1 = \frac{r\Omega}{h}\left[1 - \frac{s}{h}\left(\frac{Kr\Omega}{h}\right)^{(p-m)/m} + \cdots\right] \qquad \text{(4-B-121)}$$

A similar expression may be derived for Couette flow in Region II and both this result and Eq. (4-B-120) are then substituted into Eq. (4-B-119) for the shear stresses, which are, in turn, introduced into Eq. (4-B-76) for the torque. Mooney has integrated these expressions not only for the small slip asymptote, but also for large slippage velocities.

Mooney (157) has described how these results may be applied together with data from an instrument with a serrated disc surface to compute slip velocities.

Turner and Moore (234) (see also Hull (101) developed a pressurized biconical rheometer to characterize slippage in elastomers and their compounds. A similar instrument was later described by Montes et al. (151). Data as shown indicated the occurrence of reduced torques and apparent slippage by reducing the applied pressure.

C. ELONGATIONAL FLOW

1. Kinematics and Rheological Functions

Elongational or extensional flows are motions for which the velocity field is given by (Fig. 4-C-1)

$$\mathbf{v} = v_1(x_1)\mathbf{e}_1 + v_2(x_2)\mathbf{e}_2 + v_3(x_3)\mathbf{e}_3 \qquad (4\text{-}C\text{-}1)$$

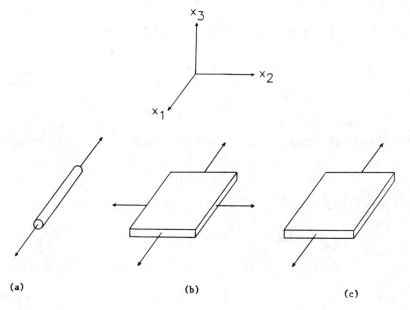

(a)   (b)   (c)

FIGURE 4-C-1. Classification biaxial elongational flow: (a) uniaxial; (b) equal biaxial; (c) planar.

The rate of deformation tensor is of the form

$$
\mathbf{d} = \begin{bmatrix} \dot{\gamma}_{e1} & 0 & 0 \\ 0 & \dot{\gamma}_{e2} & 0 \\ 0 & 0 & \dot{\gamma}_{e3} \end{bmatrix} \tag{4-C-2}
$$

where

$$
\dot{\gamma}_{ej} = \frac{dv_j}{dx_j}
$$

For an incompressible fluid, we must have

$$
\dot{\gamma}_{e2} + \dot{\gamma}_{e2} + \dot{\gamma}_{e3} = 0 \tag{4-C-3}
$$

By uniaxial extension, we mean a flow for which

$$
\dot{\gamma}_{e2} = \dot{\gamma}_{e3} = -\tfrac{1}{2}\dot{\gamma}_{e1} < 0 \tag{4-C-4}
$$

For equal *biaxial extension*

$$
\dot{\gamma}_{e1} = \dot{\gamma}_{e2} = -2\dot{\gamma}_{e3} > 0 \tag{4-C-5}
$$

For *planar extension*

$$
\dot{\gamma}_{e3} = 0, \qquad \dot{\gamma}_{e2} = -\dot{\gamma}_{e1} \tag{4-C-6}
$$

The stress response to an elongational flow will be of the form

$$
\boldsymbol{\sigma} = \begin{bmatrix} \sigma_{11} & 0 & 0 \\ 0 & \sigma_{22} & 0 \\ 0 & 0 & \sigma_{33} \end{bmatrix} \tag{4-C-7}
$$

For uniaxial extension, σ_{11} is determined by an elongational viscosity, χ, through

$$
\sigma_{11} = \chi\dot{\gamma}_{e1}; \qquad \sigma_{22} = \sigma_{33} = -p \tag{4-C-8}
$$

For equal biaxial stretching, we have a biaxial elongational viscosity

$$
\sigma_{11} = \sigma_{22} = \chi_B\dot{\gamma}_{e1}; \qquad \sigma_{33} = -p \tag{4-C-9}
$$

For planar extension

$$
\sigma_{11} = \chi_p\dot{\gamma}_{e1}, \qquad \sigma_{33} = -p \tag{4-C-10}
$$

Functions similar to χ, χ_B, and χ_p may be defined for other elongational flows.

One seeks to carry out long-duration constant-elongation rate or stress studies in order to obtain a well-defined rheological function equivalent to the shear viscosity.

Elongational flows generally occur in post-die polymer processing operations such as melt spinning, tubular film extrusion, and blow molding.

2. Historical Perspective

The origins of elongational flow measurements or rheological properties of polymer melts is in the 1906 paper of Trouton (232), who considered the uniaxial extension or rods of pitch, pitch–tar blends, and "shoemaker's wax." Trouton described four methods of measurement of (1) flow produced in a rod under traction, (2) flow under uniaxial compression, (3) flow of a freely descending stream, and (4) the rate of bending of a horizontal rod or beam under its own weight when supported only at the ends.

Following the work of Trouton, there is little in the literature dealing with elongational flow. Tamman and Jenckel (214), and others in the period 1924–1932 reported studies on elongational flow of molten glass filaments. The first reports on polymer systems appear as early as the 1940s but not careful scientific studies. The period of modern investigation begins with Ballman (14) in 1965. Ballman first reported careful uniaxial extensional measurements in which he made studies at constant deformation rate and distinguished between transient and steady-state behavior and discussed the $\chi(\dot{\gamma}_{e1})$ function. Careful studies of elongational flow of molten polymer filaments were reported during the next few years by Cogswell (41), Meissner (141), Vinogradov, Radushkevich, and Fikham (236–238), and others (71, 105, 119, 125, 138, 161, 162, 209, 210, 258).

The existence of elongational flow in the entry region of a capillary dies was first noted by various investigators in the late 1960s (42, 145). Cogswell (42) proposed and reported the use of pressure losses through orifice dies to determine elongational viscosity in polymer melts. The procedure has not gained general acceptance because of the complex nature of the flow patterns in the die entrance regions and the extent of constant elongation rate. Early studies of the use of converging flow have suggested the direction of development should be toward design of an instrument containing a die with a lubricated layer along the walls to prevent shearing motion and a position varying cross section to ensure a constant elongation rate along the die axis. A publication in the same year by Everage and Ballman (71) notes that they had developed such a rheometer in the Monsanto Textile Company's Pensacola laboratories and reports data for a polystyrene melt.

Existing methods of measurement of equal biaxial and planar extension largely involve sheet inflation. The basic experiment derives from an apparatus utilized by Treloar (231) in the 1940s to study the biaxial stretching of thin rubber sheets. Studies of this type on melts began with the work of Denson (57–59, 108) dating to 1971–1972, and involve inflation of a clamped circular or high-aspect bubble undergoes a biaxial elongational flow for circular sheets and this can be programmed to be constant in rate if the movement of fiducial marks or growth of the bubble is monitored and the inflation pressure varied to maintain this constant. Early studies with this type of instrument were carried out at room temperature using elastomers such as polyisobutylene (57, 58, 108). More recent investigations (59, 191, 259) have involved polymer melts. Here it is necessary to immerse it in a hot silicone oil reservoir and by increasing the pressure on one side, inflate the sheet and move it into the other reservoir.

There have been other efforts in recent years to measure biaxial elongational viscosity. Winter, Macosko, and Meissner have taken the lead.

Winter, Macosko, and Bennett (255) have recently developed a method for measuring equal biaxial and planar extensional viscosity using orthogonal stagnation flow. Meissner and his coworkers (143, 208) have described an equal biaxial stretching of a planar sheet to determine the biaxial elongational viscosity. Chatraei et al. (40) describe a lubricated squeezing-flow method for determination of biaxial elongational viscosity.

3. Uniaxial Extension Instruments

a. Filament Elongated by Accelerating Clamp in Uniaxial Extension

In the instrument first described by Ballman (14) and in later instruments described by Vinogradov et al. (236, 237) and Stevenson (209), an initial sample of length L_0 was stretched at a constant elongation rate $\dot{\gamma}_E$ by an accelerating clamp (Fig. 4-C-2). If we write

$$\dot{\gamma}_E = \dot{\gamma}_{e1} = \frac{1}{L}\frac{dL}{dt} \tag{4-C-11}$$

the sample must vary in length according to

$$L(t) = L_0 e^{\dot{\gamma}_E t} \tag{4-C-12}$$

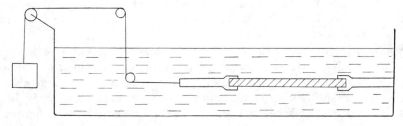

FIGURE 4-C-2. Filament elongated by accelerating clamp in uniaxial extension.

The stress on the extending filament is related to the tensile force through

$$\sigma_{11}(t) = \frac{F(t)}{A(t)} \qquad \text{(4-C-13)}$$

By incompressibility

$$\sigma_{11}(t) = \frac{F(t)}{A_0} e^{\dot{\gamma}_E t} \qquad \text{(4-C-14)}$$

where A_0 is the initial cross-sectional area.

In a variant of this instrument suggested by Cogswell (*41*), the filament is extended using a constant-stress cam (*5, 41, 3, 161, 162, 237*). This pulls the sample with a constant force over a cam that is programmed to vary with the uniformly decreasing cross-sectional area of the extending specimen.

Apparatus of this type have been developed in which the filaments are held either horizontally on the surface of a hot oil bath (*40, 237, 238*) or vertically within an air-filled chamber (*14, 161, 162, 209*).

b. Constant Filament Length Uniaxial Extension Experiments

Meissner (*140, 141*) has described an apparatus in which a filament is held horizontally between two pairs of rotating tensioned rolls (Fig. 4-C-3). Improved versions have been described by Laun and Munstedt (*119*) and Masuda et al. (*138*). Here the elongation rate is

$$\dot{\gamma}_E = \frac{V}{\frac{1}{2}L} \qquad \text{(4-C-15)}$$

where L is the distance between the rolls. The tensile stress on the filament is again given by Eq. (4-C-14).

Ide and White (*105*) have used an instrument in which a filament is floated on top of an oil bath. On one side it is clamped to a tension-measuring device and on the other side is removed onto a rotating roll with linear velocity V. The stretch

FIGURE 4-C-3. Constant filament length elongational viscosity measurement principle of Meissner (*140, 141*).

rate of the filament is given by

$$\dot{\gamma}_E = \frac{V}{L} \tag{4-C-16}$$

The tensile stress is given by Eq. (4-C-14).

Laun and Munstedt (*119*) and White and his coworkers (*210, 258*) describe an instrument where the filament is clamped at one end to a transducer and stretched by a pair of rotating rolls. The deformation rate is again given by Eq. (4-C-16).

c. Other Methods

Various methods have been discussed in the literature where uncontrolled deformation of filaments is used. They are often the only experiments that can be carried out for low-viscosity fluids which would not only yield small stresses in the instruments cited earlier, but would break up into ligaments. Fiber-spinning operations are also sometimes used (*100*). Another technique is the tubeless syphon (*8, 38*), where a stream of fluid is drawn out of a beaker by a vacuum. Force balances are used to determine stresses, diameter variations, and flow rates to obtain velocity gradiation. However, stresses and velocity gradients both vary with position.

4. Biaxial and Planar Extension Experiments

a. Sheet Inflation

In this experiment a sheet is expanded into a hemispherical shape through the injection of high-pressure fluid on one side (Fig. 4-C-4).

The first instrument to measure biaxial deformation flow behavior was developed by Denson and Gallo (*58*) and Joye et al. (*108*).

The rate of deformation is determined and controlled by observing the variation in position of a series of marks on the sheet according to

$$\dot{\gamma}_e = d_{11} = d_{22} = \frac{1}{L}\frac{dL}{dt} = \frac{d\ln L}{dt} \tag{4-C-17}$$

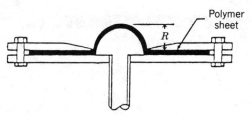

FIGURE 4-C-4. Biaxial and planar extension rheometer (sheet inflation experiments).

where L is the distance between the marks. One programs so as to make d_{jj} constant in time. The thickness $\delta(t)$ may be computed through continuity

$$d_{11} + d_{22} + d_{33} = 0$$

$$d_{33} = -(d_{11} + d_{22}) = -2d_{11} \qquad \text{(4-C-18)}$$

$$= \frac{1}{\delta}\frac{d\delta}{dt}$$

$$\delta = \delta(0)e^{-2d_{11}t}$$

For a thin shell of thickness δ and radius R, the tension $\sigma_{11} = \sigma_{22}$ in the hemispherical sheet is

$$2\pi R\delta\sigma_{11} = \pi R^2 \Delta p$$

$$\sigma_{11} = \frac{R\Delta p}{2\delta} \qquad \text{(4-C-19)}$$

$$\sigma_{11}(t) = \frac{R}{2\delta(0)}\cdot\Delta p(t)\cdot e^{2d_{11}t}$$

A limitation of this apparatus is in the total strain that can be applied before the sheet becomes too distorted (beyond a hemisphere of half cylinder). Beyond this, one will not be able to develop a constant elongation rate flow. Thus one may often not be able to achieve a steady-state elongational viscosity.

Another limitation is that the materials studied by such an apparatus must obviously have very high elongational viscosities. Denson et al.'s polymers often have viscosities in the range 10^8–10^9 pascal-second.

Denson et al. (57, 59) have described an instrument to carry out planar extensional flow. The apparatus is basically similar to that designed by Denson et al. for investigating biaxial extension and consists of the expansion of a rectangular membrane instead of a circular one. Variation of this instrument intended for thermoplastics at elevated temperature have been described by Dealy and his coworkers (191, 259).

b. Biaxial Sheet Stretching

Meissner et al. (143, 208) described a biaxial extension instrument using several rollers at varying equally distributed angles. To stretch a flat sheet, the stresses are determined from the applied tensions and velocity gradients from roll velocity and the half length of the sheet (Fig. 4-C-5a). This essentially represents a generalization of Meissner's earlier roller-based unaxial extension instrument to a two-dimension plane.

c. Squeeze Flow

Various other instruments have been proposed to measure either equal biaxial extension or planar extension.

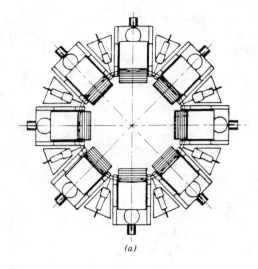

(a)

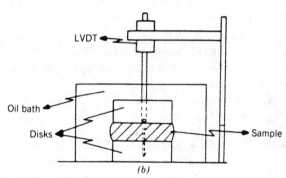

(b)

FIGURE 4-C-5. Instruments to measure biaxial elongational viscosity: (*a*) Meissner–Stephenson apparatus (*143*); (*b*) Lubricated squeezing flow instrument [see Chatraei et al. (*39*)].

A lubricated squeezing-flow instrument has been developed by Chatraei, Macosko and Winter (*40*). Here two discs move toward each other and squeeze the polymer melt sample radially outward (Fig. 4-C-5*b*). Lubrication of the ends results in slip along the disc surfaces and biaxial elongational flow.

d. Other Instruments

Winter et al. (*255*) have described an instrument using orthogonal stagnation flow to determine the biaxial elongational viscosity of polymer melts.

Recently, Williams and Williams (*254*) have developed an instrument using orthogonal stagnation flow to determine planar elongational viscosity of lower-viscosity fluids.

D. SMALL-STRAIN EXPERIMENTS

1. Kinematics and Rheological Functions

An important class of rheological experiments performed on bulk polymer systems involves studying the response of these materials to small strains. The strain is sometimes applied as step functions and the relaxation of stress following its application is measured. In other experiments a sinusoidal oscillatory deformation is applied and the stress response to it is measured. Other variants have been investigated.

These experiments, as indicated, have in common small strains of the infinitesimal strain tensor γ.

$$\gamma = \int \mathbf{d}\, dt \qquad \text{and} \qquad \gamma = \int \dot{\gamma}\, dt \qquad (4\text{-}D\text{-}1)$$

This involves for elongation and shear, respectively,

$$\gamma = \begin{bmatrix} \gamma_e & 0 & 0 \\ 0 & -\tfrac{1}{2}\gamma_e & 0 \\ 0 & 0 & -\tfrac{1}{2}\gamma_e \end{bmatrix} \qquad \gamma = \begin{bmatrix} 0 & \gamma & 0 \\ \gamma & 0 & 0 \\ 0 & 0 & 0 \end{bmatrix} \qquad (4\text{-}D\text{-}2)$$

where γ_e and γ are the strains in elongation and shear, respectively. An experiment that has been widely carried out on elastomers and other highly viscous polymer systems involves the application of a step strain to a sample in extension or shear. Here strains γ and γ_e are step strains $\dot{\gamma}_0$ and γ_{e0}. The ratio of the stress [or, experimentally, the force response $F(t)$] to the applied strain is measured. The data is interpreted in terms of a relaxation modulus. For a sample stretched in extension in direction 1 and the amount $\dot{\gamma}_{e0}$, the relaxation modulus $E(t)$ is defined as

$$E(t) = \frac{\sigma_{11}(t)}{\gamma_{e0}} \qquad (4\text{-}D\text{-}3)$$

For a measurement carried out in shear, the shear modulus, denoted by $G(t)$, is given by

$$G(t) = \frac{\sigma_{12}(t)}{\gamma_0} \qquad (4\text{-}D\text{-}4)$$

All other stress components are zero.

A common experiment performed on polymer melts is the determination of the stress response to an imposed sinusoidal oscillation

$$\gamma_{ij}(t) = \gamma_0 \sin \omega t \qquad (4\text{-}D\text{-}5)$$

The stress (generally shear stress) response is out of phase with $\gamma(t)$ by an angle δ. Considering a shear stress $\sigma_{12}(t)$, we have

$$\sigma_{12}(t) = G^*\gamma_0 \sin(\omega t + \delta) \tag{4-D-6}$$

where G^* is the complex modulus and δ is the phase angle. We may expand the sine function and rewrite Eq. (4-D-6) as

$$\sigma_{12}(t) = G'(\omega)\gamma_0 \sin \omega t + G''(\omega)\gamma_0 \cos \omega t \tag{4-D-7a}$$

with

$$G'(\omega) = G^* \cos \delta$$
$$G''(\omega) = G^* \sin \delta$$
$$\tan \delta = \frac{G''}{G'} \tag{4-D-7b}$$

where $G'(\omega)$ is called the storage modulus, $G''(\omega)$ the loss modulus, δ is referred to as the loss angle, and $\tan \delta$ is the loss tangent.

Equation (4-D-7) is often rewritten

$$\sigma_{12}(t) = G'(\omega)\gamma(t) + \eta'(\omega)\frac{d\gamma}{dt} \tag{4-D-8}$$

where $\eta'(\omega)$ is called the dynamic viscosity and is equal to $G''(\omega)/\omega$.

2. Historical Perspective

Dynamic measurements to determine rheological properties date to the nineteenth century. Clerk Maxwell (140) and others carried out sinusoidal oscillations with a rotating disc to measure the viscosity of a Newtonian fluid. Small-strain studies of the rheological behavior of polymer and related materials seem to date to the 1930s. Eisenschitz and Philippoff (67) and later Philippoff (175) described studies of sinusoidal oscillations of polymer solutions. Stress-relaxation experiments were carried out in the same period by Schofield and Scott Blair (201) on flour dough.

Measurements of this type became popular in the United States from the 1940s. The pioneer of these efforts was Herbert Leaderman (120, 121), who interpreted the responses in terms of a linear superposition integral formulation. Tobolsky and his coworkers (6) carried out extensive experiments on the stress-relaxation behavior of elastomers in the late 1940s. During the same period, research programs were initiated for using sinusoidal oscillations (137, 170, 205). By the 1960s, the measurements of $\eta'(\omega)$ and $G'(\omega)$ on polymer melts and solutions were becoming commonplace (15, 171, 172).

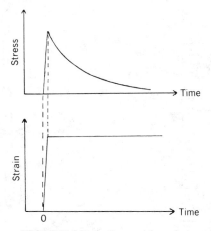

FIGURE 4-D-1. Stress relaxation.

3. Stress Relaxation

Here samples are stretched in either uniaxial extension or simple shear and the decay of the force $F(t)$ measured (Fig. 4-D-1). For uniaxial extension, the infinitesimal strain and tensile stress are given by

$$\gamma_{11} = \frac{\Delta L}{L_0}, \qquad \sigma_{11} = \frac{F(t)}{A} \qquad \text{(4-D-9a, b)}$$

where ΔL is the amount of extension, L_0 the initial length, and A the cross-sectional area. This yields $E(t)$ through Eq. (4-D-3). In shear strain

$$\gamma_{12} = \frac{\Delta L}{H}, \qquad \sigma_{12} = \frac{F(t)}{A} \qquad \text{(4-D-10a, b)}$$

where H is the sample thickness, arc obtain $G(t)$.

Care needs to be applied in both sample preparation and development and maintenance of uniform strains. Residual stresses often exist in molded specimens. These can be removed in elastomers by dissolving samples and reprecipitating them as films (e.g., on mercury), from which tensile specimens can be cut.

4. Sinusoidal Oscillations

Sinusoidal oscillation experiments (Fig. 4-D-2) have been carried out in both coaxial cylinder and cone–plate instruments. To a first approximation, one may

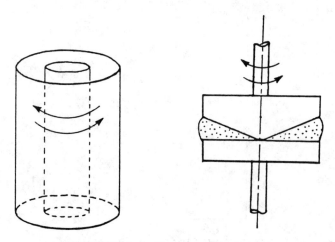

FIGURE 4-D-2. Sinusoidal oscillations.

obtained stress in the fluid from Eqs. (4-D-8) and (4-D-9a)

$$\text{(Couette)} \qquad \sigma_{12} = \frac{M(t)}{2\pi\kappa^2 R^2 L} \qquad\qquad \text{(4-D-11)}$$

$$\text{(cone–plate)} \quad \sigma_{12} = \frac{3M(t)}{2\pi R^3} \qquad\qquad \text{(4-D-12)}$$

with the strains γ_0 being given by the maximum shear strain at the torque measuring surface. However, one must in general account for instrument and fluid inertia (*131, 170, 240–243*).

5. Eccentric Rheometers

In 1965, Maxwell and Chartoff (*139*) pointed out that it is possible to obtained such dynamic measurements in steady rotational flows using eccentric parallel discs rotating at the same angular velocity. Kepes subsequently indicated that this is also possible using a rotating sphere in a rotating cavity with different axes (balance rheometer) and Abbott, Bowen, and Walters (*1*) have presented experimental studies for a number of these instruments. Basic studies of flow in these geometries have been reported by Blyler and Kurtz (*24*) and later investigators including Bird and Harris (*22*), Yamamoto (*257*), Walters (*239, 240*), Gordon and Schowalter (*87*), and Abbott and Walters (*2*). These instruments are reviewed in the monograph by Walters (*240*).

REFERENCES

1. T. N. G. Abbott and K. Walters, *J. Fluid Mech.*, **40**, 203 (1970); T. N. G. Abbott, G. W. Bowen, and K. Walters, *J. Phys. D. Appl. Phys.*, **4**, 190 (1971).

2. T. N. G. Abbott and K. Walters, *J. Fluid Mech.*, **43**, 257 (1970).

3. N. Adams and A. S. Lodge, *Phil. Trans. Roy. Soc.*, **A256**, 149 (1964).

4. J. N. C. Adamse, H. Janeschitz-Kriegl, J. L. DenOtter, and J. L. S. Wales, *J. Polym. Sci.*, **A26**, 871 (1968).

5. E. N. Andrade and B. Chalmers, *Proc. Roy. Soc.*, **A138**, 348 (1932).

6. R. D. Andrews, J. Hofman-Bang, and A. V. Tobolsky, *J. Polym. Sci.*, **3**, 669 (1948).

7. E. Ashare, R. B. Bird, and J. A. Lescaboura, *AIChE J.*, **11**, 910 (1965).

8. G. Astarita and L. Nicodemo, *Chem. Eng. J.*, **1**, 57 (1970).

9. E. B. Bagley, *J. Appl. Phys.*, **28**, 624 (1957).

10. E. B. Bagley and A. M. Birks, *J. Appl. Phys.*, **31**, 556 (1960).

11. E. D. Baily, *Trans. Soc. Rheol.*, **18**, 635 (1974).

12. T. F. Ballenger, I. J. Chen, J. W. Crowder, G. E. Hagler, D. C. Bogue, and J. L. White, *Trans. Soc. Rheol.*, **15**, 195 (1971).

13. T. F. Ballenger and J. L. White, *J. Appl. Polym. Sci.*, **15**, 1949 (1971).

14. R. L. Ballman, *Rheol. Acta*, **4**, 137 (1965).

15. R. L. Ballman and R. H. Simon, *J. Polym. Sci.*, **A2**, 3557 (1964).

16. J. J. Benbow and P. Lamb, *SPF Trans.*, **3**, 7 (1963).

17. G. A. Bialas and J. L. White, *Rubber Chem. Tech.*, **42**, 675 (1969).

18. D. M. Binding, J. F. Hutton, and K. Walters, *Rheol. Acta*, **15**, 540 (1976); D. M. Binding and K. Walters, *J. Non-Newt. Fluid Mech.*, **1**, 277 (1976).

19. E. C. Bingham, *J. Wash. Acad. Sci.*, **6**, 177 (1916); *NBS Bull.*, **13**, 309 (1916).

20. E. C. Bingham, *Fluidity and Plasticity*, McGraw-Hill, New York, E. C. Bingham and H. S. Green, *Proc. ASTM*, **19**(II), 640 (1919).

21. R. B. Bird, *SPE J.*, **11**(1), 35 (1955).

22. R. B. Bird and E. K. Harris, *AIChE J.*, **14**, 758 (1968).

23. L. L. Blyler, *Trans. Soc. Rheol.*, **13**, 39 (1969).

24. L. L. Blyler and S. J. Kurtz, *J. Appl. Polym. Sci.*, **11**, 127 (1967).

25. D. V. Boger and M. M. Denn, *J. Non-Newt. Fluid Mech.*, **6**, 163 (1980).

26. D. V. Boger, R. Gupta, and R. I. Tanner, *J. Non-Newt. Fluid Mech.*, **4**, 239 (1978).

27. D. C. Bogue, *Ind. Eng. Chem.*, **51**, 874 (1959).

28. H. C. Brinkman, *Appl. Sci. Res.*, **A2**, 120 (1951).

29. J. M. Broadbent, A. Kaye, A. S. Lodge, and D. G. Vale, *Nature*, **217**, 55 (1968).

30. J. M. Broadbent and K. Walters, *J. Phys. D. Appl. Phys.*, **4**, 1863 (1971).

31. J. G. Brodnyan, F. H. Gaskins, W. Philippoff, *Trans. Soc. Rheol.*, **1**, 107 (1957).

32. J. G. Brodnyan, F. H. Gaskins, W. Philippoff, and E. G. Lendrat, *Trans. Soc. Rheol.*, **2**, 285 (1958).

33. I. Bruker and A. S. Lodge, *J. Rheol.*, **29**, 557 (1985).

34. E. Buckingham, *Proc. ASTM*, **21**, 1154 (1921).

35. J. M. Burgers, *First Report on Viscosity and Plasticity*, Amsterdam, 1935, p. 73.

36. J. M. Castro and R. L. Fosdick, *J. Non-Newt Fluid Mech.*, **15**, 127 (1984).

37. S. Chandrasekhar, *Hydrodynamic and Hydromagnetic Stability*, Oxford, 1961.

38. K. K. M. Chao and M. C. Williams, *J. Rheol.*, **27**, 451 (1983).

39. I. J. Chen and D. C. Bogue, *Trans. Soc. Rheol.*, **16**, 59 (1972).
40. S. H. Chatraei, C. W. Macosko, and H. H. Winter, *J. Rheol.*, **25**, 433 (1981).
41. F. N. Cogswell, *Plast. Polym.*, **36**, 109 (1968).
42. F. N. Cogswell, *Rheol. Acta*, **8**, 187 (1969).
43. F. N. Cogswell, *Plast. Polym.*, **43**, 39 (1973).
44. F. N. Cogswell, *J. Non-Newt Fluid Mech.*, **4**, 9 (1978).
45. B. D. Coleman, H. Markovitz, and W. Noll, *Viscometric Flows on Non-Newtonian Fluids*, Springer, New York, 1966.
46. B. D. Coleman and W. Noll, *Arch. Rat. Mech. Anal.*, **3**, 289 (1959).
47. R. W. Connelly and J. Greener, *J. Rheol.*, **29**, 209 (1985).
48. M. Couette, *Compt. Rend.*, **107**, 388 (1888).
49. H. W. Cox and C. W. Macosko, *AIChE J.*, **20**, 785 (1974).
50. J. M. Davies, J. F. Hutton, and K. Walters, *J. Phys. D. Appl. Phys.*, **6**, 2259 (1973).
51. J. M. Davies, J. F. Hutton, and K. Walters, *J. Non. Newt. Fluid Mech.*, **3**, 141 (1977).
52. R. Darby, *J. Rheol.*, **29**, 369 (1985).
53. J. M. Dealy, Lecture at Akron Polymer Lecture Group (1988); A. J. Giacomin, T. Samurkas, and J. M. Dealy, *Polym. Eng. Sci.*, **29**, 499 (1989).
54. J. M. Dealy, R. Farber, J. Rhi Sausi, and L. A. Utracki, *Trans. Soc. Rheol.*, **20**, 445 (1976).
55. G. E. Decker and F. I. Roth, *India Rubber World*, **128**, 339 (1953).
56. M. M. Denn and J. J. Roisman, *AIChE J.*, **15**, 454 (1969).
57. C. D. Denson and D. L. Crady, *J. Appl. Polym. Sci.*, **18**, 1611 (1974).
58. C. D. Denson and R. J. Gallo, *Polym. Eng. Sci.*, **11**, 174 (1971).
59. C. D. Denson and D. Hylton, *Proc. 7th Int. Rheol. Cong.*, 386 (1976).
60. A. J. deVries and C. Bonnebat, *Polym. Eng. Sci.*, **16**, 93 (1976).
61. J. H. Dillon and N. Johnston, *Physics*, **4**, 225 (1933).
62. D. W. Dodge and A. B. Metzner, *AIChE J.*, **7**, 189 (1959); **8**, 143 (1962).
63. N. E. Dorsey, *Phys. Rev.*, **28**, 833 (1926).
64. J. L. Duda and J. S. Vrentas, *Can. J. Chem. Eng.*, **50**, 671 (1972); *Appl. Sci. Res.*, **28**, 241 (1973).
65. G. Ehrmann, *Rheol. Acta*, **15**, 8 (1976).
66. R. Eisenschitz, *Kolloid Z.*, **64**, 184 (1933).
67. R. Eisenschitz and W. Philippoff, *Naturwiss.*, **21**, 527 (1933).
68. R. Eisenschitz, B. Rabinowitsch, and K. Weissenberg, *Mitt. deutsch. Materialprulf. Sonderh.*, **9**, 9 (1929).
69. J. L. Ericksen in *Viscoelasticity, Phenomenological Aspects*, J. T. Bergen, Ed., Academic, New York, 1960.
70. A. E. Everage and R. L. Ballman, *J. Appl. Polym. Sci.*, **20**, 1137 (1976).
71. A. E. Everage and R. L. Ballman, *Nature*, **273**, 213 (1978).
72. F. D. Farrow, G. M. Lowe, and S. M. Neale, *J. Text. Instr.*, **19**, T18 (1928).
73. J. D. Ferry, *Viscoelastic Properties of Polymers*, Wiley, New York, 1969.
74. T. G. Fox and P. J. Flory, *J. Am. Chem. Soc.*, **70**, 2384 (1948).
75. A. G. Frederickson, *Principles and Applications of Rheology*, Prentice-Hall, Englewood Cliffs, NJ, 1964.

76. S. M. Freeman and K. Weissenberg, *Nature*, **161**, 334 (1948).
77. K. Funatsu and Y. Mori, *Proc. 5th Int. Rheol. Cong.*, **4**, 537 (1970).
78. I. Furuta, V. M. Lobe, and J. L. White, *J. Non-Newt. Fluid Mech.*, **1**, 207 (1976).
79. F. H. Garner and A. H. Nissan, *Nature*, **158**, 634 (1946).
80. F. H. Gaskins and W. Philippoff, *Trans. Soc. Rheol.*, **3**, 181 (1959).
81. J. Gerrard and W. Philippoff, *Proc. 4th Int. Rheol. Cong.*, **2**, 77 (1965).
82. H. Giesekus, *Proc. 4th Int. Rheol. Cong.*, **1**, 249 (1965); *Rheol. Acta*, **3**, 59 (1963) and **4**, 85 (1965).
83. R. F. Ginn and M. M. Denn, *AIChE J.*, **15**, 450 (1969).
84. R. F. Ginn and A. B. Metzner, *Proc. 4th Int. Rheol. Cong.*, **2**, 583 (1965); *Trans. Soc. Rheol.*, **13**, 429 (1969).
85. C. Goldstein, *Trans. Soc. Rheol.*, **18**, 357 (1974).
86. S. Goldstein, *Modern Developments in Fluid Dynamics*, Clarendon Press, Oxford, 1938.
87. R. J. Gordon and W. R. Schowalter, *AIChE J.*, **16**, 318 (1970).
88. H. W. Greensmith and R. S. Rivlin, *Phil. Trans. Roy. Soc.*, **A245**, 399 (1953).
89. E. Hagenbach, *Ann. Phys. Chem.*, **10**, 385 (1860).
90. C. D. Han, *J. Appl. Polym. Sci.*, **15**, 2567 (1971).
91. C. D. Han, *Trans. Soc. Rheol.*, **18**, 163 (1974).
92. C. D. Han, *Rheology in Polymer Processing*, Academic, New York, 1976.
93. C. D. Han, M. Charles, and W. Philippoff, *Trans. Soc. Rheol.*, **13**, 455 (1969).
94. E. Hatschek, *Kolloid Z.*, **12**, 238 (1913).
95. E. Hatschek and R. S. Jane, *Kolloid Z.*, **40**, 53 (1926).
96. J. W. Haynes and R. I. Tanner, *Proc. 4th Int. Rheol. Cong.*, **3**, 389 (1965).
97. W. R. Hess, *Z. Klin. Meds.*, **71**, 421 (1910); *Arch. Physiol.*, 197 (1912); *Kolloid Z.*, **27**, 154 (1920).
98. K. Higashitani and A. L. Lodge, *Trans. Soc. Rheol.*, **16**, 687 (1972).
99. K. Higashitani and W. G. Pritchard, *Trans. Soc. Rheol.*, **16**, 687 (1972).
100. N. E. Hudson, J. Ferguson, and P. Mackie, *Trans. Soc. Rheol.*, **18**, 541 (1974).
101. A. M. Hull, *J. Non-Newt. Fluid Mech.*, **12**, 175 (1983).
102. J. F. Hutton, *Proc. Roy. Soc.*, **A287**, 222 (1969).
103. J. F. Hutton, *Rheol. Acta*, **14**, 979 (1975).
104. E. Humphrey and E. Hatschek, *Proc. Phys. Soc.*, **28**, 274 (1916).
105. Y. Ide and J. L. White, *J. Appl. Polym. Sci.*, **22**, 1061 (1978).
106. A. Jobling and J. E. Roberts, *J. Polym. Sci.*, **36**, 433 (1959).
107. D. D. Joseph and R. L. Fosdick *Arc. Rat. Mech. Anal.*, **49**, 321 (1973).
108. D. D. Joye, G. W. Poehlein, and C. D. Denson, *Trans. Soc. Rheol.*, **16**, 421 (1972).
109. A. Kaye, A. S. Lodge, and D. G. Vale, *Rheol. Acta*, **7**, 368 (1968).
110. M. Keentok, A. G. Georgesru, A. A. Sherwood, and R. I. Tanner, *J. Non-Newt. Fluid Mech.*, **6**, 180 (1980).
111. R. G. King, *Rheol. Acta*, **5**, 35 (1966).
112. D. C. Kong and J. L. White, *Rubber Chem. Technol.*, **59**, 315 (1986).

113. T. Kotaka, M. Kurata, and M. Tamura, *J. Appl. Phys.*, **30**, 1075 (1959); *Bull. Chem. Soc. Japan*, **32**, 471 (1959), *Rheol. Acta*, **2**, 179 (1962).

114. T. Kotaka and K. Osaki, *J. Polym. Sci.*, **C15**, 453 (1966).

115. I. M. Krieger, *Trans. Soc. Rheol.*, **12**, 5 (1968).

116. I. M. Krieger and H. Elrod, *J. Appl. Phys.*, **24**, 143 (1953).

117. I. M. Krieger and S. H. Maron, *J. Appl. Phys.*, **23**, 147 (1952).

118. Y. Kuo and R. I. Tanner, *Rheol. Acta*, **13**, 951 (1974).

119. H. M. Laun and H. Munstedt, *Rheol. Acta*, **17**, 415 (1978).

120. H. Leaderman, *Elastic and Creep Properties of Filamentous Materials and Other High Polymers*, Textile Foundation, Washington, DC, 1943.

121. H. Leaderman, *Ind. Eng. Chem.*, **35**, 374 (1943).

122. B. L. Lee and J. L. White, *Trans. Soc. Rheol.*, **18**, 467 (1974).

123. A. I. Leonov, *Ann. N.Y. Acad. Sci.*, **491**, 106 (1987).

124. H. Lipson, *Karl Weissenberg 80th Birthday Celebration Essays*, J. Harris, Ed., East African Literature Bureau, Nairobi, 1973.

125. V. M. Lobe and J. L. White, *Polym. Eng. Sci.*, **19**, 617 (1979).

126. A. S. Lodge, *Elastic Liquids*, Academic, New York, 1964.

127. J. Lohrenz, G. W. Swift, and F. Kurata, *AIChE J.*, **6**, 415, 547 (1960).

128. C. Y. Ma, J. L. White, F. C. Weissert, A. I. Isayev, N. Nakajima, and K. Min, *Rubber Chem. Technol.*, **58**, 815 (1985).

129. C. Y. Ma, J. L. White, F. C. Weissert, and K. Min, *J. Non-Newt. Fluid Mech.*, **17**, 275 (1986); *Polym. Comp.*, **6**, 215 (1985).

130. C. W. Macosko and J. M. Starita, *SPE J.*, **27**(11), 38 (1971).

131. H. Markovitz, *J. Appl. Phys.*, **23**, 1070 (1952).

132. H. Markovitz, *Trans. Soc. Rheol.*, **1**, 37 (1957).

133. H. Markovitz, *Proc. 4th Int. Rheol. Cong.*, **1**, 189 (1965).

134. H. Markovitz, *J. Polym. Sci.*, **B3**, 3 (1965).

135. H. Markovitz, *Physics Today*, April, 23 (1968); *J. Rheol.*, **29**, 777 (1985).

136. H. Markovitz and R. B. Williamson, *Trans. Soc. Rheol.*, **1**, 25 (1957).

137. H. Markovitz, P. M. Yavorsky, R. C. Harper, L. J. Zapas, and T. W. deWitt, *Rev. Sci. Inst.*, **23**, 430 (1952).

138. T. Masuda, M. Takahashi, H. Ohno, and L. Li, *Nihon Reoroji Gakkaishi*, **16**, 111 (1988).

139. B. Maxwell and R. P. Chartoff, *Trans. Soc. Rheol.*, **9**, 41 (1965).

140. J. C. Maxwell, *Phil. Trans. Roy. Soc.*, **156** (1866).

141. J. Meissner, *Rheol. Acta*, **8**, 78 (1969); **10**, 230 (1971).

142. J. Meissner, *J. Appl. Polym. Sci.*, **16**, 2877 (1972).

143. J. Meissner, T. Raible, and S. E. Stephenson, *J. Rheol.*, **25**, 1 (1981).

144. A. B. Metzner, W. T. Houghton, R. A. Sailor, and J. L. White, *Trans. Soc. Rheol.*, **5**, 133 (1961).

145. A. B. Metzner, E. A. Uebler, and C. F. Chan Man Fong, *AIChE J.*, **15**, 750 (1969).

146. S. Middleman, *The Flow of High Polymers*, Wiley, New York, 1967.

147. S. Middleman, *Trans. Soc. Rheol.*, **13**, 123 (1969).

148. C. Miller, Ph.D., Dissertation, University of Michigan, 1967, described in J. D. Goddard, *Advances in Applied Mechanics*, Vol. 19, 1979.

149. C. E. Miller and W. H. Hoppmann, *Proc. 4th Int. Rheol. Cong.*, **2**, 619 (1965).

150. M. J. Miller and E. B. Christiansen, *AIChE J.*, **18**, 600 (1972).

151. S. Montes, J. L. White, N. Nakajima, F. C. Weissert and K. Min, *Rubber Chem. Technol.*, **61**, 698 (1988).

152. M. Mooney, *J. Rheol.*, **2**, 210 (1931).

153. M. Mooney, *Ind. Eng. Chem. Anal. Ed.*, **6**, 147 (1934).

154. M. Mooney, *Physics*, **7**, 413 (1936).

155. M. Mooney, *J. Colloid Sci.*, **2**, 69 (1947).

156. M. Mooney in *Rheology*, Vol. 2, F. R. Eirich, Ed., Academic, New York, 1958.

157. M. Mooney, Proceedings of the International Rubber Conference, Washington, DC, 1959, p. 368.

158. M. Mooney, *Rubber Chem. Technol.*, **35**, XXVII (1962).

159. M. Mooney and S. A. Black, *J. Colloid Sci.*, **7**, 204 (1952).

160. M. Mooney and R. H. Ewart, *Physics*, **5**, 350 (1934).

161. H. Munstedt, *Rheol. Acta*, **14**, 1077 (1975).

162. H. Munstedt, *Trans. Soc. Rheol.*, **23**, 421 (1979).

163. A. Y. Myers and J. T. Faucher, *Trans. Soc. Rheol.*, **12**, 183 (1968).

164. N. Nakajima and E. R. Harrell, *Rubber Chem. Technol.*, **52**, 9 (1979).

165. I. Newton, *Mathematical Principles of Natural Philosophy and System of the World*, University of California Press, 1962.

166. R. E. Nickell, R. I. Tanner, and B. Caswell, *J. Fluid Mech.*, **65**, 189 (1974).

167. Q. D. Nguyen and D. V. Boger, *J. Rheol.*, **29**, 355 (1985).

168. O. Olabisi and M. C. Williams, *Trans. Soc. Rheol.*, **16**, 727 (1972).

169. J. G. Oldroyd, *Q.J. Mech. Appl. Math.*, **4**, 271 (1951); *Proc. Roy. Soc.*, **A245**, 278 (1958).

170. J. G. Oldroyd, D. J. Strawbridge, and B. A. Toms, *Proc. Phys. Soc.*, **B64**, 44 (1951).

171. S. Onogi, T. Masuda, and T. Ibaragi, *Kolloid Z.-Z. Polym.*, **222**, 110 (1963).

172. S. Onogi, H. Kato, S. Ueki, and T. Ibaragi, *J. Polym. Sci.*, **C15**, 481 (1966).

173. F. J. Padden and T. W. deWitt, *J. Appl. Phys.*, **25**, 1086 (1954).

174. J. Pawlowski, *Kolloid Z.*, **103**, 129 (1953).

175. W. Philippoff, *Phys. Z.*, **34**, 884 (1934).

176. W. Philippoff and F. H. Gaskins, *J. Polym. Sci.*, **21**, 205 (1956).

177. W. Philippoff and F. H. Gaskins, *Trans. Soc. Rheol.*, **2**, 263 (1958).

178. G. H. Piper and J. R. Scott, *J. Sci. Inst.*, **22**, 206 (1945).

179. A. Pochettino, *Nuovo Cimento*, **8**, 77 (1914).

180. J. L. M. Poiseuille, *Mem. Pres. Div. Acad. Roy. Sci. Inst. France Sci. Math. Phys.*, **9**, 433 (1846).

181. W. F. O. Pollett, Proceedings of the 2nd International Rheology Congress, 1953, p. 85.

182. W. F. O. Pollett, *Brit. J. Appl. Phys.*, **6**, 199 (1955).

183. W. F. O. Pollett and A. H. Cross, *J. Sci. Instr.*, **27**, 209 (1950).

184. A. W. Porter and P. A. M. Rao, *Trans. Faraday Soc.*, **23**, 311 (1927).

185. B. Rabinowitsch, *Z. Phys. Chem.*, **A145**, 1 (1929).

186. M. Reiner, *Am. J. Math.*, **67**, 350 (1945).

187. M. Reiner, *Deformation and Flow*, Lewis, London, 1948.

188. M. Reiner and R. Riwlin, *Kolloid Z.*, **43**, 1 (1927).

189. O. Reynolds, *Phil. Trans. Roy. Soc.*, **174**, 935 (1883).

190. O. Reynolds, *Phil. Trans. Roy. Soc.*, **A177**, 157 (1886).

191. J. M. Rhi Sausi and J. M. Dealy, *Polym. Eng. Sci.*, **21**, 227 (1981).

192. J. E. Roberts, *Karl Weissenberg 80th Birthday Celebration Essays*, J. Harris, Ed., East African Literature Bureau, Nairobi, 1973.

193. R. Roscoe, *Phil. Mag.*, **7**, 40, 338 (1949).

194. R. J. Russell, Ph.D. Disseration, University of London, 1946.

195. R. J. Russell, *Karl Weissenberg 80th Birthday Celebration Essays*, J. Harris, Ed., East African Literature Bureau, Nairobi, 1973.

196. M. Sakai, H. Fukaya, and M. Nagasawa, *Trans. Soc. Rheol.*, **16**, 435 (1972).

197. K. Sakamoto, N. Ishida, and Y. Fukasawa, *J. Polym. Sci.A-2*, **6**, 1999 (1968).

198. B. C. Sakiadis, *AIChE J.*, **8**, 317 (1962).

199. J. G. Savins, *Soc. Pet. Eng. J.*, **4**, 203 (1964).

200. R. K. Schofield and G. W. Scott Blair, *J. Phys. Chem.*, **34**, 248 (1930).

201. R. K. Schofield and G. W. Scott Blair, *Proc. Roy. Soc.*, **A138**, 707 (1932).

202. T. Schwedoff, *J. de Phys.*, **9**, 34 (1890).

203. C. R. Shertzer and A. B. Metzner, *Proc. 4th Int. Rheol. Cong.*, **2**, 603 (1965).

204. J. C. Slattery, *J. Colloid Sci.*, **16**, 431 (1961).

205. T. L. Smith, J. D. Ferry, and F. W. Schremp, *J. Appl. Phys.*, **20**, 144 (1949).

206. R. S. Spencer and R. E. Dillon, *J. Colloid Sci.*, **3**, 163 (1948).

207. R. S. Spencer and R. E. Dillon, *J. Colloid Sci.*, **4**, 241 (1949).

208. S. E. Stephenson and J. Meissner, *Proc. 8th Int. Rheol. Cong.*, **2**, 431 (1980).

209. J. F. Stevenson, *AIChE J.*, **18**, 540 (1972).

210. Y. Suetsugu and J. L. White, *J. Appl. Polym. Sci.*, **28**, 1481 (1983).

211. G. G. Stokes, *Trans. Camb. Phil. Soc.*, **8**, 287 (1845).

212. G. G. Stokes, *Rept. Br. Assoc.*, **16**, 1 (1846); *Mathematical and Physical Papers*, **1**, 157.

213. G. G. Stokes, *Trans. Camb. Phil. Soc.*, **9**, 8 (1851).

214. G. Tamman and E. Jenckel, *Z. Anorg. Allg. Chem.*, **191**, 122 (1930).

215. R. I. Tanner, *Trans. Soc. Rheol.*, **11**, 347 (1967).

216. R. I. Tanner, *Trans. Soc. Rheol.*, **14**, 483 (1970).

217. R. I. Tanner in *Theoretical Rheology*, J. F. Hutton, J. R. A. Pearson and K. Walters, Eds., Applied Science, London, 1975.

218. R. I. Tanner and A. C. Pipkin, *Trans. Soc. Rheol.*, **13**, 471 (1969).

219. G. I. Taylor, *Phil. Trans. Roy. Soc.*, **A223**, 289 (1923).

220. R. Taylor, J. H. Fielding, and M. Mooney, Symposium in Rubber Testing, ASTM, 1947, p. 36.

221. R. H. Thomas and K. Walters, *J. Fluid Mech.*, **18**, 33 (1964); **19**, 557 (1964).

222. T. Thorpe and J. Rodgers, *Phil. Trans. Roy. Soc.*, **A185**, 397 (1894).

223. S. Toki and J. L. White, *J. Appl. Polym. Sci.*, **27**, 3171 (1982).

224. Y. Tomita, *Trans. JSME*, **25**, 938 (1959); *Bull. JSME*, **5**, 443 (1962).

225. Y. Tomita and T. Kobayoshi, *Proc. 5th Int. Rheol. Cong.*, **4**, 365 (1970).

226. H. L. Toor, *Ind. Eng. Chem.*, **48**, 922 (1956); *Trans. Soc. Rheol.*, **1**, 177 (1957).

227. J. P. Tordella, *J. Appl. Phys.*, **27**, 454 (1956).

228. J. P. Tordella, *Trans. Soc. Rheol.*, **1**, 203 (1957).

229. J. P. Tordella, *Rheol. Acta*, **1**, 216 (1958).

230. J. P. Tordella, *J. Appl. Polym. Sci.*, **7**, 215 (1963).

231. L. R. G. Treloar, *Trans. Faraday Soc.*, **40**, 59 (1944).

232. F. T. Trouton, *Proc. Roy. Soc.*, **A77**, 426 (1906).

233. R. Turian and R. B. Bird, *Chem. Eng. Sci.*, **17**, 331 (1962); **18**, 689 (1963); **20**, 771 (1965).

234. D. M. Turner and M. D. Moore, *Plast. Rubber Proc.*, **5**, 81 (1980).

235. J. Ui, Y. Ishimuru, H. Murakami, N. Kufushima, and Y. Mori, *SPE Trans.*, **4**, 295 (1964).

236. G. V. Vinogradov, V. D. Fikham, and B. V. Radushkevich, *Rheol. Acta*, **11**, 286 (1972).

237. G. V. Vinogradov, V. D. Fikham, B. V. Radushkevich, and A. Y. Malkin, *J. Polym. Sci.*, **A-2**, 8, 657 (1970).

238. G. V. Vinogradov, B. V. Radushkevich, and V. D. Fikham, *J. Polym. Sci.*, **A-2**(8) 1 (1970).

239. K. Walters, *J. Fluid Mech.*, **40**, 191 (1970).

240. K. Walters, *Rheometry*, Chapman and Hall, London, 1975.

241. K. Walters and T. E. R. Jones, *Proc. 5th Int. Rheol. Cong.*, **4**, 337 (1970).

242. K. Walters and R. A. Kemp in *Deformation and Flow in High Polymer Systems*, McMillan, London, 1972, p. 211.

243. K. Walters and N. D. Waters in *Deformation and Flow of High Polymer Systems*, McMillan, London, 1972, p. 211.

244. H. L. Weissberg, *Phys. Fluids*, **5**, 1033 (1962).

245. K. Weissenberg, *Z. Phys.*, **23**, 229 (1924).

246. K. Weissenberg, *Nature*, **159**, 310 (1947).

247. K. Weissenberg, *Proc. 1st Int. Rheol. Cong.*, **I**, 29 (1948).

248. K. Weissenberg, *Proc. 1st Int. Rheol. Cong.*, **II**, 114 (1948).

249. J. L. White, *Rubber Chem. Technol.*, **44**(5), G. 70 (1971).

250. J. L. White, *Appl. Polym. Symp.*, **20**, 155 (1973).

251. J. L. White and A. Kondo, *J. Appl. Polym. Sci.*, **21**, 2389 (1977).

252. J. L. White and A. B. Metzner, *Trans. Soc. Rheol.*, **7**, 245 (1963).

253. J. L. White and N. Tokita, *J. Appl. Polym. Sci.*, **9**, 1921 (1965).

254. P. R. Williams and R. W. Williams, *J. Non-Newt. Fluid Mech.*, **19**, 53 (1985).

255. H. H. Winter, C. W. Macosko, and K. E. Benner, *Rheol. Acta*, **18**, 323 (1979).

256. G. F. Wood, A. H. Nissan, and F. H. Garner, *J. Inst. Pet.*, **33**, 71 (1947).

257. M. Yamamoto, *Jap. J. Appl. Phys.*, **8**, 1252 (1969).

258. H. Yamane and J. L. White, *Polym. Eng. Rev.*, **2**, 167 (1982).

259. M. C. Yang and J. M. Dealy, *J. Rheology*, **31**, 113 (1987).

260. N. V. Zakharenko, F. S. Tolstukhina, and G. M. Bartenev, *Rubber Chem. Technol.*, **35**, 236 (1962).

V

EXPERIMENTAL OBSERVATIONS OF RHEOLOGICAL BEHAVIOR OF POLYMER SYSTEMS

A. INTRODUCTION

We must now turn our concerns to the characteristics of flow behavior which have been determined using the experimental methods described in Chapter 4. In this chapter we describe both qualitative observations of material response and quantitative studies of the rheological properties that we have previously defined and their dependence upon the structure of the systems investigated. Attention is given to polymer melts and the relationship of their behavior to molecular weight, its distribution, chain topology, and temperature. The characteristics of polymer solutions and the dependence of their properties on concentration and polymer structure will also be considered. Finally, we consider the major features of the rheological behavior of systems containing solid particulates. In a later chapter we will use this information to construct three-dimensional "constitutive equations" to represent the deformation and flow responses of these materials.

We begin with a critical survey of the development of the basic perspectives of the rheological characteristics of polymer systems and then proceed to discuss the behavior in turn of bulk homogeneous polymer melts, block copolymers, polymer blends, polymer solutions, liquid crystalline systems, and filled melts.

B. DEVELOPMENT OF RHEOLOGICAL CONCEPTS

1. Early Investigations and Classification of Phenomena

The concept and theory of Newtonian fluid behavior was fully established by the 1850s. Experimental studies of the rheological behavior of complex systems date to the nineteenth century. As early as 1867, James Clerk Maxwell (127) identified stress relaxation and viscoelastic behavior in various solids, and a few years later (128) he discussed the response of Canada balsam and gelatin as fluids with memory, "a sensible time of relaxation." In 1889, Schwedoff (184) measured the creep behavior of a gelatin solution and was thus apparently the first to quantitatively determine viscoelastic behavior in a near-liquid system. Schwedoff also found the existence of a yield value in shear flow, that is, a stress below which there is no flow. In 1885 Osborne Reynolds (179) described another type of fluid whose deformation behavior deviates sharply from the predictions of the Navier–Stokes equations. In studying the deformation behavior of an assemblage of rigid particles whose interstices were filled with water, Reynolds found that the system expanded volumetrically during shearing deformation. Reynolds introduced the term *dilatancy* (from dilation–volumetric enlargment) to represent this new type of behavior.

We now come to the opening years of the twentieth century. This was the era of a great effort to study the properties of colloidal systems. Bingham (18) points out that the early colloidal chemists considered that the viscometer was to colloid chemistry what the galvanometer was to electrical science. While the work of Schwedoff was apparently forgotten, colloid chemists noted that their materials seemed to show viscosities varying with duration of shear and flow. The existence of a viscosity that decreased with shear rate was conclusively shown by the experiments of Hatschek (75) and by Bingham and Green (17, 18). Bingham found that the suspensions he studied were actually soft solids possessing yield values, and thus the term Bingham plastics. Beyond the yield value, the deformation rate would, according to Bingham, vary proportionally with the difference between the applied stress and the yield stress. However, more extensive experimental data, especially on polymer solutions, for example, by Farrow, Lowe, and Neale (42), and Ostwald (162, 163), showed that Bingham's attempt to maintain linearity was not generally true. The phenomenon of a viscosity decreasing with shear has been called *strukturviskositat* by Ostwald (162), *pseudoplasticity* by Williamson, and *non-Newtonian* by Reiner (177, 178).

Studies of elastic recovery following flow in colloidal suspensions and polymer solutions were reported by various investigators in the early 1920s (160). Herzog and Weissenberg (81) carefully discussed elastic energy in flow and the development of elastic recovery during flow. Eisenschitz (39) argued that non-Newtonian viscosity may be traced to viscoelastic behavior. Elastic recovery phenomena in rubber following flow was measured by Mooney, and studies of the stress response of fluids to sinusoidally oscillating flows were initiated by Eisenschitz and Philippoff (40, 167).

Other phenomena were noted in suspensions in this period. One effect, which Freundlich (51, 52, 54) called *thixotropy* (*thixis*, touch or movement; *trepo*, to change), consists of fluid behavior in which the apparent viscosity decreased with increasing extent of deformation of an initially virgin material. When the material is not being deformed, the viscosity builds up to its initial value. Freundlich and Juliusberger (53) found a second class of fluid behavior in which the viscosity of a colloidal system increases during deformation, and this phenomenon they denoted *rheopexy* (*rheo*, to flow; *pectous*, solidified). Freundlich and Roder (54) extended Reynold's observations (179) on dilatant suspensions, noting that the mechanism proposed by Reynolds would imply that the viscosity increases with shear rate. Freundlich and his coworkers (53, 54) have studied the occurrence of and relationship between thixotropy, yield values, and dilatancy. Systems with small particles were noted to exhibit both thixotropy and yield values. Dilatancy was observed in suspensions with large particles.

2. Beginnings of Quantitative Experimental Rheological Research on Polymers (1925–present)

In the late 1920s and during the 1930s there was a considerable program to comprehensively study the non-Newtonian viscosity of polymer systems and suspensions. The variation of viscosity of polymer solutions over a wide range (decades) of shear rate was investigated. The general dependence was first found by Ostwald (162) in 1925 and more extensive results were given in succeeding years by Reiner (177, 178) and Philippoff (168). At very low shear rates, the viscosity has a value of η_0, which remains constant up to a characteristic shear rate range and then steadily decreases in a second characteristic shear rate range, above which it maintains a constant value, η_∞.

The first melt viscosity data on bulk polymers appeared in this same period. During the 1920s and 1930s several investigators (15, 34–36, 41, 61) repeated similar extrusion studies of natural rubber and its compounds, which was the major industrially important bulk polymer material of that period. These showed non-Newtonian viscous character. Dillon and Johnston (35) suggested the existence of yield values in rubber–carbon black compounds. The possibility of die-wall slippage and shear viscosity of natural rubber are first found in the work of Mooney (141), published in the same period. Mooney (142) later described the perspectives and concerns of rheological studies of rubber in this period.

Although the rheological studies of the 1920s and 1930s focused on non-Newtonian shear viscosity and thixotropic behavior, attention was also given to the elastic properties, especially recoil following flow (75, 140, 163). Quantitative studies of the dynamic viscosity response, η', to small oscillations were initiated by Philippoff (167–169) in the 1930s and showed behavior similar to that of the $\eta(\dot\gamma)$ of polymer solutions.

The modern period of experimental rheological research on polymer systems dates to about 1940. Flory (43), and later Fox and Flory (46–48), from 1948 investigated the relationship between the zero shear viscosity and molecular weight in polymers. These studies resulted in the 3.4 power-law relationship. Quantitative studies of the small-strain viscoelastic behavior were initiated by Leaderman (107, 108) and then carried out extensively on polyisobutylene samples by Tobolsky and his coworkers (4, 198).

The 1950s saw the beginnings of normal-stress measurements on polymer solutions and melts. The early work of Pollett (170, 171), Greensmith and Rivlin (61), Markovitz and Williamson (114), and Brodnyan, Gaskins, and Philippoff (22, 23) are noteworthy. There was concern over both normal-stress functions N_1 and N_2, the first of which was easily established to be positive. Correlations of N_1 with molecular weight and its distribution were developed (138, 151, 197, 213). In time it came to be generally accepted that N_2/N_1 was negative and of magnitude 0.1–0.3 (21, 38, 55, 57, 91, 96, 105, 134, 152, 176, 180, 196).

Since the mid-1960s attention has turned to the influence of molecular-weight distribution on the shear rate dependence of the reduced viscosity η/η_0 (59, 138, 158, 172, 203). Various plots of η/η_0 versus reduced shear rate or stress were proposed. With broadening distribution in linear polymers, η/η_0 fell off more rapidly.

Elongational viscosity measurements on polymer melts only date to the late 1960s (11, 28, 129). This subject has been a topic of considerable attention for a generation. It is only since about 1980 that there have been studies of molecular-weight distribution effects on this property (138, 224).

Measurements on more complex polymer melt systems, such as suspensions of particles on polymer melts, block copolymers, and liquid crystalline systems, have only been of concern since the 1970s.

C. POLYMER MELTS

1. Shear Viscosity Function

The shear viscosity function η_0 is found in polymer melts to be a constant η_0 at low shear rates and to decrease at higher shear rates $\dot{\gamma}$. This viscosity decreases with increasing temperature and increases with molecular weight, and the non-Newtonian character becomes more pronounced. This is shown in Figure 5-C-1. If we plot η_0 versus molecular weight for linear polymers, such as vinyl polymers and polydienes, we obtain a slope of about unity in the low-molecular-weight region and about 3.4 or 3.5 in the higher-molecular-weight region, that is,

$$\eta_0 = k'_\eta M, \qquad M > M_c \qquad (5\text{-}C\text{-}1a)$$

$$\eta_0 = k''_\eta M^{3.5}, \qquad M > M_c \qquad (5\text{-}C\text{-}1b)$$

This result is from Fox and Flory (*46–48*) and has been confirmed by numerous later investigations (*14, 49, 50, 62, 101, 138, 173*). Generally, it is found for polydisperse systems that the molecular weight which should be used is a weight-average molecular weight M_w.

The transition marked by M_c in Eq. (5-C-1) represents in some sense that between low-molecular-weight and high-molecular-weight polymers. It is sometimes called the entanglement transition and is considered to present the molecular weight at which entangling of the flexible polymer chains occurs. The value of M_c varies considerably from polymer to polymer. It is 4000 for polyethylene and 40,000 for polystyrene. The value of M_c would seem to correspond to a number of chain atoms Z_c in vinyl polymers that is independent of polymer type, where Z_c would appear to be about 350.

Early studies of the viscosity of branched polymers by Schaefgen and Flory (*183*) on polyamides indicated that η_0 at a fixed molecular weight was reduced by the presence of long-chain branching. This was confirmed by Masuda et al. (*120, 121*) for narrow-molecular-weight distribution star-branched polystyrenes

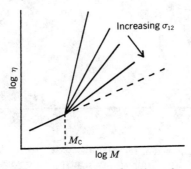

FIGURE 5-C-1. Shear viscosity as a function of molecular weight.

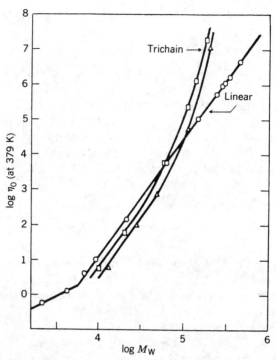

FIGURE 5-C-2. Dependence of Newtonian viscosity of branched polybutadienes on molecular weight: (○) linear; (□) trichain; (△) tetrachain [after Kraus and Gruver (*102*) with permission of publisher].

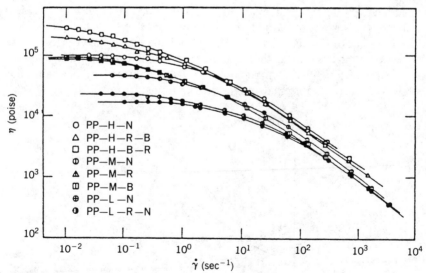

FIGURE 5-C-3. Typical viscosity–shear rate data for polypropylene melts [after Minoshima et al. (*138*)]. The symbols represent polypropylenes of varying molecular weight and distribution.

with four and six branches. They found that η_0 increased with about the 4.5 power of molecular weight. A different result was found by Kraus and Gruver (102) using narrow-molecular-weight distribution tri- and tetra-chain polybutadienes. They found that η_0 is decreased by branching in the low-molecular-weight region, but increases more rapidly at higher molecular weights. At about 60,000 for the tri-chain polymer and 100,000 for the tetra-chain polymer, the viscosity crosses the linear data and becomes higher (Fig. 5-C-2). They suggest the discrepancy with observations on other polymers is due to the short chain lengths, Z, of the polyamides and polystyrenes. The observations of Kraus and Gruver have been confirmed by Folt (45).

The influence of molecular weight and its distribution on viscosity–shear rate behavior has received considerable study. Below M_c, only Newtonian behavior is observed. Above M_c and the entanglement transition, strongly non-Newtonian viscosity is found (Fig. 5-C-3). Generally, broadening molecular-weight distribution increases deviations from Newtonian flow at low shear rates (138, 172). To quantitatively test the effect of molecular-weight distribution on linear polymers, a method of data reduction is required. Porter et al. (172) proposed reduced plots of η/η_0 as a function of shear stress σ_{12} multiplied by absolute temperature and reduced by concentration (for solutions). Later, Vinogradov and Malkin (203) suggested representing shear viscosity data in terms of plots of η/η_0 versus $\eta_0\dot{\gamma}$. This procedure gives a temperature-independent universal plot. It is found that experimental data so plotted take the form shown in Figure 5-C-4 (taken from Minoshima et al. ref. 138) with η/η_0 for broader molecular-weight distribution linear polymers falling off more rapidly than for narrower distribution polymers. The narrower molecular-weight distribution polymers show, after first exhibiting a slow decrease in η/η_0, a sudden rapid decrease.

The approach of Minoshima et al. (138) described above has been calibrated and quantified by Yamane and White (220). Knowledge of η/η_0 at specific $\eta_0\dot{\gamma}$ allows the estimation of M_w/M_n at least for polymers of similar molecular-weight distribution (see Table 5-C-1). They applied this to study the narrowing of the

TABLE 5-C-1. Non-Newtonian Shear Viscosity $\eta/\eta_0(\eta_0\dot{\gamma} = 10^5$ Pa $= 10^6$ dyn/cm^2) as a function of Molecular Weight Distribution Breadth[a]

M_w/M_n	$\eta/\eta_0(\eta_0\dot{\gamma} = 10^6\,\text{dyn/cm}^2)$
1.06	0.515
2.6	0.375
3.5	0.380
4.6	0.270
6.7	0.255
9.0	0.180
17.0	0.173

[a]From Yamane and White (220).

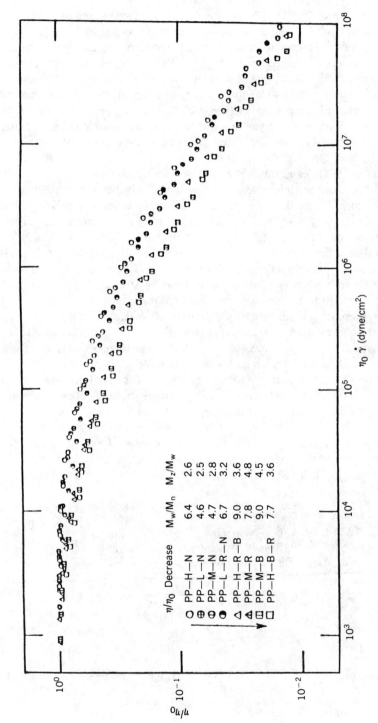

FIGURE 5-C-4. Vinogradov–Malkin reduced viscosity plot showing effect of molecular weight distribution [after Minoshima et al. (138)]. Data is polypropylene.

molecular-weight distribution of polypropylene and polybutene-1 due to thermal degradation.

A similar plotting procedure has been proposed by Graessley (59), which involves plotting η/η_0 versus $\tau_G\dot{\gamma}$, where τ_G is a characteristic time defined in terms of the reciprocal of the shear rate at which η/η_0 begins to decrease. Thus, $\tau_G\dot{\gamma}$ is really a $\dot{\gamma}/\dot{\gamma}_G$ where $\dot{\gamma}_G$ is a characteristic shear rate. It is found in Graessley plots as in Vinogradov–Malkin plots that η/η_0 falls off more rapidly for broader molecular-weight-distribution polymers (158).

Viscosity decreases with increasing temperature. The specific temperature dependence of the zero shear is generally expressed in terms of the Williams–Landel–Ferry (WLF) (215) parameter. Specifically,

$$\eta_0(T) = a_T\eta_0(T_s)$$

or

$$\eta_0(T) = a_T\frac{\rho T}{\rho_s T_s}\eta_0(T_s) \tag{5-C-2}$$

where

$$\log a_T = -\frac{C_1(T - T_s)}{C_2 + (T - T_s)} \tag{5-C-3}$$

where T_s is a characteristic temperature and C_1 and C_2 are constants. T_s may be taken as T_g, the glass transition temperature, in which case C_1 and C_2 have the values 17.4 and 51.6. This expression holds best within 100°C of T_g. Clearly the closer T is to the T_g, the larger the variations of a_T with T. Polystyrene with a T_g of 100°C shows a much greater temperature dependence than polypropylene with an effective T_g of about $-15°C$. The lowest temperature dependence of the viscosity is found with polyethylene, which has an effective T_g of $-100°C$.

It is often convenient to represent the temperature dependence of viscosity through an activation energy E of viscous flow.

$$\eta_0 = Ae^{-E/RT} \tag{5-C-4}$$

Generally E increases with increasing T_g, as might be expected from Eq. (5-C-3), and is thus larger for polystyrene than polyethylene when comparisons are made at the same temperature. In the non-Newtonian viscosity range E is best determined at constant stress.

The viscosity $\eta(\dot{\gamma})$ is also pressure-dependent, generally increasing with increasing pressure (29, 166).

2. Principal Normal Stress Difference

Measurements of the principal normal stress difference N_1 for polymer melts were first initiated by Pollett and Cross (170, 171) in the 1950s using a Weissenberg rheogoniometer. A new generation of investigations began with the work of King (97) in 1965, which was followed by the growth of a moderately extensive literature $(2, 10, 65, 67, 84, 109, 138, 214)$. We limit our citations here to papers using the cone–plate geometry, which yields N_1, and exclude parallel-disc instruments, which give a combination of N_1 and N_2. These studies indicated that N_1 was a monotonic, positive increasing function of shear rate. At low shear rates, the normal stress coefficient Ψ_1 seems to go asymptocally to a constant value. We should also cite the slit rheometer studies of Han $(63, 65–71)$, which give higher shear-rate values of N_1.

Investigations of the influence of molecular weight and its distribution on N_1 and Ψ_1 have been reported in papers by the author and his coworkers $(138, 150, 214)$. White and Kondo (213) report that Ψ_1 varies with about the sixth power of molecular weight for polystyrene at a shear rate of 1 sec^{-1}. They surmise that at very low shear rates both Ψ_1 and η are constant at Ψ_{10} and η_0. From the theory of second-order fluids (Section 7-F-6),

$$\lim_{\dot{\gamma} \to 0} \Psi_1 = \Psi_{10} = K_{\psi\eta}\eta_0^2 \qquad (5\text{-}C\text{-}5a)$$

where $K_{\psi\eta}$ is independent of molecular weight. This suggests that Ψ_{10} depends on about the seventh power. Thus

$$\Psi_{10} = K'_\psi(M^{3.5}) = K'_\psi M^7 \qquad (5\text{-}C\text{-}5b)$$

Han and Yu $(65, 72)$ have found that the exit-pressure normal-stress data are independent of temperature if they are plotted as a function of shear stress. A more extensive use of $N_1 - \sigma_{12}$ plots has been proposed by Oda et al. (151), who found data for polystyrenes to be not only independent of temperature, but also of molecular weight, as long as the molecular-weight distribution was not varied. However, variation of the breadth of the molecular-weight distribution changes the position of the data on the plot. Broadening the distribution increases N_1 at fixed σ_{12}. This is shown in Figure 5-C-5. It was also found that increasing the breadth of the distribution decreases the slope of the plot. This was confirmed for polypropylenes by Minoshima et al. (138).

If we note that Eq. (5-C-5a) is equivalent to

$$N_1 = K_{\psi\eta}\sigma_{12}^2 \qquad (5\text{-}C\text{-}6)$$

the plots of the paragraph above indicate that $K_{\psi\eta}$ is dependent upon molecular-weight distribution. Minoshima et al. (138) note, on the basis of available data in the literature, that

$$K_{\psi\eta} = K'\left(\frac{M_z}{M_w}\right)^{3.5}, \qquad \Psi_{10} = K'M_z^{3.5} M_w^{3.5} \qquad (5\text{-}C\text{-}7a,b)$$

As the shear rate increases, the dependence of Ψ_1 upon molecular weight decreases, as in the case of the viscosity function.

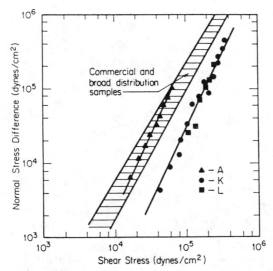

FIGURE 5-C-5. Plot of N_1 versus σ_{12} showing molecular weight distribution effect [after Oda et al. (*151*)].

It may be seen that the magnitude of the principal normal stress difference, properly interpreted, is a sensitive index of the presence of high-molecular-weight fractions in linear polymers.

Generally, N_1 is found to be increased at fixed σ_{12} for long-chain branched polymer melts in comparison to linear melts at the same molecular-weight distribution. This has been described by IUPAC working party investigators for polyesters (*215*) and polyethylenes (*216*).

3. Second Normal Stress Difference

There have been few studies of the second normal stress difference N_2 for polymer melts. Those studies that do exist, by Sakamoto and Porter (*182*), Lee and White (*109*), Hayashida and Kato (*77*) and Ehrmann (*38*), generally confirm the more extensive investigations on polymer solutions (Section F-3 of this chapter), which we will describe in another section. N_2 is found to be about an order of magnitude less than N_1 and to be negative in sign.

4. Elongational Flow

Uniaxial elongational flow experiments on polymer melts often have difficulties which prevent the determination of steady-state data. The greatest problem is the development of nonuniformities in the diameter of the filament, as described in the papers of Ide and White (*88*) and Minoshima et al. (*138*). This leads to neck development and filament failure (Fig. 5-C-6). The polymers that have consist-

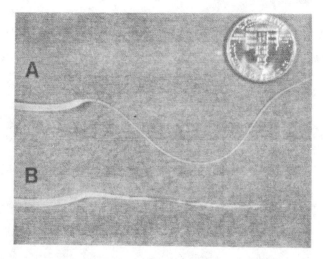

FIGURE 5-C-6. Development of necks in uniaxial elongational flow [after Minoshima et al. (*138*)]. Material B (polypropylene) develops necks and fails. Material A (a narrow molecular-weight distribution polypropylene) draws out more uniformly.

ently given this difficulty are high-density (linear) polyethylenes (HDPE) and polypropylenes (PP). Minoshima et al. argue on the basis of a series of polypropylenes that this difficulty is associated with breadth of molecular-weight distribution in linear polymer systems. This seems consistent with other linear polymer systems for which experimental data is available. It is confirmed in studies by Yamane and White (*220*), who investigated the characteristics of thermally degraded polypropylene and polybutene-1. Degradation narrows the molecular-weight distribution and produces filaments that elongate more uniformly. Long-chain branching also stabilizes filaments (*88*).

Experimental studies of uniaxial elongational flow behavior of polymer melts in which stable elongational flow is found generally show that at low stretch rates $\dot{\gamma}_E$ the elongational viscosity χ goes asymptotically to $3\eta_0$ (*11, 28, 29, 88, 106, 129, 189, 202, 205*)

$$\lim_{\dot{\gamma}_E \to 0} \chi = 3\eta_0 \tag{5-C-8}$$

This is identical to Trouton's result for Newtonian fluids (*201*). The polymers exhibiting this behavior have included polystyrene (PS) (*88, 144, 202*), polyisobutylene (PIB) (*189, 205*), low-density polyethylene (LDPE) (*28, 88, 129*), and certain polypropylenes (PP) (*138, 220*). There have been difficulties with high-density polyethylenes (HDPE) and polypropylenes in verifying this, which seems to be due in large part to necking in the filaments. Usually, decreasing

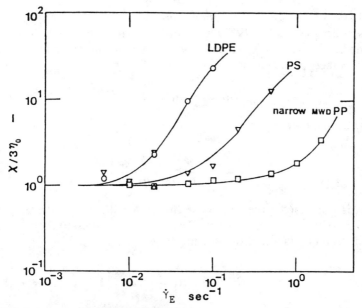

FIGURE 5-C-7. Uniaxial elongational viscosity as a function of deformation rate [after Ide and White (*88*), Minoshima et al. (*138*)].

elongational viscosity functions are found when these functions are estimated.

We restrict further considerations to melts that do achieve a steady state and show the $3\eta_0$ asymptote. Various research groups have reported that $\chi(\dot{\gamma}_E)$ increases with deformation rate for LDPE (*28, 88, 129*), PS (*88, 144, 202*), PIB (*205*), and certain PPs (*138*) (see Fig. 5-C-7). Laun and Munstedt (*106*) found in their experiments for LDPE that $\chi(\dot{\gamma}_E)$ increases from $3\eta_0$, goes through a maximum, and then decreases. Munstedt (*144*) interprets his experiments on PS in a similar manner.

Experimental observations for more complex flows are rather limited. Winter et al. (*216*) found for planar elongational flow that χ_P at low $\dot{\gamma}_E$ is $4\eta_0$, as would be expected for a Newtonian fluid. Denson and his coworkers find both χ_P (*33*) and χ_B (*93*) for PIB to be decreasing functions of stretch rate.

5. Small-Strain Behavior

Generally two classes of small-strain experiments on polymer melts have received extensive attention. These are stress relaxation and sinusoidal oscillation experiments, which we will discuss in turn. The results of these two different experiments can be described in a consistent manner using Boltzmann's superposition theory of linear viscoelasticity. We shall, however, delay consideration of this point to Chapter 7.

The basic studies of stress relaxation in polymers were carried out by Tobolsky and his colleagues (4, 198, 199) on PIB (4) and later PS (199). They found, following Leaderman (107), that data obtained at different temperatures could be shifted together to form a master curve. Specifically

$$G(\underline{t}, T) = G(t/a_T, T_0) \tag{5-C-9a}$$

or, as it is often written,

$$G(t, T) = \frac{\rho T}{\rho_0 T_0} G\left(\frac{t}{a_T}, T_0\right) \tag{5-C-9b}$$

where a_T is a shift factor, the same WLF shift factor that defines the temperature dependence of the zero shear viscosity. Tobolsky et al. expressed $G(t)$ in terms of a spectrum $H(\tau)$ of relaxation times defined by

$$G(t) = \int_0^\infty H(\tau) \frac{e^{-t/\tau}}{\tau} d\tau \tag{5-C-10}$$

The spectrum $H(\tau)$ was found for relatively narrow molecular-weight distribution PIBs to have the form (Fig. 5-C-8)

$$H(\tau) = \frac{A}{\sqrt{\tau}}, \qquad \tau < \tau_1 \tag{5-C-11a}$$

$$H(\tau) = H_0, \qquad \tau_2 < \tau < \tau_m \tag{5-C-11b}$$

Here τ_m is known as the maximum relaxation time. It was found that A, H_0, τ_1, and τ_2 were independent of molecular weight, but τ_m exhibited a strong

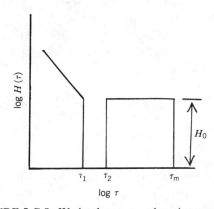

FIGURE 5-C-8. Wedge-box type relaxation spectrum.

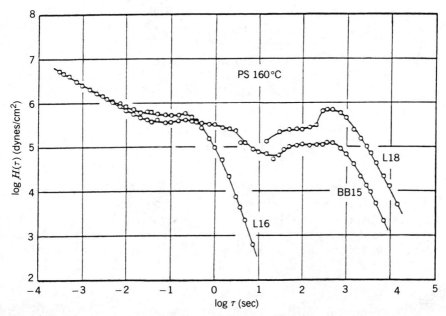

FIGURE 5-C-9. Relaxation spectrum for narrow and broad in molecular-weight-distribution polymers [after Masuda et al. (*117*). With permission of the publisher.]

dependence, which may be expressed as

$$\tau_m = K_\tau M^{3.5} \tag{5-C-12}$$

These results have largely been confirmed by later investigators. Masuda et al. (*117*) found a wedge-box form, but with a spike at the end of the box for narrow distribution PS (Fig. 5-C-9).

The influence of molecular-weight distribution on $H(\tau)$ has been considered by Ninomiya (*149*), Bogue et al. (*20*), and others (*59,60*). These involve linear and quadratic mixing rules. For binary blends

Ninomiya (linear)

$$H_b(\tau) = \phi_1 H_1\left(\frac{\tau}{\lambda_{11}}\right) + \phi_2 H_2\left(\frac{\tau}{\lambda_{22}}\right) \tag{5-C-13a}$$

Bogue–Masuda–Einaga–Onogi (quadratic)

$$H_b(\tau) = \phi_1^2 H_1\left(\frac{\tau}{\lambda_{11}}\right) + 2\phi_1\phi_2 H_{12}\left(\frac{\tau}{\lambda_{12}}\right) + \phi_2^2 H_2\left(\frac{\tau}{\lambda_{22}}\right) \tag{5-C-13b}$$

where ϕ_j corresponds to the volume fraction of component i, $H_j(\tau)$ is the spectrum of component J, λ_{jj} are shift factors due to mixing, and H_{12} is a cross term due to mixing. These and other blending rules are critically discussed by Masuda et al. (*122*).

The rheological behavior of polymer melts subjected to sinusoidal oscillations has been extensively investigated in the literature, with the most notable studies being those of Onogi and Masuda (*117, 118, 120, 154, 155*). Generally, it is observed that

$$\lim_{\omega \to 0} \eta'(\omega) = \eta_0, \qquad \lim_{\omega \to 0} G'(\omega) = 0 \qquad \text{(5-C-14a,b)}$$

The $G'(\omega)$ curve extends to lower frequencies in broad molecular-weight distribution samples than in narrower distribution polymers.

Cox and Merz (*31*) have found an empirical relationship between the shear viscosity function $\eta(\dot{\gamma})$ and dynamic properties $\eta'(\omega)$ and $G(\omega)$ which is valid for flexible-chain polymer melts in the lower and intermediate $\dot{\gamma}$ and ω range. Specifically,

$$\eta(\dot{\gamma}) = \sqrt{(\eta'(\omega))^2 + \left(\frac{G'(\omega)}{\omega}\right)^2}\bigg|_{\omega = \dot{\gamma}} \qquad \text{(5-C-15a)}$$

$$\eta(\dot{\gamma}) = \sqrt{(\eta'(\omega))^2 + (\eta''(\omega))^2}\big|_{\omega = \dot{\gamma}} \qquad \text{(5-C-15b)}$$

This is valid in a wide range of polymer melts (Fig. 5-C-10). However, it tends to fail at higher $\dot{\gamma}$.

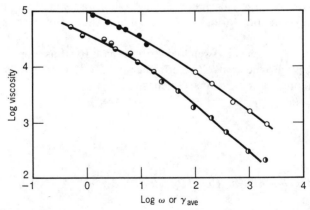

FIGURE 5-C-10. Cox–Merz relation between dynamic and steady shear viscosity [after Cox and Merz (*31*) with permission of the publisher]. Typical flow data for polystyrene B. Complex dynamic viscosity, $|\eta^*|$ (●), compared to steady-flow apparent viscosity, η_a(○); and dynamic viscosity, η'(◐)compared to steady flow consistency, η_c(◑).

6. Reversible Elasticity Reductions in Branched Polymer Melts

Various investigators (74, 84, 175, 181) have noted that continuous shear of long-chain branched polymers substantially reduces the viscoelastic characteristics in a reversible manner, while not affecting the melt viscosity or polymer molecular weight. This phenomenon has been described in both LDPE (74, 84, 181) and branched polyacetals (175). In most cases, elasticity is measured by extrudate swell (74, 84, 181) rather than basic scientific measurements. However, the author has been told of unpublished investigations involving reductions in normal stresses (and normal stress overshoot) following continuous shearing flow in a cone–plate geometry, while viscosity remains the same.

Hanson (74) and later investigators have interpreted this behavior in terms of a disentangling of branched polymer chains. When the melt is allowed to sit quiescently, it is hypothesized that the chains reentangle.

D. BLOCK COPOLYMERS

It has been found that block copolymers of the form AB and ABA tend to segregate into separate regions of A and B segments (104). Most studies have involved butadiene(B)–styrene(S) copolymers with an SBS structure. These separate into a range of morphologies that depend upon the relative polystyrene and polybutadiene contents. Studies of the shear-flow melt viscosity of SBS blocks have been given by Kraus and Gruver (101), Holden et al. (83), and Ghijsels and Raadson (56) representing an IUPAC Working Party program involving a large number of laboratories. The shear viscosity data at low shear rates, as measured by the cone–plate instrument, was highly sensitive to shear history. Ghijsels and Raadson (56) find such data not to be reproducible between laboratories. The material does not exhibit a zero shear viscosity, but increases indefinitely as the shear rate is lowered.

The behavior described above appears to be due to the segregation of the styrene segments in separate regions. Flow involves pulling styrene segments through the butadiene regions. It would appear that the work required to break up the two-phase region to induce flow results in the material having a yield value or at least the appearance of one.

Studies of branched butadiene–styrene block copolymers are described by Kraus et al. (104). Branching decreased the viscosity of all the block copolymer systems at equal M_w.

E. RUBBER-MODIFIED PLASTICS

Basic investigations of the rheological properties of rubber-modified molten plastics have been reported by Lee (111), Zosel (222), Aoki (6), Tanaka and White (194), Munstedt (145) and Saito (182), among others. These are two-phase systems formed during polymerization, often produced by dissolving an elastomer in a monomer which is subsequently polymerized. The most important cases are

HIPS and acrylonitrile–butadiene–styrene (ABS) resins. In most (if not all) cases, the rubber particles are cross-linked and contain graft copolymer on their surface, which makes them more compatible with the matrix polymer.

Studies using large grafted particles, especially at lower loadings (HIPS), indicate behavior similar to homogeneous polymer melts (194). Newtonian low-deformation-rate viscosities are observed. At higher shear rates, the viscosity decreases. Substantial normal stresses are observed, but these appear to be reduced relative to the matrix phase. The Cox–Merz (31) relation between $\eta'(\omega)$ and $\eta(\dot{\gamma})$ is observed.

Rubber-modified plastics with high-volume fractions of smaller ($< 1\,\mu\text{m}$) particles, such as ABS resins, exhibit an unbounded buildup of viscosity at low deformation rates (6, 111, 145, 222), as shown in Figure 5-E-1. This appears to be due to the agglomeration of the rubber particles. The dynamic viscosity $\eta'(\omega)$, which responds to the rubber particle gel, is much greater than $\eta(\dot{\gamma})$ at equivalent deformation rates and the Cox–Merz (31) relation is invalid. The elongational flow behavior is studied by Saito (182).

F. POLYMER BLENDS

Experimental measurements on the rheological properties of polymer melt blends are largely limited to shear viscosity studies. The phase morphology, which can be complex and dependent on variables such as the interfacial tension (110, 136), is usually not measured.

Most studies are limited to a single instrument, usually a capillary rheometer which only yields data in a high shear rate range. Han et al. (66), Lee and White (110), Liang et al. (112), and Min et al. (135) present both cone–plate and extrusion rheometer data for a series of blends (PS/PMMA, PS/PE) which they find to be consistent. At low shear rates, η is a constant η_0. The shear viscosity decreases at higher shear rates.

The dependence of viscosity on composition has in many cases been found to be monotonic in character (3, 77, 110). However, systems exist in which striking minima are observed (1, 68, 72, 110, 135, 136), and in some cases maxima are found (90, 206). Both maxima and minima have also been observed in the same viscosity–composition curves (68, 135). The mechanisms leading to this behavior are not clear. Hypotheses such as that proposed by Hayashida et al. (78), which presume the fluid to flow in parallel lamellae, suggest that the viscosity should change monotonically with composition and be intermediate between the viscosities of the individual components. Presumably, the viscosity extremes with composition are associated with changes in phase morphology with composition.

There are few measurements of normal stresses in polymer blend systems. They are almost entirely by Han and his coworkers (63, 68, 69, 73), though we note the work of Liang et al. (112) and Min et al. (135). Han's results are largely obtained by exit-pressure measurements using a slit rheometer. Maxima and

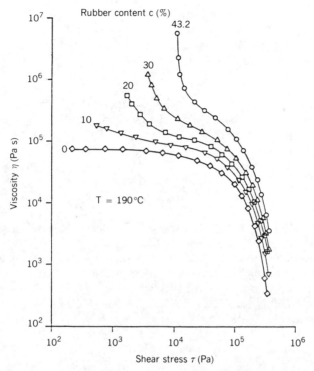

FIGURE 5-E-1. Viscosity as a function of shear stress for ABS with different rubber contents (*145*) (with permission of the publisher).

minima as a function of composition are found. These are certainly attributable to variations in phase morphology.

G. POLYMER (INCLUDING BLOCK COPOLYMER) SOLUTIONS

1. Shear Viscosity

There is a very large literature on the rheological properties of solutions of polymers. The studies involve a range of polymers, including flexible hydrocarbon polymer chains such as PS, PIB, and polydienes which are fully soluble in hydrocarbon solvents, as well as partially soluble polar and rigid polymers. Our understanding of the behavior of the flexible hydrocarbon chains is the greatest and we shall place our major emphasis there.

The most extensive studies of rheological behavior have been for shear viscosity. In dilute solutions, the viscosity increases with concentration according to (*44, 195*)

$$\eta = \eta_0(1 + [\eta]c + k'[\eta]^2c^2 + \cdots) \tag{5-G-1}$$

where $[\eta]$ is the intrinsic viscosity, c is the concentration, and k' is known as the Huggins constant. The intrinsic viscosity is related to molecular weight for monodisperse systems through

$$[\eta] = KM^a \qquad (5\text{-G-}2)$$

which is known as the Mark–Houwink equation. The value of a ranges from a low of about 0.4 for branched flexible chain polymers, to 0.5–0.8 for linear flexible chains (44, 189), to 1.8 for rigid macromolecules (37). For polydisperse systems, the viscosity-average molecular weight

$$M_\eta = \left[\frac{\sum N_i M_i^{a+1}}{\sum N_i M_i} \right]^{1/a} \qquad (5\text{-G-}3)$$

is used in Eq. (5-G-2). The form of Eq. (5-G-3) may be derived from Eqs. (5-G-1) and (5-G-2).

At higher concentrations, the solutions become non-Newtonian. The low shear-rate viscosity η_0 has been found to be empirically related to concentration through a fifth-power relation for PIB (92), PS (82, 87, 99, 116), polydienes (210), and flexible-chain polyamides (5). If we combine this with Fox and Flory's 3.5 power law, we obtain (see Fig. 5-G-1)

$$\eta_0 = K_s M^{3.5} c^5 \qquad (5\text{-G-}4)$$

Onogi and Masuda (153) have proposed that the zero shear viscosity–concentration behavior of polymer solutions may be expressed as

$$\eta = K_s' c^\alpha M^\beta = K'[M^{\beta/\alpha} c]^\alpha \qquad (5\text{-G-}5)$$

They argue that β/α is the same exponent a in the Mark–Houwink equation. Perhaps more generally,

$$\eta_0 = F[M^a c] \qquad (5\text{-G-}6)$$

With increasing shear rate, the viscosity is found to decrease in a manner similar to polymer melts (22, 87, 99, 162, 163). At very high shear rates, the shear viscosity η decreases to a high shear-rate asymptote η_∞ (22, 162, 163).

Masuda and his coworkers (120) have studied the viscosity of solutions of branched polystyrenes. They find that the viscosity of solutions of polystyrene is reduced by long-chain branching at fixed molecular weight and concentration.

The viscosity of solutions of block polymers has been investigated by various authors. Kotaka and White (100) found in solutions of SBS block copolymers that the viscosity exhibits substantial increases with decreasing solvent quality for the styrene blocks. This was accomplished using mixed solvents. This results in the creation of two-phase structures in the solution. For SBS blocks, an elastic gel seems to be produced. For SB blocks, thixotropic time-dependent viscosity

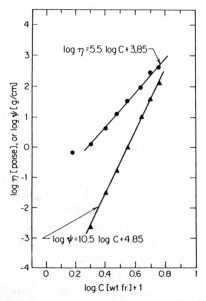

FIGURE 5-G-1. Shear viscosity–concentration behavior of polymer solutions [after Ide and White (87)]. Zero-shear viscosity η and zero-shear principal normal stress coefficient ψ plotted against concentration for the polystyrene–styrene solutions.

behavior is observed. Watanabe and Kotaka (208) have studied the SB block copolymer–polybutadiene–solvent ternary system.

2. Principal Normal Stress Difference

Most investigations of normal stresses in solutions of flexible polymer chains are concerned with the characteristics of individual systems or comparison to constitutive equations and not with dependence upon concentration. An early noteworthy exception is in the work of Brodnyan, Gaskins, and Philippoff (22), who investigated polyisobutylene solutions. They concluded that

$$N_1 = \frac{1}{G}\sigma_{12}^2 \qquad (5\text{-}G\text{-}7)$$

where G is a constant.

Tanner (196) has correlated the extensive PIB solution data in the literature with both concentration and molecular weight. His correlation is of the form

$$N_1 = 1.2 \times 10^{-5}\left(\frac{cM}{10^6}\right)^{0.13}\frac{\sigma_{12}^{1.9}}{c^2} \qquad (5\text{-}G\text{-}8)$$

This does not differ significantly from the hypothesis of Brodnyan et al.

If we assume that the zero shear viscosity depends upon concentration and molecular weight through Eq. (5-G-4), it follows that the low-shear-rate principal stress difference coefficient Ψ_{10} may be expressed through

$$\Psi_{10} = K'_{\psi s} M^{7.0} c^{10} \text{ (Brodynan et al.)} \tag{5-G-9a}$$

$$\Psi_{10} = K''_{\psi s} M^{6.8} c^{7.6} \text{ (Tanner)} \tag{5-G-9b}$$

Kotaka et al. (99) find that Ψ_1 for polystyrene solutions is of the form

$$\Psi_{10} = K'_s M^{6.8} c^b \tag{5-G-9c}$$

where b is between 7.5 and 8.0. Ide and White (87) find a 10.5 power for b in polystyrene–styrene solutions. In any case, the dependence of Ψ_{10} on concentration is much greater than for the viscosity.

Kotaka et al. (99) suggest that there are strong molecular-weight distribution effects on the $N_1-\sigma_{12}$ relationship, similar to those described earlier for melts.

Kotaka and White (100) have described measurements of the principal normal stress difference in solutions of polybutadiene as a function of solvent quality. Using mixed decalin–decane solvent, N_1 generally follows the trend of σ_{12} and η, so that the $N_1-\sigma_{12}$ relationship is independent of solvent quality.

Measurements of N_1 for solutions of SB and SBS block copolymers have been presented by Kotaka and White (100). For the SBS block copolymers, N_1 varies with σ_{12} so as to be independent of solvent quality. It was found in SB blocks that N_1 increased more rapidly than σ_{12} as solvent quality for the S segment deteriorated.

3. Second Normal Stress Difference

There have been extensive studies of the second normal stress difference on polymer solutions dating to the investigations of Roberts (180) in the 1950s. Early researchers divided into two schools, one following the views of Weissenberg (209) that N_2 was zero, the second following Greensmith and Rivlin (61) and Markovitz and Williamson (114) that N_2 is large and positive. More recent investigations (21, 57, 91, 96, 105, 134, 149, 197) have shown that N_2 for hydrocarbon and polar polymer solutions is negative and ranges from 0 to 20% of N_1.

Christiansen and his coworkers et al. (55) report that N_2 depends on molecular-weight distribution in the same manner as N_1 for concentrated polystyrene solutions.

Ramachandran and Christiansen (176) studied N_2 for solutions of BS block copolymers. The ratio N_2/N_1 was found to have a value of about -0.2.

H. LIQUID CRYSTALLINE SYSTEMS

Liquid crystalline polymer systems exhibit structural order in a state of rest, as may be seen through optical retardation studies using a polarized light microscope (5, 7, 101) (see Chapter 6). These macromolecules, in general, possess largely rigid structures. The classical system of this type are solutions of poly-γ-benzyl glutamate, which is representative of polypeptides. Internal hydrogen bonding makes this polymer form a helix in solution. Liquid crystalline character occurs in both concentrated solutions and melts.

Unbounded viscosity buildup at low shear rates in liquid crystalline bulk polymers such as polyesters (191, 217–220) and cellulose ethers (187, 192) have been described in the literature. These studies indicate yield values in these materials where $\eta'(\omega)$ becomes infinite as ω goes to zero and $G'(\omega)$ remains positive. The Cox–Merz relation of Eq. (5-C-15) is not obeyed with $\eta'(\omega)$ exceeding $\eta(\dot{\gamma})$. This is certainly also associated with the yield value.

For at least one cellulose ether (192), it has been observed that the viscosity exhibits a sharp increase at high $\dot{\gamma}$ and loss of the yield value when a transformation to an isotropic phase occurs by increasing temperature.

Hermans (80) has found that the viscosity of poly-γ-benzyl glutamate solutions first increases with concentration, reaches a maximum, and then decreases. This has been confirmed by later investigators (5, 89). Similar behavior has been found in solutions of p-linked aromatic polyamides (5, 8, 165) (Fig. 5-H-1). The viscosity decrease is associated with the formation of the ordered liquid crystalline phase at this concentration. Similar behavior is found in solutions of other aromatic polycondensates (221).

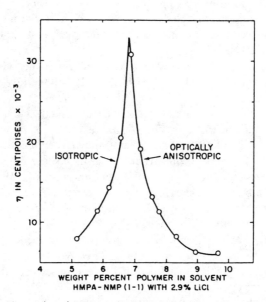

FIGURE 5-H-1. Shear viscosity–concentration behavior of liquid crystalline polymer solution [after P. W. Morgan, *Macromolecules*, **10**, 1381 (1977) with permission of the publisher].

The viscosity–shear rate behavior of poly-γ-benzyl glutamate solutions generally is constant at low shear rates and then decreases with increasing shear rate (5, 86, 89, 98), very much like other polymer solutions. The p-linked aromatic polyamides have been found by some investigators to exhibit yield values in shear flow (5, 165).

Normal stress measurements in liquid crystalline polymer solutions as a function of shear rate have been reported by Aoki et al. (5), Kiss and Porter (98), and Baird et al. (9) among others (219). Kiss and Porter (98) report negative principal normal stress differences. Prasadorao et al. (174) report negative normal stresses in liquid crystalline polyesters. Problems exist in measurement of N_1 because the liquid crystalline systems exhibit yield values.

I. PARTICLE-FILLED POLYMER MELTS AND SOLUTIONS

1. Viscosity–Shear Rate Behavior

Freundlich and Jones (52) have contrasted the rheological behavior of a wide range of concentrated particle suspensions. They note that suspensions of small particles differ from large particles in exhibiting significant aggregation which gives rise to yield values of low shear rates and time-dependent, that is, thixotropic, viscosities.

Particles of varying size and shape have been added to polymer melts and solutions. It is convenient to begin our discussions with large isotropic particles and then proceed to anisotropic particles and finally to very small particles. The characteristics of large particles in suspensions in Newtonian liquids and presumably non-Newtonian matrices may be handled reasonably rigorously by hydrodynamic analyses (12, 13).

The influence of glass spheres of diameter 10–60 μm on melt viscosity has been investigated by various researchers (148, 212). It is generally found that the viscosity is constant at low shear rates and then decreases with increasing shear rates. The η–$\dot{\gamma}$ behavior resembles that of the pure melt.

Studies of the influence of glass fibers of diameter 10–15 μm on the viscosity of polymer melts and solutions have been reported by various investigators (25, 32, 133, 164). Most have been carried out at high shear rates using capillary rheometers. The studies of Chan et al. (25) and Czarnecki and White (32) indicate that the melt exhibits a low-shear-rate Newtonian viscosity and then decreases with increasing shear rates (Fig. 5-I-1). Similar observations were made for cellulose and aramid fiber-filled melts (32, 212).

Czarnecki and White (32) found that viscosity data for fiber-filled polymer melts may be represented through a Vinogradov–Malkin master curve of η/η_0 versus $\eta_0\dot{\gamma}$. The data for the fiber-filled melts superposes with that for the pure melt.

Most studies of the shear viscosity of small-particle-filled polymer melts are only single-point measurements indicating viscosity increase or capillary rheometer investigations in a high shear-rate range, which give little additional basic

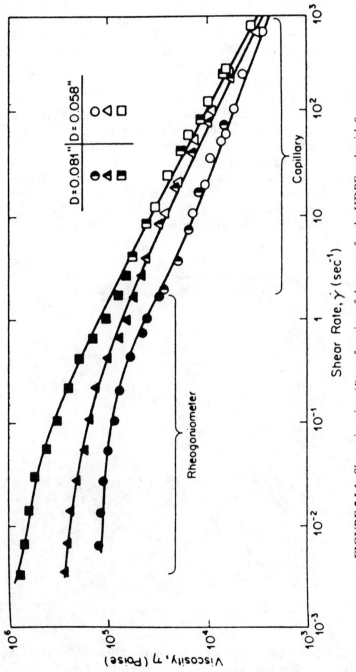

FIGURE 5-I-1. Shear viscosity $\eta(\dot\gamma)$ as a function of shear rate for the HDPE melt with 0 (circles) 20 (triangles), and 40 (squares) wt% glass fibers; $T = 180°C$ [after Chan et al. (25)].

information. More basic studies exist. A shear viscosity that decreases with time over long periods, coupled with a structure–viscosity buildup at rest, was found in suspensions of carbon black in rubber by Mullins and Whorlow (*143*). Similar studies have been reported by Montes et al. (*140*). This indicates the occurrence of thixotropy similar to that observed by Freundlich for suspensions in low-viscosity liquids (*51–54*). A wide range of rheological measurements on rubber–carbon black compounds has been published by Nakajima et al. (*146*), who show the lack of validity of the Cox–Merz relation.

As early as 1933, Dillon and Johnson (*35*) interpreted capillary rheometer viscosity–shear rate behavior of rubber compounds as indicating yield values. It was only in the 1960s that very low shear rate–viscosity behavior of a carbon-black-filled elastomer was shown by Zakharenko et al. (*221*) to exhibit a yield value. The observation of carbon black inducing a yield value in elastomers was extended and confirmed by Vinogradov et al. (*204*) and by White and his coworkers (*139, 188, 200, 214*). More recently, the occurrence of yield values in carbon-black-filled polystyrenes has been noted (*113, 193*) (Fig. 5-I-2). Matsu-

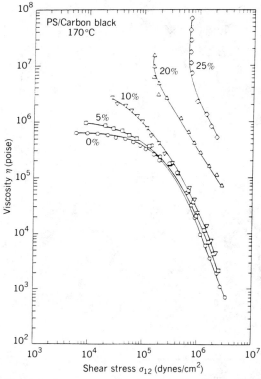

FIGURE 5-I-2. Viscosity–shear stress behavior of carbon-black-filled melts showing the influence of carbon black at various volume fractional percentages [after Lobe and White (*113*)].

moto and Onogi (123–126) have found particle–filled polymer solutions to exhibit yield values.

White et al. (214) have been able to fit rubber–carbon black compound data with empirical equations of the form

$$\sigma_{12} = Y + K\dot{\gamma}^n, \qquad \sigma_{12} = Y + \frac{A\dot{\gamma}}{1 + B\dot{\gamma}^{1-n}} \qquad \text{(5-I-1a,b)}$$

Many other particulates produce compounds with polymer melts that exhibit yield values. This has been found by Chapman and Lee (26) for talc, by Minagawa and White (137) and Tanaka and White (193) for titanium dioxide, and by Kataoka et al. (94, 95), Tanaka and White (193), and Suetsugu and White (190) for calcium carbonate (Fig. 5-I-3). Typical yield values for different particles are shown in Figure 5-I-4.

Generally, the magnitude of the yield value increases with particle loading and with decreasing particle size. Suetsugu and White (190) find that the yield Y varies inversely with the particle diameter in calcium-carbonate-filled polystyrene (Fig. 5-I-4).

The yield value can be strongly influenced by additives. Thus, adding stearic acid to calcium carbonate substantially reduces the yield value of molten polymer

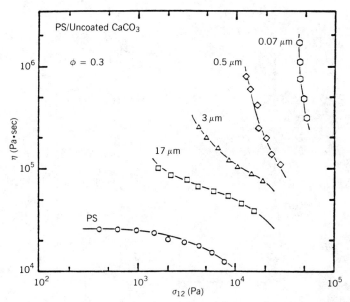

FIGURE 5-I-3. Shear viscosity of polystyrene–calcium carbonate compounds as a function of shear stress with different particle size with $\phi = 0.3$ [after Suetsugu and White (190)].

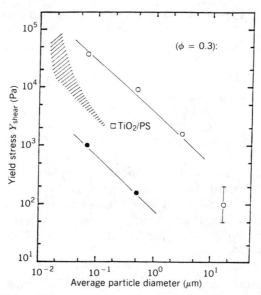

FIGURE 5-I-4. The yield value of 30 volume percent polystyrene–calcium carbonate compounds (○) as a function of particle size; stearic acid treated $CaCO_3$ (●). The shading is for carbon black compounds. Polystyrene–titanium dioxide is also shown.

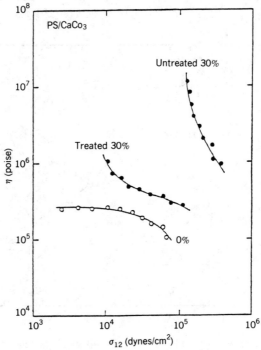

FIGURE 5-I-5. Viscosity–shear stress behavior of $CaCO_3$-filled melts [after Tanaka and White (*193*)]. The treated $CaCO_3$ has been coated with stearic acid.

compounds (*190, 193*) (Figs 5-I-5 and 5-I-4). A similar effect has been reported for addition of titanate coupling agents (*70*).

The mechanism of the yield value seems to be particle–particle attraction, which leads to aggregation and the formation of gel structures. The addition of surface additives reduces particle–particle attraction and the tendency to form a gel structure.

2. Normal Stresses

Chan et al. (*25*) and Czarnecki and White (*32, 212*) have shown that the presence of fibers greatly increases the value of the principal normal stress difference. In terms of an N_1–σ_{12} plot, it has been found that the addition of spherical particles has little influence on the correlation for the pure melt. However, glass, aramid, and cellulose fibers increase N_1 at fixed σ_{12} with the extent of increase depending upon the aspect ratio and modulus of the fibers. Masticating these compounds and degrading the fibers reduces N_1. This indicates that high-volume contents of long rigid fibers give the greatest N_1 (Figure 5-I-6). This phenomenon is completely hydrodynamic in character. Weissenberg effects have been described by several authors (*115, 131, 147*) for suspensions of fibers in Newtonian fluids.

Hutton (*86*) was the first to be concerned with normal stress measurements in polymer systems with large quantities of small particles in his studies of greases.

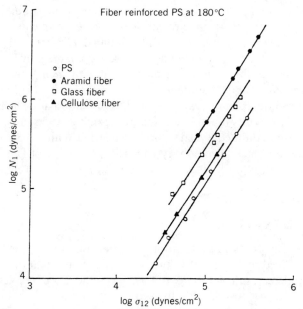

FIGURE 5-I-6. Principal normal stress difference–shear stress data for small particle and fiber-filled polymer melts [after White et al. (*212*)].

Hutton points to the difficulty of measuring normal stresses in materials with yield values in cone–plate rheometers. This eventually led him to the design of special instruments for this purpose (16). Similar behavior is found for melts filled with small particles (113, 190). If we can account for such an effect, measurements in polymer melts filled with small particles seem to indicate substantial reduction in normal stresses. The level of decrease increases with extent of loading. These results are confirmed by the exit-pressure measurements of Han (64) on calcium carbonate compounds.

3. Elongational Viscosity

The elongational viscosity of polymer melts containing large glass spheres has been found to be three times the shear viscosity (147). Chan et al. (25, 203) presented a basic study of elongational flow of fiber-filled polymer melts. The filaments are unstable and readily develop necks. The elongational viscosity function seems to be a decreasing function of extension rate. They explained this in terms of a mechanism developed by Batchelor (13) and expanded by Goddard (58). In elongational flow, the fibers are parallel to each other. They move past each other as the filament is stretched, creating a shearing flow. The tensile force on the filament is stretched, creating a shearing flow between the parallel fibers, which are gliding past each other. As the melt is non-Newtonian in shear flow, the elongational viscosity takes on a form similar to the shear flow except that the magnitude is amplified. White and Czarnecki (211) contrast experimental observations with the predictions of Goddard's theory.

These studies are supported by investigations on Newtonian fluids. Mewis and Metzner (131) have described elongational flow experiments in fiber-filled Newtonian oils. They found a greatly amplified elongational viscosity as predicted from the theory of Batchelor and described in the previous paragraph.

The elongational flow of polymer melts filled with small particles has been investigated by White and his students (113, 187, 193). In studies of carbon-black- (113, 193), calcium-carbonate-(187, 193), and titanium-dioxide-filled (187) melts, the elongational viscosity function was found to exhibit a yield value Y_e whenever the shear viscosity exhibited a yield value Y_s. Lobe and White (113) found Y_e is 1.2 Y_s for their carbon-black-filled polystyrene melt. The studies of Tanaka (193) and Suetsugu (187) with White find that Y_e is about 1.8 Y_s for several compounds that they studied.

4. Small-Strain Experiments

Experimental studies of stress relaxation of molten polystyrene filled with carbon black are described by Lobe and White (113). They found that the carbon black substantially increases the modulus and decreases the rate of relaxation. At high loadings, the stresses do not relax to zero.

The response of a carbon-black-filled polystyrene melt to sinusoidal oscill-ations has been investigated by Lobe and White (113). They found that as frequency ω went to zero, $\eta''(\omega)$ goes to infinity and $G''(\omega)$ goes to a finite value. This behavior is typical of a solid and not a fluid. With pure melts, $G''(\omega)$ goes to zero as ω goes to zero and $\eta''(\omega)$ to the zero shear viscosity η_0. Nakajima et al. (146) found that the Cox–Merz relation (31) of Eq. (5-C-15) is not applicable to rubber–carbon black. The shear viscosity $\eta(\dot{\gamma})$ is much lower than the corresponding dynamic quantity.

Dynamic measurements on polymer solutions filled with small particles have been reported by Matsumoto and Onogi (123, 124, 132, 133, 159, 160). They found that $G''(\omega)$ went to a nonzero value as ω went to zero. This is again the response of a solid.

REFERENCES

1. T. I. Ablazova, M. V. Tsebrenki, A. V. Yudin, G. V. Vinogradov, and B. V. Yarlykov, *J. Appl. Polym. Sci.*, **19**, 1781 (1975).

2. J. W. C. Adamse, H. Janeschitz-Kriegl, J. L. denOtter, and J. L. S. Wales, *J. Polym. Sci.*, **A-2**(6), 871 (1968).

3. N. Alle and J. Lyngaae,-Jorgenson, *Rheol. Acta*, **19**, 94 (1980).

4. R. D. Andrews, N. Hofman-Bang, and A. V. Tobolsky, *J. Polym. Sci.*, **3**, 669 (1948).

5. H. Aoki, J. L. White, and J. F. Fellers, *J. Appl. Polym. Sci.*, **23**, 2293 (1979.)

6. Y. Aoki, *Nihon Reoroji Gakkaishi (J. Soc. Rheology Japan)*, **7**, 20 (1979).

7. T. Asada, H. Muramatsu, and S. Onogi, *Nihon Reoroji Gakkaishi (J. Soc. Rheol. Japan)*, **6**, 130 (1978).

8. D. C. Baird, *Trans. Soc. Rheol.*, **24**, 465 (1980).

9. D. G. Baird, R. L. Ballman, and A. E. Everage, *Rheol. Acta*, **19**, 183 (1980).

10. T. F. Ballenger, I. H. Chen, J. W. Crowder, G. E. Hagler, D. C. Bogue, and J. L. White, *Trans. Soc. Rheol.*, **15**, 195 (1971).

11. R. L. Ballman, *Rheol. Acta*, **4**, 137 (1965).

12. G. K. Batchelor, *J. Fluid Mech.*, **46**, 813 (1971); *G. K. Batchelor and J. T. Green, J. Fluid Mech.*, **56**, 401 (1972).

13. G. K. Batchelor, *J. Fluid Mech.*, **46**, 813 (1971).

14. G. F. Baumann and S. Steingiser, *J. Polym. Sci.*, **A1**, 3395 (1963).

15. J. Behre, *Kautschuk*, **8**, 2 (1932).

16. D. M. Binding, J. F. Hutton, and K. Walters, *Rheol. Acta*, **15**, 540 (1976).

17. E. C. Bingham, *J. Wash. Acad. Sci.*, **6**, 177 (1916).

18. E. C. Bingham, *Fluidity and Plasticity*, McGraw-Hill, New York, 1922.

19. E. C. Bingham and H. Green, *Proc. ASTM*, **19**, 640 (1919).

20. D. C. Bogue, T. Masuda, Y. Einaga, and S. Onogi, *Polym. J.*, **1**, 563 (1970).

21. J. Brindley and J. M. Broadbent, *Rheol. Acta*, **12**, 48 (1973).

22. J. G. Brodnyan, F. H. Gaskins, and W. Philippoff, *Trans. Soc. Rheol.*, **1**, 109 (1957).

23. J. G. Brodnyan, F. G. Gaskins, W. Philippoff, and E. G. Lendrat, *Trans. Soc. Rheol.*, **2**, 285 (1958).

24. W. F. Busse and R. Longworth, *Trans. Soc. Rheol.*, **6**, 179 (1962).

25. Y. Chan, J. L. White, and Y. Oyanagi, *Trans. Soc. Rheol.*, **22**, 507 (1978).

26. F. M. Chapman and T. S. Lee, *SPE J.*, **26**(1), 37 (1970).

27. J. M. Charrier and J. M. Rieger, *Fiber Sci. Technol.*, **7**, 191 (1974).

28. F. N. Cogswell, *Plast. Polym.*, **36**, 109 (1968); *Rheol. Acta*, **8**, 187 (1969).

29. F. N. Cogswell, *Plast. Polym.*, **41**, 39 (1973).

30. F. N. Cogswell, *Appl. Polym. Symp.*, **27**, 3 (1975).

31. W. P. Cox and E. H. Merz, *J. Polym. Sci.*, **28**, 619 (1958).

32. L. Czarnecki and J. L. White, *J. Appl. Polym. Sci.*, **25**, 1217 (1980).

33. C. D. Denson and D. L. Crady, *J. Appl. Polym. Sci.*, **18**, 1611 (1974).

34. J. H. Dillon, *Physics*, **7**, 73 (1936).

35. J. H. Dillon and N. Johnson, *Physics*, **4**, 225 (1933).

36. J. H. Dillon and P. M. Torrance, *Physics*, **6**, 53 (1935).

37. P. Doty, J. H. Bradbury, and A. M. Hotzer, *J. Am. Chem. Soc.*, **78**, 947 (1956).

38. G. Ehrmann, *Rheol. Acta*, **15**, 8 (1976).

39. R. Eisenschitz, *Kolloid Z.*, **54**, 184 (1933).

40. R. Eisenschitz and W. Philippoff, *Naturwiss.*, **28**, 527 (1933).

41. F. Elliott, *Trans. Inst. Rubber Ind.*, **3**, 468 (1927–8).

42. F. D. Farrow, G. M. Lowe, and S. M. Neale, *J. Text. Inst.*, **19**, T18 (1928).

43. P. J. Flory, *J. Am. Chem. Soc.*, **62**, 1057 (1940).

44. P. J. Flory, *Principles of Polymer Chemistry*, Cornell University Press, Ithaca, New York, 1953.

45. V. M. Folt, *Rubber Chem. Technol.*, **42**, 1294 (1969).

46. T. G. Fox and P. J. Flory, *J. Am. Chem. Soc.*, **70**, 2384 (1948).

47. T. G. Fox and P. J. Flory, *J. Appl. Phys.*, **21**, 581 (1950).

48. T. G. Fox and P. J. Flory, *J. Phys. Chem.*, **55**, 221 (1951).

49. T. G. Fox, S. Gratch, and S. Loshaek in *Rheology*, Vol. 1, F. R. Eirich Ed., Academic, New York, 1956.

50. T. G. Fox and S. Loshaek, *J. Polym. Sci.*, **15**, 371 (1955).

51. H. Freundlich and L. L. Bircumshaw, *Kolloid Z.*, **40**, 19 (1926).

52. H. Freundlich and A. D. Jones, *J. Phys. Chem.*, **40**, 1217 (1936).

53. H. Freundlich and F. Juliusberger, *Trans. Faraday Soc.*, **31**, 920 (1935).

54. H. Freundlich and H. L. Roder, *Trans. Faraday Soc.*, **34**, 308 (1938).

55. H. W. Gao, S. Ramachandran, and E. B. Christiansen, *J. Rheol.*, **25**, 213 (1981).

56. A. Ghijsels and J. Raadsen, *Pure Appl. Chem.*, **52**, 1359 (1980).

57. R. F. Ginn and A. B. Metzner, *Proc. 4th Int. Rheol. Cong.*, **2**, 583 (1965); *Trans. Soc. Rheol.*, **13**, 429 (1969).

58. J. D. Goddard, *J. Non-Newt. Fluid Mech.*, **1**, 1 (1976).

59. W. W. Graessley, *J. Chem. Phys.*, **43**, 2696 (1965).

60. W. W. Graessley, *J. Chem. Phys.*, **47**, 1942 (1967).

61. H. W. Greensmith and R. S. Rivlin, *Phil. Trans. Roy. Soc.*, **A245**, 399 (1953).

62. J. T. Gruver and G. Kraus, *J. Polym. Sci.*, **A2**, 797 (1964).

63. C. D. Han, *Trans. Soc. Rheol.*, **18**, 163 (1974).

64. C. D. Han, *J. Appl. Polym. Sci.*, **18**, 821 (1974).
65. C. D. Han, *Rheology in Polymer Processing, Academic, New York,* 1976.
66. C. D. Han, K. U. Kim, J. Parker, N. Siskovic, and C. R. Huang, *Appl. Polym. Symp.*, **17**, 95 (1973).
67. C. D. Han, K. U. Kim, N. Siskovic, and C. R. Huang, *Appl. Polym. Symp.*, **17**, 95 (1973).
68. C. D. Han and Y. W. Kim, *Trans. Soc. Rheol.*, **19**, 245 (1975).
69. C. D. Han, Y. M. Kim, and S. J. Chen, *J. Appl. Polym. Sci.*, **19**, 2831 (1975).
70. C. D. Han, C. Sanford, and H. J. Yoo, *Polym. Eng. Sci.*, **18**, 849 (1978).
71. C. D. Han and C. A. Villamizar, *J. Appl. Polym. Sci.*, **22**, 1677 (1978).
72. C. D. Han and T. C. Yu, *Rheol. Acta*, **10**, 398 (1971).
73. C. D. Han and T. C. Yu, *J. Appl. Polym. Sci.*, **15**, 1163 (1971).
74. D. E. Hanson, *Polym. Eng. Sci.*, **9**, 405 (1969).
75. E. Hatschek, *Kolloid Z.*, **7**, 301 (1913); E. Humphrey and E. Hatschek, *Proc. Phys. Soc.*, **28**, 274 (1916).
76. K. Hayashida and M. Katsuyama, *Bull. Faculty Text. Sci. Kyoto Univ. Ind. Arts Text. Fibers*, **8**, 91 (1977).
77. K. Hayashida and J. Kato, *Kobunshi Ronbonshu*, **31**, 406 (1974).
78. K. Hayashida, J. Takahashi, and M. Matsui, *Proc. 5th Int. Rheol. Cong.*, **4**, 525 (1970).
79. K. Hayashida and S. Yonei, *Rheol. Acta*, **14**, 158 (1975).
80. J. Hermans, *J. Colloid Sci.*, **17**, 638 (1962).
81. R. O. Herzog and K. Weissenberg, *Kolloid Z.*, **46**, 277 (1928).
82. M. Hirose, E. Oshima, and H. Inoue, *J. Appl. Polym. Sci.*, **12**, 9 (1968).
83. G. Holden, E. T. Bishop, and N. R. Legge, *J. Polym. Sci.*, **C26**, 37 (1969).
84. E. R. Howells and J. J. Benbow, *Trans. J. Plast. Inst.*, **30**, 240 (1962).
85. D. C. Huang and J. L. White, *Polym. Eng. Sci.*, **19**, 609 (1979).
86. J. F. Hutton, *Rheol. Acta*, **14**, 979 (1975).
87. Y. Ide and J. L. White, *J. Appl. Polym. Sci.*, **18**, 2997 (1974).
88. Y. Ide and J. L. White, *J. Appl. Polym. Sci.*, **22**, 1067 (1978).
89. E. Iizuka, *Mol. Cryst. Liq. Cryst.*, **25**, 284 (1974).
90. K. Iwakura and T. Fujimura, *J. Appl. Polym. Sci.*, **19**, 1427 (1975).
91. R. F. Jackson and A. Kaye, *Br. J. Appl. Phys.*, **17**, 1355 (1966).
92. M. F. Johnson, W. W. Evans, I. Jordan, and J. D. Ferry, *J. Colloid Sci.*, **7**, 498 (1952).
93. D. D. Joye, G. W. Poehlein, and C. D. Denson, *Trans. Soc. Rheol.*, **16**, 421 (1972).
94. T. Kataoka, T. Kitano, M. Sasahara, and K. Nishijima, *Rheol. Acta*, **17**, 149 (1978).
95. T. Kataoka, T. Kitano, Y. Oyanagi, and M. Sasahara, *Rheol. Acta*, **18**, 635 (1979).
96. A. Kaye, A. S. Lodge, and D. G. Vale, *Rheol. Acta*, **7**, 368 (1968).
97. R. G. King, *Rheol. Acta*, **5**, 35 (1965).
98. G. Kiss and R. S. Porter, *J. Polym. Sci. Polym. Phys.*, **18**, 361 (1980).
99. T. Kotaka, M. Kurata, and M. Tamura, *Rheol. Acta*, **2**, 179 (1962).
100. T. Kotaka and J. L. White, *Trans. Soc. Rheol.*, **17**, 587 (1973).
101. G. Kraus and J. T. Gruver, *J. Appl. Polym. Sci.*, **9**, 739 (1965).

102. G. Kraus and J. T. Gruver, *J. Polym. Sci.*, **P-A**, **3**, 102 (1965).

103. G. Kraus and J. T. Gruver, *J. Appl. Polym. Sci.*, **11**, 2121 (1967).

104. G. Kraus, F. E. Naylor, and K. W. Rollman, *J. Polym. Sci.*, **A2**, 9 (1971).

105. Y. Koo and R. I. Tanner, *Rheol. Acta*, **13**, 951 (1974); A. Keentok, A. G. Georgescu, A. A. Sherwood, and R. I. Tanner, *J. Non-Newt. Fluid Mech.*, **6**, 303 (1980).

106. H. M. Laun and H. Munstedt, *Rheol. Acta*, **17**, 415 (1978).

107. H. Leaderman, *Elastic and Creep Properties of Filamentous Materials*, Textile Foundation, 1943.

108. H. Leaderman, *Ind. Eng. Chem.*, **35**, 374 (1943).

109. B. L. Lee and J. L. White, *Trans. Soc. Rheol.*, **18**, 467 (1974).

110. B. L. Lee and J. L. White, *Trans. Soc. Rheol.*, **19**, 481 (1975).

111. T. S. Lee, *Proc. 5th Int. Rheol. Cong.*, **4**, 421 (1970).

112. B. Liang, J. L. White, J. E. Spruiell, and B. C. Goswami, *J. Appl. Polym. Sci.*, **28**, 2011 (1983).

113. V. M. Lobe and J. L. White, *Polym. Eng. Sci.*, **19**, 617 (1979).

114. H. Markovitz and R. B. Williamson, *Trans. Soc. Rheol.*, **1**, 25 (1957).

115. R. O. Maschmeyer and C. T. Hill, *Adv. Chem. Ser.*, **134**, 95 (1974).

116. T. Masuda, Ph.D. Dissertation, Kyoto University, Kyoto, Japan, 1973.

117. T. Masuda, K. Kitagawa, T. Inoue, and S. Onogi, *Macromolecules*, **3**, 116 (1970).

118. T. Masuda, K. Kitagawa, and S. Onogi, *Polym. J.*, **1**, 418 (1970).

119. T. Masuda, Y. Nakagawa, Y. Ohta, and S. Onogi, *Polym. J.*, **3**, 92 (1972).

120. T. Masuda, Y. Ohta, and S. Onogi, *Macromolecules*, **4**, 763 (1971).

121. T. Masuda, Y. Saito, Y. Ohta, and S. Onogi, *J. Soc. Mater. Sci. Japan*, **22**, 438 (1973).

122. T. Masuda, M. Takahashi, and S. Onogi, *Appl. Polym. Symp.*, **20**, 49 (1973).

123. T. Matsumoto, Ph.D. Dissertation, Kyoto University, Kyoto Japan, 1978.

124. T. Matsumoto, C. Hitomi, and S. Onogi, *Trans. Soc. Rheol.*, **19**, 541 (1975).

125. T. Matsumoto, T. Masuda, K. Tsutsui, and S. Onogi, *Nippon Kagaku Zasshi*, **90**, 360 (1969).

126. T. Matsumoto, A. Takashima, T. Masuda, and S. Onogi, *Trans. Soc. Rheol.*, **14**, 617 (1970).

127. J. C. Maxwell, *Phil. Trans. Roy. Soc.*, **157**, **249** (1867).

128. J. C. Maxwell, *Proc. Roy. Soc.*, **22**, 46 (1873).

129. J. Meissner, *Rheol. Acta*, **10**, 230 (1971).

130. J. Mewis and R. de Bleyser, *Rheol. Acta*, **14**, 721 (1975).

131. J. Mewis and A. B. Metzner, *J. Fluid Mech.*, **62**, 593 (1974).

132. Y. Mikami, T. Matsumoto, and S. Onogi, *Nihon Reoroji Gakkaishi* (*J. Soc. Rheol. Japan*), **4**, 86 (1976).

133. Y. Mikami, T. Matsumoto, and S. Onogi, *Nihon Reoroji Gokkaishi* (*J. Soc. Rheol. Japan*), **4**, 149 (1976).

134. M. J. Miller and E. B. Christiansen, *AIChE J.*, **18**, 600 (1972).

135. K. Min, J. L. White, and J. F. Fellers, *J. Appl. Polym. Sci.*, **29**, 2117 (1984).

136. K. Min, J. L. White, and J. F. Fellers, *Polym. Eng. Sci.*, **24**, 1327 (1984).

137. N. Minagawa and J. L. White, *J. Appl. Polym. Sci.*, **20**, 510 (1976).

138. W. Minoshima, J. L. White, and J. E. Spruiell, *Polym. Eng. Sci.*, **20**, 1166 (1980).

139. S. Montes and J. L. White, *Rubber Chem. Technol.*, **55**, 1354 (1982).

140. S. Montes, J. L. White, and N. Nakajima, *J. Non-Newt. Fluid Mech.*, **28**, 183 (1988).

141. M. Mooney, *Physics*, **7**, 413 (1936).

142. M. Mooney, *Rubber Chem. Technol.*, **35**(5), xxvii (1962).

143. L. Mullins, *J. Phys. Colloid Chem.*, **54**, 639 (1950); L. Mullins and R. W. Whorlow, *Trans. IRI.* **27**, 55 (1951).

144. H. Munstedt, *Rheol. Acta*, **14**, 1077 (1975); *J. Rheol.*, **23**, 421 (1979).

145. H. Munstedt, *Polym. Eng. Sci.*, **21**, 259 (1981).

146. N. Nakajima, H. H. Bowerman, and E. A. Collins, *J. Appl. Polym. Sci.*, **21**, 3063 (1977).

147. M. A. Nawab and S. G. Mason, *J. Phys. Chem.*, **62**, 1249 (1958).

148. F. Nazem and C. T. Hill, *Trans. Soc. Rheol.*, **18**, 87 (1974).

149. K. Ninomiya, *J. Colloid Sci.*, **14**, 49 (1959); **17**, 759 (1962).

150. N. Nishida, D.G. Salladay, and J. L. White, *J. Appl. Polym. Sci.*, **15**, 1181 (1971).

151. K. Oda, J. L. White, and E. S. Clark, *Polym. Eng. Sci.*, **18**, 25 (1978).

152. O. Olabisi and M. C. Williams, *Trans. Soc. Rheol.*, **9**, 365 (1972).

153. S. Onogi and T. Masuda, *J. Polym. Sci.*, **A2**(5), 899 (1967).

154. S. Onogi, T. Masuda, and T. Ibaragi, *Kolloid Z.-Z. Polym.*, **222**, 100 (1968).

155. S. Onogi, T. Masuda, and K. Kitagawa, *Macromolecules*, **3**, 109 (1970).

156. S. Onogi, T. Masuda, and T. Matsumoto, *Nippon Kagaku Zasshi*, **89**, 464 (1968).

157. S. Onogi, T. Masuda, and T. Matsumoto, *Trans. Soc. Rheol.*, **14**, 275 (1970).

158. S. Onogi, T. Masuda, I. Shiga, and F. M. Costaschuk, *Appl. Polym. Symp.*, **20**, 37 (1973).

159. S. Onogi, T. Matsumoto, and Y. Warashina, *Trans. Soc. Rheol.*, **17**, 175 (1973).

160. S. Onogi, Y. Mikami, and T. Matsumoto, *Polym. Eng. Sci.*, **17**, 1 (1977).

161. Y. Onogi, J. L. White, and J. F. Fellers, *J. Polym. Sci. Polym. Phys.*, **18**, 663 (1980); *J. Non-Newt. Fluid Mech.*, **7**, 121 (1980).

162. W. Ostwald, *Kolloid Z.*, **36**, 99 (1925).

163. W. Ostwald and R. Auerbach, *Kolloid Z.*, **38**, 261 (1926).

164. Y. Oyanagi and Y. Yamaguchi, *Nihon Reoroji Gakkaishi* (*J. Soc. Rheol. Japan*), **3**, 64 (1975).

165. S. P. Papkov, V. G. Kulichikhin, V. D. Kalmykova, and A. Y. Malkin, *J. Polym. Sci., Polym. Phys.*, **12**, 1753 (1974).

166. R. C. Penwell and R. S. Porter, *J. Polym. Sci.*, **A-2**(9), 731 (1971).

167. W. Philippoff, *Phys. Z.*, **35**, 844 (1934).

168. W. Philippoff, "Viskositat der Kolloide", Edwards Bros., Ann Arbor, MI, 1943.

169. W. Philippoff and K. Hess, *Z. Phys. Chem.*, **B31**, 237 (1937).

170. W. F. O. Pollett, *Br. J. Appl. Phys.*, **6**, 199 (1955).

171. W. F. O. Pollett and A. H. Cross, *J. Sci. Inst.*, **27**, 209 (1950); *Proc. 2nd Int. Rheol. Cong.*, 85 (1955).

172. R. S. Porter, M. J. R. Cantow, and J. F. Johnson, *Proc. 4th Int. Rheol. Cong.*, **2**, 479 (1965).

173. R. S. Porter and J. F. Johnson, *Proc. 4th Int. Rheol. Cong.*, **2**, 467 (1965).

174. M. Prasadarao, E. M. Pearce, and C. D. Han, *J. Appl. Polym. Sci.*, **27**, 1343 (1982).

175. J. H. Pritchard and K. F. Wissbrun, *J. Appl. Polym. Sci.*, **13**, 233 (1969).

176. S. Ramachandran and E. B. Christiansen, *J. Non-Newt. Fluid Mech.*, **13**, 21 (1983).

177. M. Reiner, *J. Rheol.*, **1**, 11 (1929).

178. M. Reiner, *Deformation and Flow*, Lewid, London, 1948.

179. O. Reynolds, *Phil. Mag.*, **20**(5), 469 (1885).

180. J. E. Roberts, *Proc. 2nd Int. Rheol. Cong.*, (1954).

181. M. Rokudai, *Nihon Reoroji Gakkaishi*, **8**, 161 (1980); *J. Appl. Polym. Sci.*, **23**, 463 (1979).

182. Y. Saito, *Nihon Reoroji Gakkaishi*, **10**, 123, 128, 135 (1982).

183. K. Sakamoto and R. S. Porter, *J. Polym. Sci.*, **B8**, 171 (1970).

184. J. R. Schaefgen and P. J. Flory, *J. Am. Chem. Soc.*, **70**, 2709 (1948).

185. T. G. Schwedoff, *J. Phys.*, (2)**8**, 341 (1889); **9**, 34 (1890).

186. M. Shida and L. V. Cancio, *J. Appl. Polym. Sci.*, **14**, 3083 (1970).

187. K. Shimamura, J. L. White, and J. F. Fellers, *J. Appl. Polym. Sci.*, **26**, 2165 (1981).

188. H. Song, J. L. White, K. Min, N. Nakajima, and F. C. Weissert, *Adv. Polym. Technol.*, **8**, 431 (1988).

189. J. F. Stevenson, *AIChE J.*, **18**, 540 (1972).

190. Y. Suetsugu and J. L. White, *J. Appl. Polym. Sci.*, **28**, 1481 (1983).

191. H. Sugiyama, D. N. Lewis, J. L. White, and J. F. Fellers, *J. Appl. Polym. Sci.*, **30**, 2324 (1985).

192. S. Suto, J. L. White, and J. F. Fellers, *Rheol. Acta*, **21**, 62 (1982).

193. H. Tanaka and J. L. White, *Polym. Eng. Sci.*, **20**, 949 (1980).

194. H. Tanaka and J. L. White, *Polym. Eng. Rev.*, **1**, 89 (1981).

195. C. Tanford, *Physical Chemistry of Macromolecules*, Wiley, New York, 1960.

196. R. I. Tanner, *Trans. Soc. Rheol.*, **14**, 483 (1970).

197. R. I. Tanner, *Trans. Soc. Rheol.*, **17**, 365 (1973).

198. A. V. Tobolsky, *Properties and Structure of Polymers*, Wiley, New York, 1960.

199. A. V. Tobolsky and K. Murakami, *J. Polym. Sci.*, **40**, 443 (1959).

200. S. Toki and J. L. White, *J. Appl. Polym. Sci.*, **27**, 3171 (1982).

201. F. T. Trouton, *Proc. Roy. Soc.*, **A77**, 426 (1906).

202. G. V. Vinogradov, V. D. Fikham, B. V. Radushkevich, and A. Y. Malkin, *J. Polym. Sci.*, **A-2**(8), 1 (1970).

203. G. V. Vinogradov and A. Y. Malkin, *J. Polym. Sci.*, **A-24**, 135 (1966).

204. G. V. Vinogradov, A. K. Malkin, E. P. Plotnikova, O. Y. Sabsai, and N. E. Nikolayeva, *Int. J. Polym. Mater.*, **2**, 1 (170)

205. G. V. Vinogradov, B. V. Radushkevich, and V. D. Fikham, *J. Polym. Sci., Polym. Phys.*, **A-2**(8), 1 (1970).

206. G. V. Vinogradov, B. V. Yarlokov, M. V. Tsebrenko, A. V. Yudin, and T. I. Ablakova, *Polymer*, **16**, 609 (1975).

207. H. L. Wagner and K. F. Wissbrun, *SPE Trans.*, **2**, 222 (1962).

208. H. Watanabe and T. Kotaka, *J. Rheol.*, **27**, 223 (1983).

209. K. Weissenberg, *Proc. 2nd Int. Rheol. Cong.*, I-24 (1948).

210. J. L. White, S. Chankraiphon, and Y. Ide, *Trans. Soc. Rheol.*, **21**, 1 (1977).

211. J. L. White and L. Czarnecki, *J. Rheol.*, **24**, 501 (1980).

212. J. L. White and L. Czarnecki, and H. Tanaka, *Rubber Chem. Technol.*, **53**, 823 (1980).

213. J. L. White and A. Kondo, *J. Non-Newt. Fluid Mech.*, **3**, 41 (1977).

214. J. L. White, Y. Wang, A. I. Isayev, N. Nakajima, F. C. Weissert, and K. Min, *Rubber Chem. Technol.*, **60**, 337 (1987).

215. M. L. Williams, R. F. Landel, and J. D. Ferry, *J. Am. Chem. Soc.*, **77**, 3701 (1955).

216. H. H. Winter, C. W. Macosko, and K. E. Bennett, *Rheol. Acta*, **18**, 323 (1979).

217. K. F. Wissbrun, *Br. Polym. J.*, 163 (1980).

218. K. F. Wissbrun, *J. Rheol.*, **25**, 619 (1981).

219. C. P. Wong, H. Ohnuma, and G. C. Berry, *J. Polym. Sci., Polym. Symp.*, **65**, 173 (1978).

220. H. Yamane and J. L. White, *Polym. Eng. Rev.*, **2**, 167 (1982).

221. N. V. Zakharenko, F. S. Tolstukhina, and G. M. Bartenev, *Rubber Chem. Technol.*, **35**, 326 (1962).

222. K. Zosel, *Rheol. Acta*, **11**, 229 (1972).

VI

RHEO-OPTICS AND FLOW BIREFRINGENCE

A. INTRODUCTION

One of the more striking experimental observations in the flow of polymer solutions and melts is the development of optical anisotropy associated with molecular orientation during flow. This area is known as rheo-optics as distinguished from rheology, which deals with mechanical responses. The refractive indices in different directions in the flow field change as the result of flow, and the magnitude of the difference depends in many flexible chain melts and solutions on the stress field. This optical anisotropy results in retardation of polarized light passing through the fluid. Birefringent fields are observed which allow calculation of the detailed local stress field. It is our purpose in this chapter to describe the phenomenon of rheo-optics, its relationship to stress fields, and structure in polymer fluids. We begin with a historical survey of research in this area. We then turn to a classification of optical properties of polymer fluid and a development of its basis in terms of the electromagnetic theory of light. The relationship between birefringence and molecular orientation is developed. The discussion then turns to the measurement of birefringence during flow. Experimental results and in particular the rheo-optical law between birefringence and stress are described. The relationship of the parameters arising in the rheo-optical law, specifically the stress optical constant, to molecular structure is considered. The chapter closes with a discussion of the interpretation of birefringence in polymers solidified under stress.

B. HISTORICAL BACKGROUND

The phenomenon of materials with optical anisotropy, that is, different refractive indices in different directions, had been recognized as occurring naturally in certain classes of crystals by Huyghens (40) in the seventeenth century. These observations were extended by various investigators, most notably Brewster (8–10) in the early nineteenth century, who noted that birefringence could be induced into certain optically isotropic material, such as organic gels and glass, by the action of applied stresses. Clerk Maxwell (51) carried this one step further by presuming the indicated birefringence varied linearly and isotropically with applied stresses. He showed that the birefringent fields developed in polarized light could be interpreted as stress fields. This gave rise to the development of the area of photoelastic stress analysis. Clerk Maxwell (53) later observed the development of birefringence during the flow of Canada balsam, a naturally occurring polymer solution, after having failed in attempts with sugar solutions. In this same period Clerk Maxwell (52, 54) formulated the electromagnetic theory of light and related the phenomenon of birefringence to the dielectric constant becoming a second-order tensor. Experimental proof of Maxwell's theory was forthcoming from the spark-coil experiments of Hertz (35), who later gave a more general presentation (36) of the electromagnetic theory.

During the first part of the present century, there were extensive investigations of the flow birefringent characteristics of suspensions (6, 39, 62, 67) and polymer solutions (20, 22, 23). The modern perspective of birefringence in flow of polymer fluids dates to the 1940s and 1950s. Kuhn and Grun (44) showed that Maxwell's birefringence–stress relation derived naturally from the molecular theory of networks of flexible chains. In stress relaxation experiments, Stein and Tobolsky (86, 87) found that birefringence decayed proportionally to stress. In 1956 Lodge (46) argued that Kuhn and Grun's arguments could be extended to a network of entangled chains and hence to polymer solutions and melts. This was subsequently verified by Philippoff and his coworkers (12, 13, 21, 63–66) for a range of homogeneous polymer fuids. The first quantitative investigations of birefringence in the flow of polymer melts were later presented by Wales and his coworkers (2, 90, 91). Quantitative verification of the linear birefringence–stress relationship was obtained in various experiments by Janeschitz–Kriegl and his coworkers (25, 88). The method has been extensively used to characterize the stress field in complex flows by Bogue and his students for polymer solutions (1), by Han et al. (27–30, 95), and by Arai and Hatta (3, 5) for polymer melts.

Birefringence has long been realized to be related to the orientation of polymer chains (15, 68). The concept was first quantified by P. H. Hermans and his coworkers (32–34) in the late 1930s and early and mid-1940s for uniaxial orientation. Muller (55) considered the relationship of birefringence to the distribution of chain orientation. Subsequent investigators have extended the perspective to multiaxial orientation (17, 59, 60, 82, 93).

There have been various investigations in recent years on nonisothermal birefringent characteristics. Matsumoto and Bogue (50) found that the birefringence–stress relationship maintains its linearity in polystyrene under nonisothermal conditions. The birefringence "frozen in" quenched polystyrene and polycarbonate is similarly related to the stresses at vitrification (17, 43, 61).

C. CLASSIFICATION OF OPTICAL CHARACTERISTICS OF POLYMER SYSTEMS

Many materials such as common gases, liquids, and glasses are optically isotropic and homogeneous. The interaction of these materials with light waves is characterized by a single constant, a refractive index n defined by

$$n = \frac{c}{v} \qquad (6\text{-C-}1)$$

where c is the velocity of light in a vacuum and v in the medium in question.

In anisotropic materials such as single crystals or oriented polymer systems, the optical characteristics seem to be determined by a refractive index tensor with components

$$\mathbf{n} = \begin{bmatrix} n_{11} & n_{12} & n_{13} \\ n_{12} & n_{22} & n_{23} \\ n_{13} & n_{23} & n_{33} \end{bmatrix} \qquad (6\text{-C-}2)$$

This is generally considered to be a symmetric tensor with n_{ij} equal to n_{ji}. As will be seen later, $\mathbf{n}$ is related to the dielectric tensor.

The refractive index tensor has often been represented as an ellipsoid (44, 69) in a manner similar to the stress, deformation rate, and strain tensors, and this approach will be followed here.

D. BASIC CONCEPTS OF ELECTROMAGNETIC THEORY/OPTICS

To understand the interaction of light through polymeric fluids, it is necessary to concisely describe the electromagnetic theory of light. The interactions of electric and magnetic fields are described by Maxwell's equations (7, 36, 52, 54, 77):

$$\nabla \times \mathbf{E} = -\frac{\partial \mathbf{B}}{\partial t}, \quad \nabla \times \mathbf{H} = \mathbf{J} + \frac{\partial \mathbf{D}}{\partial t} \qquad \text{(6-D-1a–d)}$$

$$\nabla \cdot \mathbf{D} = \rho, \qquad \nabla \cdot \mathbf{B} = 0$$

Here $\mathbf{E}$ is the electrical field strength, $\mathbf{D}$ is Maxwell's displacement vector, $\mathbf{J}$ is the electrical current density, $\mathbf{B}$ is the magnetic field strength, $\mathbf{H}$ is the magnetic excitation, and ρ is the charge density.

In an anisotropic dielectric medium, Maxwell's displacement vector $\mathbf{D}$ is related to the electrical field strength $\mathbf{E}$ through

$$\mathbf{D} = \boldsymbol{\varepsilon} \cdot \mathbf{E}, \qquad D_i = \varepsilon_{ij} E_j \qquad \text{(6-D-2)}$$

Here ε is the dielectric constant tensor. The quantity ε_{ij} has been represented as an ellipsoid and is generally called Fresnel's ellipsoid (78). The electric current flux $\mathbf{J}$ is zero in a dielectric medium. The magnetic behavior is generally linear and isotropic, that is,

$$\mathbf{B} = \mu \mathbf{H} \qquad \text{(6-D-3)}$$

Substituting Eq. (6-D-3) into Eq. (6-D-1a) and taking the curl yields:

$$\nabla \times \nabla \times \mathbf{E} = -\mu \frac{\partial}{\partial t} \nabla \times \mathbf{H} \qquad \text{(6-D-4)}$$

From Eq. (6-D-1b):

$$\nabla \times \nabla \times \mathbf{E} = \mu \boldsymbol{\varepsilon} \cdot \frac{\partial^2 \mathbf{E}}{\partial t^2} \qquad \text{(6-D-5)}$$

Expanding $\nabla \times \nabla \times \mathbf{E}$ leads to

$$\nabla^2 \mathbf{E} - \nabla(\nabla \cdot \mathbf{E}) = \mu \boldsymbol{\varepsilon} \cdot \frac{\partial^2 \mathbf{E}}{\partial t^2} \qquad \text{(6-D-6)}$$

For the special case of an isotropic medium

$$\nabla^2 \mathbf{E} = \mu \varepsilon \frac{\partial^2 \mathbf{E}}{\partial t^2} \qquad \text{(6-D-7)}$$

This may be recognized as a vector wave equation with velocity $1/\sqrt{\mu\varepsilon}$. This result is from Clerk Maxwell and represents the basis of the electromagnetic theory of light. From Eq. (6-C-1) the refractive index is

$$n = \sqrt{\frac{\mu\varepsilon}{\mu_0\varepsilon_0}}, \qquad n^2 = \frac{\mu}{\mu_0}\frac{\varepsilon}{\varepsilon_0} \tag{6-D-8a}$$

where the subscript zero represents values in a vaccum. For dielectrics, including polymers, μ is close to μ_0

$$n^2 \cong \frac{\varepsilon}{\varepsilon_0} \tag{6-D-8b}$$

Certain characteristics of the solution of Eq. (6-D-7) deserve discussion. From Eq. (6-D-1), it is readily shown that vectors $\mathbf{E}$ and $\mathbf{H}$ are perpendicular both to the direction of propagation of the wave and to each other. It is thus a *transverse wave*. This follows by taking $\mathbf{E}$ to depend upon direction x_1 only and noting that Eq. (6-D-1a–d) reduces to

$$\mathbf{e}_2\left(-\frac{\partial E_3}{\partial x_1}\right) + \mathbf{e}_3\left(\frac{\partial E_2}{\partial x_1}\right) = -\mu\frac{\partial \mathbf{H}}{\partial t}$$

$$\mathbf{e}_2\left(-\frac{\partial H_3}{\partial x_1}\right) + \mathbf{e}_3\left(\frac{\partial H_2}{\partial x_1}\right) = \varepsilon\frac{\partial \mathbf{E}}{\partial t} \tag{6-D-9a–d}$$

$$\frac{\partial E_1}{\partial x_1} = 0, \qquad \frac{\partial H_1}{\partial x_1} = 0$$

If E_3 is taken to zero, H_2 must be zero and $\mathbf{E}$ and $\mathbf{H}$ are perpendicular. This is shown in Figure 6-D-1.

The energy flux of an electromagnetic wave is given by the Poynting vector (7, 78).

$$\mathbf{S} = \mathbf{E} \times \mathbf{H} \tag{6-D-10}$$

An electromagnetic wave may consist of (1) a wave as defined by Eq. (6-D-9), where the $\mathbf{E}$ and $\mathbf{H}$ vectors will maintain their directions in space, or (2) a superposition of waves out of phase with each other and whose $\mathbf{E}$ and $\mathbf{H}$ vectors are at various angles to each other, though the direction of propagation is the same. The former simpler wave is said to be *plane-polarized* (or often simply polarized). The second case, where there is a superposition or out-of-phase waves, leads to $\mathbf{E}$ and $\mathbf{H}$ vectors, which rotate in time about the direction of propagation to define the shape of an ellipse. One thus speaks of *elliptically polarized* light. If the principal axes of the ellipse are equal, the electromagnetic wave is said to be *circularly polarized*.

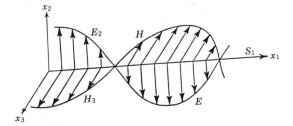

FIGURE 6-D-1. Polarized electromagnetic wave.

An electromagnetic wave may be expressed

$$\mathbf{E} = \mathbf{E}_0 e^{i(kx_1 - \omega t)} = \mathbf{E}_0 e^{ik(x_1 - ct)} \qquad (6\text{-}D\text{-}11)$$

where ω is the frequency and k is the wave number. These are interrelated by

$$k = \frac{\omega}{c} = \frac{2\pi}{\lambda} \qquad (6\text{-}D\text{-}12)$$

The quantity λ is known as the wavelength and is the most common characterization of a wave. Visible light is an electromagnetic wave in the range 4000–7000 Å, with the shortest wavelengths being violet and the larger wavelengths red. Shorter wavelengths include ultraviolet (40–3500 Å), hard x-rays (0.06–1.2 Å) and beyond these γ rays of radioactive substance. Longer wavelengths are described as infrared. Still longer wavelengths include microwaves and radiowaves.

E. PROPAGATION OF WAVES IN ANISOTROPIC MEDIA

Equation (6-D-1) represents the starting point for analysis of wave propagation in anisotropic media (7, 36, 54, 78). If we take 1 as the direction of propagation, Eqs. (6-D-1a, b, c, d) yield

$$\mathbf{e}_2\left(-\frac{\partial E_3}{\partial x_1}\right) + \mathbf{e}_3\left(\frac{\partial E_2}{\partial x_1}\right) = -\mu\frac{\partial \mathbf{H}}{\partial t}$$

$$\mathbf{e}_2\left(-\frac{\partial H_3}{\partial x_1}\right) + \mathbf{e}_3\left(\frac{\partial H_2}{\partial x_1}\right) = \frac{\partial \mathbf{D}}{\partial t} \qquad (6\text{-}E\text{-}1a\text{–}d)$$

$$\frac{\partial D_1}{\partial x_1} = 0, \qquad \frac{\partial H_1}{\partial x_1} = 0$$

The vectors $\mathbf{E}$ and $\mathbf{H}$ are perpendicular, as are $\mathbf{D}$ and $\mathbf{H}$. However, $\mathbf{D}$ and $\mathbf{E}$, as indicated by Eq. (6-D-2) are not parallel as in isotropic media. $\mathbf{D}$ and $\mathbf{H}$ may be identified with the 2 and 3 directions. $\mathbf{E}$ now lies in the 1–2 plane normal to $\mathbf{H}$.

We may write the vectors $\mathbf{E}$, $\mathbf{D}$, and $\mathbf{H}$ as

$$
\begin{aligned}
\mathbf{E} &= \mathbf{E}_0 e^{i(kx_1 - \omega t)} = \mathbf{E}_0 e^{ik(x_1 - vt)} \\
\mathbf{D} &= \mathbf{D}_0 e^{i(kx_1 - \omega t)} = \mathbf{D}_0 e^{ik(x_1 - vt)} \\
\mathbf{H} &= \mathbf{H}_0 e^{i(kx_1 - \omega t)} = \mathbf{H}_0 e^{ik(x_1 - vt)}
\end{aligned}
\tag{6-E-2}
$$

It follows from Eq. (6-E-1) that

$$
H_{03} = vD_{02}, \qquad E_{02} = \mu v H_{03}, \qquad E_{02} = \mu v^2 D_{02}
\tag{6-E-3}
$$

The relation between $\mathbf{D}$ and $\mathbf{E}$ may be expressed equivalently as

$$
\mathbf{D} = \varepsilon \cdot \mathbf{E} = \frac{1}{\mu v^2}[\mathbf{E} - E_1 \mathbf{e}_1] = \frac{1}{\mu v^2}[\mathbf{E} - (\mathbf{E}\cdot\mathbf{s})\mathbf{s}]
\tag{6-E-4}
$$

where $\mathbf{s}$ is the direction of propagation.

It is convenient to reformulate this problem in terms of a coordinate system based on the principal axes of the ε_{ij} tensor. This allows us to write

$$
\mathbf{D} = \varepsilon_1 E_1 \mathbf{e}_1 + \varepsilon_2 E_2 \mathbf{e}_2 + \varepsilon_3 E_3 \mathbf{e}_3
\tag{6-E-5}
$$

and Eq. (6-E-4) as

$$
\begin{aligned}
D_i &= \frac{1}{\mu v^2}\left[\frac{1}{\varepsilon_i}D_i - (\mathbf{E}\cdot\mathbf{s})s_i\right] \\
&= \frac{1}{\mu v^2}\left[\frac{1}{\varepsilon_i}D_i - \left(\sum_m \frac{1}{\varepsilon_m}D_m s_m\right)s_i\right]
\end{aligned}
\tag{6-E-6}
$$

This has the solution

$$
D_i = \frac{s_i}{v^2 - 1/\mu\varepsilon_i}\left(\sum_m \frac{1}{\mu\varepsilon_m}D_m s_m\right)
\tag{6-E-7}
$$

As the vector $\mathbf{D}$ is normal to the vector $\mathbf{s}$,

$$
\mathbf{s}\cdot\mathbf{D} = 0
\tag{6-E-8}
$$

Substitution of Eq. (6-E-7) into Eq. (6-E-8) leads to

$$\frac{s_1^2}{v^2 - 1/\varepsilon_1\mu} + \frac{s_2^2}{v^2 - 1/\varepsilon_2\mu} + \frac{s_3^2}{v^2 - 1/\varepsilon_3\mu} = 0 \qquad (6\text{-E-}9a)$$

$$s_1^2\left(v^2 - \frac{1}{\varepsilon_2\mu}\right)\left(v^2 - \frac{1}{\varepsilon_3\mu}\right) + s_2^2\left(v^2 - \frac{1}{\varepsilon_1\mu}\right)\left(v^2 - \frac{1}{\varepsilon_3\mu}\right)$$

$$+ s_3^2\left(v^2 - \frac{1}{\varepsilon_1\mu}\right)\left(v^2 - \frac{1}{\varepsilon_2\mu}\right) = 0. \qquad (6\text{-E-}9b)$$

This is Fresnel's equation of wave normals. This was derived by Fresnel before the development of the electromagnetic theory, using the analog of propagation of waves in an anisotropic elastic solid. It was shown by Clerk Maxwell to also arise in the propagation of electromagnetic waves in an anisotropic medium.

For each direction **s** in space (i.e., the set s_1, s_2, s_3), there correspond two velocities. For **s** parallel to one of the principal axes of the dielectric tensor, say 1, these are

$$v' = \frac{1}{\sqrt{\varepsilon_2\mu}}, \qquad v'' = \frac{1}{\sqrt{\varepsilon_3\mu}} \qquad (6\text{-E-}10)$$

The vectors **D'** and **D''** associated with these waves are given by Eqs. (6-E-7). The orientation of **D'** and **D''** in space relative to each other may be investigated by forming a scalar product.

$$\mathbf{D'\cdot D''} = K\sum_j \frac{s_j^2}{[v'^2 - 1/\mu\varepsilon_j][v''^2 - 1/\mu\varepsilon_j]} \qquad (6\text{-E-}11)$$

which is equivalent to

$$\mathbf{D'\cdot D''} = \frac{K}{v'^2 - v''^2}\left[\sum_j \frac{s_j^2}{v'^2 - 1/\mu\varepsilon_j} - \sum_k \frac{s_k^2}{v''^2 - 1/\mu\varepsilon_k}\right] \qquad (6\text{-E-}12)$$

It follows from Fresnel's equation of wave normals, Eq. (6-E-9), that we have

$$\mathbf{D'\cdot D''} = 0 \qquad (6\text{-E-}13)$$

The **D'** vector associated with the wave of velocity v' is perpendicular to the **D''** vector associated with the wave of velocity v''. The corresponding **H'** and **H''** vectors will also be perpendicular.

Equations (6-E-9) and (6-E-10) suggest a fundamental relationship between the refractive index and dielectric tensors. The principal values are interrelated through

$$n_i = \frac{c}{v_i} = c\sqrt{\mu\varepsilon_i} \tag{6-E-14}$$

Fresnel's ellipsoid

$$\varepsilon_1 x_1^2 + \varepsilon_2 x_2^2 + \varepsilon_3 x_3^2 = C' \tag{6-E-15a}$$

is equivalent to

$$n_1^2 x_1^2 + n_2^2 x_2^2 + n_3^2 x_3^2 = C'' \tag{6-E-15b}$$

The reciprocals of the lengths of the principal axes of Fresnel's ellipsoid are proportional to the principal refractive indices (see Fig. 6-E-1).

We need to develop these ideas further. We may define an *optically uniaxial crystal* as an anisotropic material for which

$$\varepsilon_1 = \varepsilon_2 \neq \varepsilon_3 \qquad \text{or} \qquad n_1 = n_2 \neq n_3 \tag{6-E-16a}$$

The wave for which $v_1 = v_2$ is referred to as the ordinary wave and the third

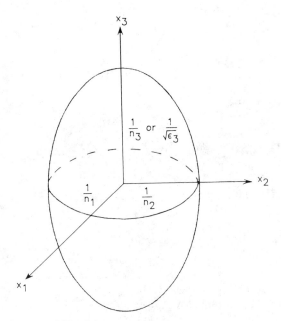

FIGURE 6-E-1. Fresnel's ellipsoid.

component v_3 as the *extraordinary wave*. If

$$\varepsilon_1 \neq \varepsilon_2 \neq \varepsilon_3 \qquad \text{or} \qquad n_1 \neq n_2 \neq n_3 \tag{6-E-16b}$$

the anisotropic material is referred to as an *optically biaxial crystal*.

In summary, an electromagnetic wave in an anisotropic medium consists of two waves of different velocities v' and v'', whose planes of polarization are normal to each other. This key result is the basis of flow birefringence investigations.

F. BIREFRINGENCE AND MOLECULAR ORIENTATION

There is a basic relationship between molecular orientation and the optical properties of homogeneous polymers. The displacement vector of Eq. (6-D-2) is related to the polarization vector[†]

$$\mathbf{D} = \boldsymbol{\varepsilon} \cdot \mathbf{E} = \varepsilon_0 \mathbf{E} + \mathbf{P} = \varepsilon_0 \mathbf{E} + N' \varepsilon_0 \boldsymbol{\alpha} \cdot \mathbf{E} \tag{6-F-1}$$

where $\mathbf{P}$ is the polarization vector and $\boldsymbol{\alpha}$ the polarizability tensor. If we consider the medium to be isotropic and the polarization vector to linearly depend upon the field strength $\mathbf{E}$, we have

$$\mathbf{D} = \varepsilon \mathbf{E} = \varepsilon_0 \mathbf{E} + N' \alpha \varepsilon_0 \mathbf{E}, \qquad \varepsilon - \varepsilon_0 = N' \alpha \varepsilon_0 \tag{6-F-2a, b}$$

From Eq. (6-D-8b) we have:

$$n^2 - 1 = N' \alpha \tag{6-F-3}$$

Generally, $\mathbf{P}$ is considered to depend upon an effective field which is the sum of $\mathbf{E}$ and contributions from dipoles. Taking this into consideration, Lorentz (47, 48) and Lorenz (49) found the relationship between refractive index and polarizability for isotropic materials to be

$$\frac{n^2 - 1}{n^2 + 2} = \frac{N' \alpha}{3} \tag{6-F-4}$$

Muller (55) and Kuhn and Grun (44) have proposed that if a material is oriented, Eq. (6-F-1) may be applied to the principal axes of the α_{ij} and n_{ij} tensors, that is,

$$\frac{n_i^2 - 1}{n_i^2 + 2} = \frac{N' \alpha}{3} \tag{6-F-5}$$

This cannot be strictly true because of the isotropy restriction in the arguments of Lorentz and Lorenz.

[†]The reader should note that we are using the same symbol $\mathbf{P}$ for the polarization vector as we have used elsewhere for extra stress [Eq. (4-F-4)]. We will not use this symbol to represent stress in this chapter.

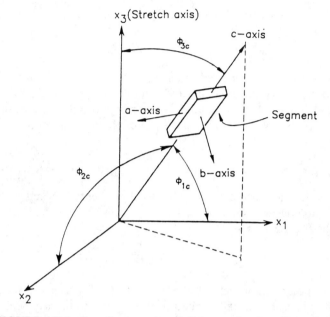

FIGURE 6-F-1. Coordinate system defining angles ϕ_{1c}, ϕ_{2c}, and ϕ_{3c}.

It is possible to express orientation in terms of the anisotropy of polarizability. This has been formulated in various ways. It is possible to accomplish this using spherical coordinated (Euler's angles) (59, 60) or with cartesian coordinates (92, 93). Consider an oriented segment of an individual macromolecule which is represented in an **a**, **b**, and **c** (chain axis) coordinate system. The **c** axis is oriented relative to laboratory coordinate axes 1, 2, 3 in terms of angles ϕ_{1c}, ϕ_{2c}, ϕ_{3c}. It has polarizability $\alpha_\parallel$ along the chain and $\alpha_\perp$ perpendicular to the chain. The polarizability components α_{ij} along the 1, 2, and 3 coordinate axes are (see Fig. 6-F-1)

$$\alpha_{11} = \alpha_\parallel \cos^2 \phi_c + \alpha_\perp \cos^2 \phi_{1a} + \alpha_\perp \cos^2 \phi_{1b}$$

$$\alpha_{22} = \alpha_\parallel \cos^2 \phi_{2c} + \alpha_\perp \cos^2 \phi_{2a} + \alpha_\perp \cos^2 \phi_{2b} \qquad \text{(6-F-6a–c)}$$

$$\alpha_{33} = \alpha_\parallel \cos^2 \phi_{3c} + \alpha_\perp \cos^2 \phi_{3a} + \alpha_\perp \cos^2 \phi_{3b}$$

The differences in polarizability may be expressed

$$\alpha_{11} - \alpha_{33} = \alpha_\parallel (\cos^2 \phi_{1c} - \cos^2 \phi_{3c})$$
$$+ \alpha_\perp (\cos^2 \phi_{1a} + \cos^2 \phi_{1b} - \cos^2 \phi_{3a} - \cos^2 \phi_{3b}) \qquad \text{(6-F-7)}$$

$$\alpha_{22} - \alpha_{33} = \alpha_\parallel (\cos^2 \phi_{2c} - \cos^2 \phi_{3c})$$
$$+ \alpha_\perp (\cos^2 \phi_{2a} + \cos^2 \phi_{2b} - \cos^2 \phi_{3a} - \cos^2 \phi_{3b}) \qquad \text{(6-F-8)}$$

and using the Pythagorean theorem

$$\cos^2 \phi_{ja} + \cos^2 \phi_{jb} + \cos^2 \phi_{jc} = 0 \qquad (6\text{-F-}9)$$

we may write

$$\alpha_{11} - \alpha_{33} = (\alpha_{\parallel} - \alpha_{\perp})(\cos^2 \phi_{1c} - \cos^2 \phi_{3c}) \qquad (6\text{-F-}10a)$$

$$\alpha_{22} - \alpha_{33} = (\alpha_{\parallel} - \alpha_{\perp})(\cos^2 \phi_{2c} - \cos^2 \phi_{3c}) \qquad (6\text{-F-}10b)$$

Introducing the Pythagorean relationship

$$\cos^2 \phi_{1c} + \cos^2 \phi_{2c} + \cos^3 \phi_{3c} = 1 \qquad (6\text{-F-}11)$$

we may write

$$\frac{\alpha_{11} - \alpha_{33}}{\alpha_{\parallel} - \alpha_{\perp}} = 2 \cos^2 \phi_{1c} + \cos^2 \phi_{2c} - 1 \qquad (6\text{-F-}12a)$$

$$\frac{\alpha_{22} - \alpha_{33}}{\alpha_{\parallel} - \alpha_{\perp}} = 2 \cos^2 \phi_{2c} + \cos^2 \phi_{1c} - 1 \qquad (6\text{-F-}12b)$$

The above expression was derive for an individual macromolecule, we may average over all of the macromolecules present to yield (93)

$$f_1^B = \frac{\bar{\alpha}_{11} - \bar{\alpha}_{33}}{\alpha_{\parallel} - \alpha_{\perp}} = \overline{2 \cos^2 \phi_{1c}} + \overline{\cos^2 \phi_{2c}} - 1 \qquad (6\text{-F-}13a)$$

$$f_2^B = \frac{\bar{\alpha}_{22} - \bar{\alpha}_{33}}{\alpha_{\parallel} - \alpha_{\perp}} = \overline{2 \cos^2 \phi_{2c}} + \overline{\cos^2 \phi_{1c}} - 1 \qquad (6\text{-F-}13b)$$

where f_1^B and f_2^B are considered as biaxial orientation factors.

We shall express the differences in mean molecular polarizability in terms of refractive indices using the modified Lorentz–Lorenz equation, Eq. (6-F-5), specifically noting that

$$\frac{n_i^2 - 1}{n_i^2 + 2} - \frac{n_j^2 - 1}{n_j^2 + 2} = \frac{2(n_i^2 - n_j^2)}{(n_i^2 + 2)(n_j^2 + 2)} = \frac{2(n_i + n_j)}{(n_i^2 + 2)(n_j^2 + 2)}(n_i - n_j)$$

$$\cong \frac{12\bar{n}}{(\bar{n}^2 + 2)^2}(n_i - n_j) \qquad (6\text{-F-}14)$$

where $\bar{n}$ is the mean refractive index. This leads to

$$\bar{\alpha}_{ii} - \bar{\alpha}_{jj} = \frac{12\bar{n}}{N(\bar{n}^2 + 2)^2}(n_i - n_j) \qquad (6\text{-F-}15)$$

As the Lorentz–Lorenz equation presumes isotropy, this approach may be criticized especially for highly oriented systems as shown by Stein (37, 85). However, the proportionality of $(\bar{\alpha}_{ii} - \bar{\alpha}_{jj})$ to $(n_i - n_j)$ is a weaker approximation.

In any case, substitution of Eq. (6-F-15) into Eq. (6-F-13) leads to (94)

$$\frac{n_1 - n_3}{\Delta^\circ} = 2\overline{\cos^2 \phi_{1c}} + \overline{\cos^2 \phi_{2c}} - 1 = f_1^B$$

$$\frac{n_2 - n_3}{\Delta^\circ} = 2\overline{\cos^2 \phi_{2c}} + \overline{\cos^2 \phi_{1c}} - 1 = f_2^B$$

(6-F-16a, b)

where Δ° is the maximum or *intrinsic* birefringence.

For an isotropic material

$$f_1^B = 0, \qquad f_2^B = 0$$

While for uniaxial orientation in the 1 or 2 directions,

$$f_1^B = 1, \qquad f_2^B = 0 \qquad \text{or} \qquad f_1^B = 0, \qquad f_2^B = 1$$

For equal biaxial orientation

$$f_1^B = f_2^B$$

The quantitative representation of orientation by polarizability was originally developed by P. H. Hermans and his coworkers (32, 33) for uniaxial orientation [compare Muller (55)]. We may obtained their results from Eqs. (6-F-13) and (6-F-16) by setting

$$\bar{\alpha}_{22} = \alpha_{33} \qquad \text{or} \qquad n_2 = n_3 \qquad \text{(6-F-17)}$$

This leads to

$$\overline{\cos^2 \phi_{2c}} = \tfrac{1}{2}[1 - \overline{\cos^2 \phi_{1c}}] \qquad \text{(6-F-18)}$$

and

$$\frac{\bar{\alpha}_{11} - \bar{\alpha}_{22}}{\alpha_\parallel - \alpha_\perp} = \frac{3\overline{\cos^2 \phi_{1c}} - 1}{2} = f_H \qquad \text{(6-F-19)}$$

For optically homogeneous materials

$$\frac{n_1 - n_2}{\Delta^\circ} = \frac{3\overline{\cos^2 \phi_{1c}} - 1}{2} = f_H \qquad \text{(6-F-20)}$$

The quantity f_H is known as the Hermans orientation factor.

This may also be applied to the two-dimensional case. Here we have

$$\overline{\cos^2 \phi_{3c}} = 0, \qquad \overline{\cos^2 \phi_{2c}} = 1 - \overline{\cos^2 \phi_{1c}} \qquad \text{(6-F-21)}$$

$$\frac{\alpha_{11} - \alpha_{22}}{\alpha_\parallel - \alpha_\perp} = \overline{\cos^2 \phi_{1c}} - \overline{\cos^2 \phi_{2c}} = 2\overline{\cos^2 \phi_{1c}} - 1 \qquad \text{(6-F-22)}$$

This development for chain-axis orientation may be generalized for crystallographic axes. This was first described by Stein (82) for uniaxial orientation and has been similarly applied to biaxial orientation (94).

It is possible to express orientation in terms of an orthogonal polynomial expansion. This was first done with Legendre polynomials for uniaxial orientation by Muller (55). This was expanded to more general orientations by Roe (69) and by Nomura and Kawai (60).

G. METHODS OF PRODUCING POLARIZED LIGHT

Monochromatic light as usually produced is elliptically polarized, that is, the directions of the electrical vectors **E** (and **D**) and the magnetic vector **H** vary in direction and amplitude with time. To study the birefringent characteristics of materials it is best to have linearly polarized light, where the directions of **E** and **H** remain unchanged. An apparatus that accomplishes this is known as a polariscope.

Various methods have been used through the years to produce polarized light (Fig. 6-G-1). These are summarized in monographs and treatises on optics (7, 79, 95). It was discovered in the early nineteenth century that elliptically polarized light reflected off a surface at certain angles is linearly polarized. The specific angle of impingement [Brewster's angle (8)] varies with the refractive index of the medium, being 57° (relative to the normal) for glass and 53° for water.

It is also possible to produce polarized light by refraction. The light transmitted when a beam of monochromatic light is impinged at Brewster's angle is largely polarized. However, its extent of polarization is generally less under practical circumstances than the reflected light. However, using bundles of glass plates placed at Brewster's angle to the impinging beam, one may produce polarized light. One may conceive of a polariscope consisting of a tube filled with numerous pieces of cleaned glass slides mounted at an angle of 33° to the axis of a tube.

Double refraction in birefringent media may also be used to produce linearly polarized light. A beam of elliptically polarized light impinging on a birefringent material, for example, a crystal, divides into two polarized waves which travel at different velocities in different directions. The directions of **D** and **H** of the individual waves are defined by the principal axes of the dielectric or polarizability tensor of the material.

Two different classes of devices have been developed on this basis. In the

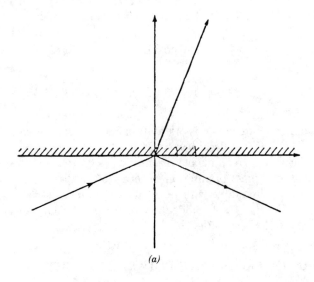

(a)

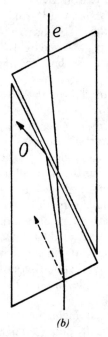

(b)

FIGURE 6-G-1. Methods of producing polarized light: (*a*) reflection at Brewster's angle; (*b*) Nicol prism—e is the extraordinary ray passing through the prism and O is the totally reflected ordinary ray; (*c*) sheet polarizers—anisotropic (dichroic) absorption.

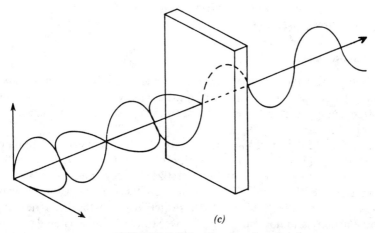

(c)

FIGURE 6-G-1. (*Continued*).

Nicol prism (57), a crystal of Iceland spar is used. The two end faces of a natural rhombus of spar are cut down sufficiently to reduce the normal angle of 71° between the faces to 68°. The crystal is then cut in two along a diagonal opposite to the 68° angles. After polishing the surfaces, they are cemented together with Canada balsam, which has a refractive index intermediate between the indices of the spar for the ordinary and extraordinary rays. When elliptically polarized light is impinged on the prism, the extraordinary ray strikes the balsam film at an angle greater than the critical angle for total reflection of light moving from a medium of high refractive index to low refractive index. This light is reflected to one side and adsorbed by a coating of black paint. The ordinary ray passes through and is the source of polarized light.

Another class of polariscopes based on birefringent media uses the phenomenon of *dichroism* (7, 45). Here light is divided by the briefringent medium into two polarized beams, but now one of these is absorbed by the medium. Crystals of tourmaline have the remarkable property of absorbing the ordinary and transmitting the extraordinary beam. Land (45) has developed sheets of polymeric films with dichroic character. Some of these consist of uniaxially oriented polyvinyl alcohol containing iodine.

One may also convert elliptically polarized light into circularly polarized light. An apparatus that accomplishes this is known as a *circular polariscope*. This is commonly done using what are called *quarter-wave plates*. These are plates of mica of such thickness that the path difference (see the next section) between the ordinary and extraordinary rays is one quarter of a wavelength. For yellow light, this is 0.0032 cm or 32 μm. One may produce circularly polarized light from elliptically polarized light by a two-layer sheet. One layer consists of Land's dichroic polaroid® sheet and the second a birefringent sheet which introduces a specified path difference on retardation.

H. RETARDATION CHARACTERIZATION OF BIREFRINGENT SHEETS

1. General Remarks

The basic apparatus for birefringence measurements consists of a monochromatic light source, polarizer, analyzer, and intensity recording apparatus, as summarized in Fig. 6-H-1. Optical benches and polarized light microscopes have this configuration.

Consider the Maxwell displacement vector **D**, associated with a nonpolarized, or elliptically polarized light. The direction of **D** rotates about the direction of propagation. Upon passing through a polarizer the components of **D** in all but one direction are absorbed. The wave now enters a point in the birefringent sample with **D** generally at an angle to the principal axes of the dielectric ellipsoid. In photoelasticity and flow birefringence, we study two-dimensional samples so we may represent the dielectric ellipsoid by an ellipse with two axes, 1 and 2. The components of the displacement vector ending the sample may be expressed

$$D_1 = D_0 \cos \alpha \cos pt, \qquad D_2 = D_0 \sin \alpha \sin pt \qquad \text{(6-H-1a, b)}$$

where α is the angle between the polarized wave component and the principal axis of the Fresnel or refractive index ellipsoid (or better ellipse in this two-

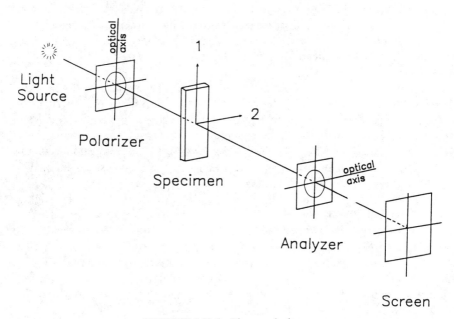

FIGURE 6-H-1. Plane polariscope.

dimensional material). The polarized wave component along the 1 axis will be propagated at a velocity v_1 and that along the 2 axis at a velocity v_2. The components D_1 and D_2, when exiting the sample, will be out of phase and represented through the expressions

$$D_1 = D_0 \cos \alpha \cos p(t - t_1) = D_0 \cos \alpha \cos pt' \qquad \text{(6-H-2a)}$$

$$D_2 = D_0 \sin \alpha \cos p(t - t_2) = D_0 \sin \alpha \cos (pt' - \Delta) \qquad \text{(6-H-2b)}$$

where, for a sample of thickness h,

$$t_1 = \frac{h}{v_1}, \qquad t_2 = \frac{h}{v_2}$$

On leaving the sample, these components have a phase difference

$$\Delta = \omega(t_2 - t_1) = \frac{2\pi h}{\lambda} \left(\frac{1}{v_2} - \frac{1}{v_1} \right)$$

$$= \frac{2\pi h}{\lambda c} (n_2 - n_1) \qquad \text{(6-H-3)}$$

where

$$n_1 = \frac{c}{v_1}, \qquad n_2 = \frac{c}{v_2} \qquad \text{(6-H-4)}$$

The analyzer transmits only vibrations in its own plane of polarization. If this is at right angles to the plane of polarization of the polarizer, and if the birefringent specimen is removed, no light is transmitted. We now consider what occurs when the sample is present. The waves arriving at the analyzer may be represented by Eq. (6-H-2a, b). If the analyzer is at right angles to the polarizer, we may write the remaining component of **D** at right angles to the initial polarized wave as

$$D = D_1 \sin \alpha - D_2 \cos \alpha$$

$$= D_0 \cos \alpha \sin \alpha [\cos \omega t' - \cos (\omega t' - \Delta)] \qquad \text{(6-H-5a)}$$

$$D = -D_0 \sin 2\alpha \sin \frac{\Delta}{2} \sin \left(\omega t' - \frac{\Delta}{2} \right) \qquad \text{(6-H-5b)}$$

The factor $\sin(\omega t - \Delta/2)$ represents a simple harmonic variation with time. The amplitude is

$$D_0 \sin 2\alpha \sin \frac{\Delta}{2}$$

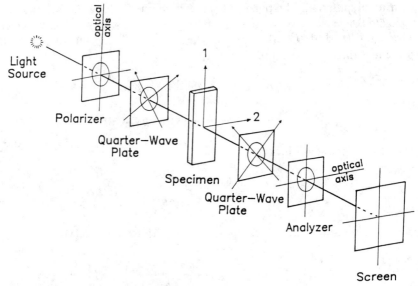

FIGURE 6-H-2. Circular polariscope.

It follows that light reaches the screen unless either

$$\sin 2\alpha = 0, \qquad \sin\frac{\Delta}{2} = 0 \qquad \text{(6-H-6a, b)}$$

Equation (6-H-6a) means that if the principal refractive indices are parallel to the directions of polarization of P and A, all rays passing through such points will be extinguished and no light appears on the screen in back of the analyzer. These points form a curve which is indicated by a dark band. This band is called an *isoclinic*. Equation (6-H-6b) means that

$$\frac{\Delta}{2} = N\pi, \qquad \frac{h}{\lambda c}(n_2 - n_1) = N \qquad \text{(6-H-7)}$$

where N is zero or an integer. The lines connecting the points on the screen where this occurs again form a dark fringe. When N is unity, this is called a first-order fringe, when N is 2, we have a second-order fringe, and so on. These fringes are called *isochromatics*, since when white light is used, they correspond to extinction of a certain wavelength and therefore to a color band. To evaluate the magnitude of the local birefringence it is necessary to know the order of the fringes. In photoelastic stress analysis, this is commonly done by studying the gradual loading of the solid object being investigated.

2. Measurement of Birefringence

When one knows the directions of the principal axes of the refractive index and dielectric tensor, it is only necessary to be concerned with Eqs. (6-H-3) and

(6-H-7). If the retardation may be measured, the birefringence can be determined.

There are two methods of measuring the retardation. One is to measure the intensity of light I passing through the object, which is given by

$$I \sim |\mathbf{E} \times \mathbf{H}| \sim D_0^2 \sin^2 2\alpha \sin^2 \frac{\Delta}{2} \qquad (6\text{-H-}8)$$

Measurement of I yields $\Delta/2$ and, from Eq. (6-H-3), the birefringence $n_1 - n_2$.

The second procedure is to use a compensator (7, 42, 79). The most widely used compensator is that of Babinet, which consists of two quartz wedges whose optic axes are aligned perpendicular to each other. By sliding these relative to each other so that one has thickness x_1 and the other x_2, retradation of magnitude

$$\frac{2\pi}{\lambda}(n_e - n_0)(x_1 - x_2)$$

may be introduced. The Berek compensator uses a rotating crystal plate.

3. Measurement of Isochromatics and Isoclinics

The device described above involves the use of a *plane polariscope*: it does not allow the separation of isochromatic and isoclinic lines. Birefringence cannot be measured if the directions of the principal axes are not known. This is achieved in the *circular polariscope*, which introduces optical devices known as quarter-wave plates between the polarizer and the sample and between the sample and the analyzer (Fig. 6-H-2). A quarter-wave plate is a crystal (mica) plate that introduces a retardation equivalent to a phase difference of $\pi/2$ or a quarter wave. Thus the plate is stationed relative to the polarizer, so that the angle between the direction of polarization and the principal axes of the mica is 45°.

We now treat the optics of this apparatus. From Eq. (6-H-2a, b) it follows that the components of the displacement vector leaving the mica in the quarter-wave plate principal axis system are

$$D_1^Q = D_0 \cos 45° \cos \psi = \frac{D_0}{\sqrt{2}} \cos \psi \qquad (6\text{-H-}9a)$$

$$D_2^Q = D_0 \sin 45° \cos\left(\psi + \frac{\pi}{2}\right) = \frac{D_0}{\sqrt{2}} \sin \psi \qquad (6\text{-H-}9b)$$

where ψ is $\omega t + $ constant. A vector moving with these components rotates in a circle and is thus circularly polarized.

The circularly polarized light now enters the sample. If β is the angle between the principal axes of the quarter-wave plate and the sample, the components in the sample principal axis system are

$$D_1 = \frac{D_0}{\sqrt{2}}[\cos\beta\cos\psi - \sin\beta\sin\psi]$$

$$= \frac{D_0}{\sqrt{2}}\cos(\psi + \beta) = \frac{D_0}{\sqrt{2}}\cos\psi'$$

$$(6\text{-H-}10a,b)$$

$$D_2 = \frac{D_0}{\sqrt{2}}[\sin\beta\cos(\psi - \Delta) + \cos\beta\sin(\psi - \Delta)]$$

$$= \frac{D_0}{\sqrt{2}}\sin(\psi + \beta - \Delta) = \frac{D_0}{\sqrt{2}}\sin(\psi' - \Delta)$$

If we write ψ' as $\psi'' + \Delta/2$, we may express Eq. (6-H-10) as

$$D_1 = \frac{D_0}{\sqrt{2}}\cos\left(\psi'' + \frac{\Delta}{2}\right) = \frac{D_0}{\sqrt{2}}\left(\cos\frac{\Delta}{2}\cos\psi'' - \sin\frac{\Delta}{2}\sin\psi''\right)$$

$$(6\text{-H-}11a,b)$$

$$D_2 = \frac{D_0}{\sqrt{2}}\cos\left(\psi'' - \frac{\Delta}{2}\right) = \frac{D_0}{\sqrt{2}}\left(\cos\frac{\Delta}{2}\sin\psi'' - \sin\frac{\Delta}{2}\cos\psi''\right)$$

The wave consists of the sum of two circularly polarized beams; one clockwise and the other counterclockwise of amplitudes

$$\frac{D_0}{\sqrt{2}}\cos\frac{\Delta}{2}, \quad \frac{D_0}{\sqrt{2}}\sin\frac{\Delta}{2}$$

The light leaving the sample contains a retardation Δ in it, but the angle α defining the position of the principal axes of the sample relative to the direction of polarization is lost. We no longer have the source of the isoclinic lines in our analysis.

The second quarter-wave plate introduces a retardation $\Delta/2$ in the direction of the principal axes of the quarter-wave plate. The motion in this quarter-wave plate system may be represented by Eq. (6-H-10), replacing ψ' by ψ'''. The light exiting the quarter-wave plate in this frame is expressed as

$$D_1^Q = \frac{D_0}{\sqrt{2}}\cos\frac{\Delta}{2}\cos\psi''' - \frac{D_0}{\sqrt{2}}\sin\frac{\Delta}{2}\sin\left(\psi''' - \frac{\pi}{2}\right)$$

$$(6\text{-H-}12a,b)$$

$$D_2^Q = \frac{D_0}{\sqrt{2}}\cos\frac{\Delta}{2}\sin\left(\psi''' - \frac{\pi}{2}\right) - \frac{D_0}{\sqrt{2}}\sin\frac{\Delta}{2}\sin\psi'''$$

The light exiting the second quarter-wave plate now goes through a polarizer whose principal axis is at 45° to the analyzer. The components of the light in $\cos \Delta/2$ are cancelled out, leaving the emerging wave from the analyzer to have the component

$$D^{A} = \sqrt{2} D_{0} \sin \frac{\Delta}{2} \sin \psi'''$$ (6-H-13)

The intensity of light on a screen placed behind the analyzer will be

$$I \sim (D^{A})^{2} \sim \sin^{2} \frac{\Delta}{2}$$ (6-H-14)

Only isochromatic lines will appear.

I. RHEO-OPTICAL LAW

There is a long history of experimental studies of the birefringence developed in stretched elastomers and flowing polymer melts and solutions. Various investigators inspired by Kuhn and Grun's model of the optical properties of deformed networks of polymer chains have sought to contrast the levels of birefringence developed with those of the principal stresses. Typical experimental results are shown in Figure 6-I-1. The behavior may be seen to be generally linear in character except at very high stress levels.

The results described above suggest a linear isotropic relationship between birefringence and stress. This is most simply stated in terms of the refractive index

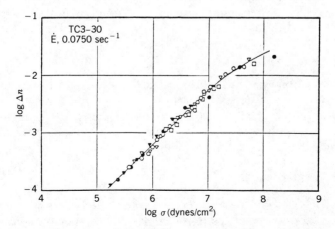

FIGURE 6-I-1. Experimental relationship between birefringence and stress in a polystyrene melt [Matsumoto and Bogue (50), with permission of the publisher]. The data were obtained under nonisothermal conditions.

tensor $\mathbf{n}$ as

$$\mathbf{n} = \tfrac{1}{3}(\operatorname{tr}\mathbf{n})\mathbf{I} + C\mathbf{T} \tag{6-I-1a}$$

where $\mathbf{T}$ is the deviatoric stress tensor and C the stress-optical constant. The rheo-optical law is more commonly stated in terms of principal values, that is,

$$n_i - n_j = C(\sigma_i - \sigma_j) = C(T_i - T_j) \tag{6-I-1b}$$

Generally the stress optical constant, C, is a temperature dependent quantity. For polystyrene as shown by Matsumoto and Bogue (50) C is independent of temperature in the melt. This appears to be true of many but not all thermoplastics.

The rheo-optical law for polymer melts actually represents a more general relationship between stress and orientation. From Eqs. (6-F-13) and (6-F-16), it follows that not only is stress anisotropy related to anisotropy of the polarizability tensor, but that Eq. (6-J-1) may be expressed as

$$f_1^{\mathrm{B}} = \frac{C}{\Delta^{\circ}}(\sigma_1 - \sigma_3), \qquad f_2^{\mathrm{B}} = \frac{C}{\Delta^{\circ}}(\sigma_2 - \sigma_3) \tag{6-I-2}$$

The biaxial orientation factors, and the Hermans orientation factor, are linearly related to stress in flowing melts.

J. MEASUREMENTS OF STRESSES

It follows from Eq. (6-I-1) that isoclinics and isochromatics characterize stress fields in these materials as well as birefringence fields. Isoclinics represent directions of the principal axes of the stress tensor and isochromatics where the difference in principal stresses has the magnitude from Eqs. (6-I-1) and (6-H-7) of

$$\frac{h}{\lambda c}C(\sigma_1 - \sigma_2) = N, \qquad \sigma_1 - \sigma_2 = \frac{N\lambda c}{hC} \tag{6-J-1}$$

Various investigators (1, 3–5, 27–30) have used this technique to investigate stress fields in polymer-melt processing operations.

As noted above, flow birefringence has been widely applied to determine stress fields in the flow of polymer melts. Isochromatic lines for fully developed flow in a slit are shown in Figure 6-J-1. Here we know the shear stress is given by

$$\sigma_{12}(x_2) = \frac{x_2}{2}\frac{dp}{dx_1} \tag{6-J-2}$$

and the fringes correspond to Eqs. (6-H-7) and (6-J-1). These allow us to compute the stress-optical constant if we know $N_1(\sigma_{12})$ by noting that

$$\sigma_1 - \sigma_2 = \sqrt{N_1^2 + 4\sigma_{12}^2} \tag{6-J-3}$$

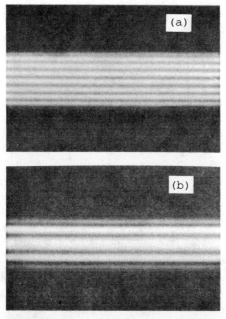

FIGURE 6-J-1. Isochromatic fringe for flow of a birefringent polymer melt in a slit [from Han and Drexler (29)].

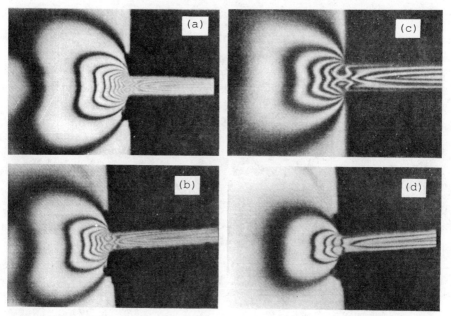

FIGURE 6-J-2. Isochromatic lines for flow of a polymer melt into the entrance of a slit die [from Han and Drexler (29)].

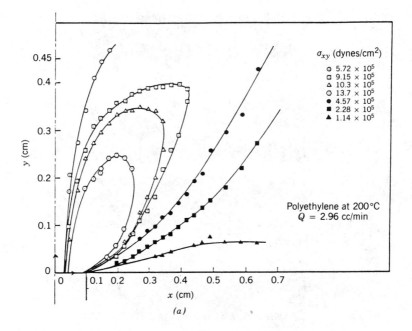

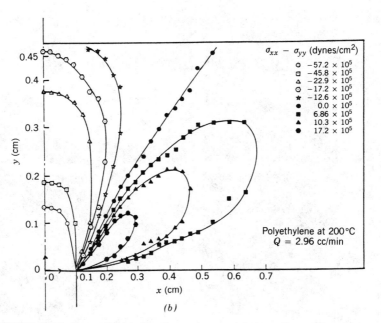

FIGURE 6-J-3. Stress field in flowing polymer melt in a die entrance [from Han and Drexler (29)].

Flow birefringence is also a technique which may be used to measure rheological properties, notably normal stress differences. If σ_{12} and isoclinics and isochromatics are known, N_1 may be computed. Such procedures were pioneered by Philippoff (*12, 13, 63–66*), growing out of the hypothesis of Lodge (*46*).

Let us now look at the application to some complex problems. Quantitative flow birefringence studies in melts at the entrance of a die have been described by Han and Drexler (*29*), Han (*27*), and Vinogradov, Isayev, and their coworkers (*11*). Flow birefringence isochromatic lines from Han and Drexler (*29*) for flow of HDPE at the entrance of a slit die are shown in Figure 6-J-2 at different extrusion rates. These have been, together with the isoclinics, converted in Figure 6-J-3a and b to principal normal stress differences.

Vinogradov, Isayev, and coworkers (*89*) have investigated the flow birefringence of polymer melts in materials subjected to sinusoidal oscillations.

K. OPTICAL PROPERTIES OF POLYMER MELTS

In this section we turn to the material properties of polymer melt systems. There are three optical properties of general concern to us. These are the mean refractive index $\bar{n}$

$$\bar{n} = \tfrac{1}{3}(n_1 + n_2 + n_3) \tag{6-K-1}$$

the intrinsic birefringence Δ°, and the stress optical constant C.

The refractive indices of important organic liquids are summarized in Table 6-K-1. The numbers are in the range 1.28–1.75. Aliphatic hydrocarbon liquids have values of order 1.39. Sample alcohols and acids are slightly lower. Substantial reductions are obtained by introducing fluorine. Benzene and toluene have values of 1.498 and 1.494. Introduction of chlorine, bromine, sulfur, and iodine into organic compounds raises the refractive index. Generally, polymers of similar structure have higher refractive indices then their low-molecular-weight-homologs.

On a molecular level, the refractive indices are related to molecular polarizabilities through the Lorentz–Lorenz equation, Eq. (6-F-4)

$$\frac{n^2 - 1}{n^2 + 2} = \frac{N'}{3}\alpha \tag{6-K-2}$$

The ordering of the refractive indices in Table 6-K-1 may be noted to vary with polarizability of the atoms present, generally increasing with those that have larger electron clouds and higher polarizabilities. It has been argued by Denbigh (*19*) that the refractive indices of a wide range of compounds can be calculated from the molecular polarizability computed by an additive scheme where definite increments are associated with each bond present.

TABLE 6-K-1   Refractive Indices[a]

Material	Refractive Index n or $\bar{n}$[b]
Heptane	1.385
Octane	1.395
Benzene	1.498
Toluene	1.494
Bromobenzene	1.557
Carbon disulfide	1.628
Ethanol	1.359
Trifluoroethanol	1.290
Polyethylene	1.51–1.54
Polypropylene (atactic)	1.473
Polybutene-1 (isotactic)	1.512
Polybutadiene	1.517
cis-1,4-Polyisoprene	1.519–1.520
Poly(acrylonitrile)	1.520
Polystyrene	1.595
Polyvinyl acetate	1.466
Polyoxymethylene	1.48
Polymethylmethacrylate	1.489
Polyvinyl alcohol	1.49–1.53
Polyvinylidene fluoride	1.42
Polytetrafluoroethylene	1.35–1.38
Polytrifluorochloroethylene	1.42–1.43

[a] From J. Brandrup and E. H. Immergut, Eds., *Polymer Handbook*, Wiley, New York, 1975.
[b] Measured at sodium D wavelength.

The intrinsic birefringence of polymers is related to the difference in polarizabilities along the chain and perpendicular to a fully oriented chain, $\alpha_\parallel - \alpha_\perp$, and to the mean refractive index $\bar{n}$. This may be seen by Eq. (6-F-15) to be

$$\Delta^\circ = \frac{(\bar{n}^2 + 2)^2}{12\bar{n}} N'(\alpha_\parallel - \alpha_\perp) \qquad (6\text{-}K\text{-}3)$$

It is possible to estimate Δ° in many manners, ranging from calculations based on the polarizability of extended chains to studies of highly oriented materials (26, 37, 70, 80, 84, 85). There are serious problems in predictions of Δ° from first principles. Such analyses involve uncertain knowledge of individual bond polarizabilities and handling of internal field effects. Observations based on experimental results indicate values of Δ° for polyethylene decrease as increasingly polarizable substituents are introduced. For a phenyl group, the effect is so large that Δ° becomes strongly negative (see Table 6-K-2).

TABLE 6-K-2   Intrinsic Birefringences for Different Polymers

Material	Δ°	Source	Remarks
Polyethylene	0.0615	Bunn and Daubeny (14)	
	0.0585	Hoshino et al. (38)	
Polypropylene	0.0331	Samuels (71)	
Polyvinyl alcohol	0.0443		Crystalline
	0.0404	Nomura and Kawai (58)	Amorphous
Polyvinyl chloride	0.027	Shindo et al. (77)	Amorphous at 80°C
cis-1,4-Polyisoprene	0.13	Hashiyama et al. (31)	
Polystyrene	−0.16	Stein (84)	

TABLE 6-K-3   Stress-Optical Constants for Polymer Melts

Material	T (°C)	$C \times 10^{-9}$ pascal (brewsters)	Source
Polyethylene	190	1810	Den Otter (quoted in (29))
	190	1600	Wales and Philippoff (91)
	200	1230	Han and Drexler (29)
Polypropylene	210	900	Adamse et al. (2)
	180	605	Han and Drexler (29)
Polybutadiene	30	2200	Furukawa et al. (24)
cis-1,4-Polyisoprene	20	2330	Treloar (87)
Hevea	20	1880	Saunders (73)
	20	1640	Shimomura et al. (76)
Guayule	20	2380	Shimomura et al. (76)
Synthetic	20	1870–2640	Shimomura et al. (76)
Polystyrene	140	−4000	Furukawa et al. (24)
	196	−4890	Den Otter (quoted in (29))
	190	−4100	Wales and Philippoff (91)
	210	−4950	Han and Drexler (29)
	120–157	−6100	Matsumoto and Bogue (50)
Butadiene–styrene copolymer			
13% Styrene	30	1870	Furukawa et al. (24)
38.8% Styrene	80	280	Furukawa et al. (24)
48.0% Styrene	80	−280	Furukawa et al. (24)
84.0% Styrene	120	−2900	Furukawa et al. (24)

TABLE 6-K-4   Approximate Values of the Ratio $C/\Delta°$

Material	C (average)	$C/\Delta°$ (average)(Brewsters)
Polyethylene	1550	26,000
Polypropylene	750	23,000
cis-1, 4-Polyisoprene	2000	16,000
Polystyrene	4650	28,000

Experimental stress optical constants for different polymer melts are summarized in Table 6-K-3. The numbers are in the range $600–5000 \times 10^{-12} Vm^2/N$ and may be positive or negative. The dependence seems similar to $\Delta°$, with C decreasing with introductions of substituents. Like $\Delta°$, C becomes negative with polystyrene.

Kuhn and Grun (*44*) have shown on the basis of the kinetic theory of deformation of flexible-chain networks that the stress-optical constant C is given by

$$C = \frac{1}{90kT} \frac{(\bar{n}+2)}{\bar{n}} (\alpha_\| - \alpha_\perp) \tag{6-K-4}$$

The discussion given above indicates that C and $\Delta°$ are related by

$$\frac{C}{\Delta°} = \frac{2}{15kTN} \tag{6-K-5}$$

One would expect that $C/\Delta°$ should be a constant roughly independent of material. We summarize values of $C/\Delta°$ for different polymer melts in Table 6-K-4.

L. FROZEN-IN ORIENTATION AND ITS INTERPRETATION

It has been shown by Oda et al. (*61*) Choi et al. (*18*), and Kang and White (*43*) that birefringence developed during melt flow of polystyrene is "frozen in" when the polymer vitrifies into a glass. Quantitatively, the birefringence is linearly related to the principal stresses acting at the point of vitrification, that is, through the rheo-optical law. Oda et al.'s experiments included uniaxial extension, melt spinning, and simple shear (see Fig. 6-L-1) and Choi et al.'s tubular film extrusion (Fig. 6-L-2).

It is believed that the frozen-in orientation result is general for vitrified systems, but the relation with birefringence is limited to cases where the stress-optical coefficient in the melt is considerably larger than that in the glass and is largely temperature independent. This is because of the existence of large

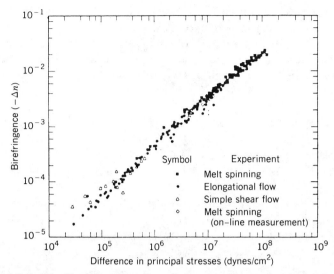

FIGURE 6-L-1. Birefringence of tensile and shear specimens as a function of principal stress difference $\sigma_1 - \sigma_2$ [from Oda et al. (*61*)].

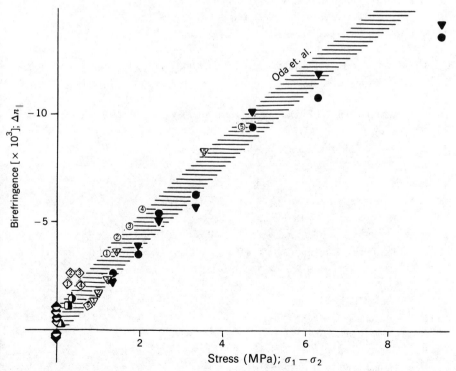

FIGURE 6-L-2. Birefringence of polystyrene films due to Choi et al. (*18*) as a function of applied stress at vitrification for tubular film extension [comparison with correlation of Oda et al. (*61*)].

thermal quench stresses in the vitrified product, which may well be even larger than that in the flowing melt [compare Isayev (*41*)].

Evidence also exists that the state of orientation in polymer melts crystallized under stress is also quantitatively related to the state of stress at the position of solidification in melt-spinning of fibers (*18,56*) and tubular film extrusion (*16*).

REFERENCES

1. E. B. Adams, J. C. Whitehead, and D. C. Bogue, *AIChE J.*, **11**, 1026 (1965).
2. J. W. C. Adamse, H. Janeschitz-Kriegl, J. L. Den Otter, and J. L. S. Wales, *J. Polym. Sci.*, **A-2**(6), 871 (1968).
3. T. Arai and H. Hatta, *Nihon Reoroji Gakkaishi* (*J. Soc. Rheol. Japan*) **8**, 67 (1980).
4. T. Arai and H. Hatta, *Nihon Reoroji Gakkaishi* (*J. Soc. Rheol. Japan*) **7**, 173 (1979).
5. T. Arai, H. Ishikawa, and H. Hatta, *Kobunshi Ronbunshu*, **38**, 29 (1981).
6. D. C. Bogue and F. N. Peebles, *Trans. Soc. Rheol.*, **6**, 317 (1962).
7. M. Born and E. Wolf, *Principles of Optics*, 4th ed., Pergamon, Oxford, 1970.
8. D. Brewster, *Phil. Trans. Roy. Soc.*, **103**, 101 (1813).
9. D. Brewster, *Phil. Trans. Roy. Soc.*, **105**, 60 (1815).
10. D. Brewster, *Phil. Trans. Roy. Soc.*, **106**, 156 (1816).
11. V. I. Brizitsky, G. V. Vinogradov, A. I. Isayev, and V. Y. Podolsky, *J. Appl. Polym. Sci.*, **22**, 751 (1978).
12. J. G. Brodynan, F. H. Gaskins, and W. Philippoff, *Trans. Soc. Rheol.*, **1**, 109 (1957).
13. J. G. Brodnyan, F. H. Gaskins, W. Philippoff, and E. G. Lendrat, *Trans. Soc. Rheol.*, **2**, 285 (1958).
14. C. W. Bunn and R. de P. Daubeny, *Trans. Faraday Soc.*, **50**, 1173 (1954).
15. W. H. Carothers and J. W. Hill, *J. Am. Chem. Soc.*, **54**, 1579 (1932).
16. K. J. Choi, J. E. Spruiell, and J. L. White, *J. Polym. Sci., Polym. Phys.*, **20**, 27 (1982).
17. K. J. Choi, J. L. White, and J. E. Spruiell, *J. Appl. Polym. Sci.*, **25**, 2777 (1980).
18. J. R. Dees and J. E. Spruiell, *J. Appl. Polym. Sci.*, **18**, 1053 (1974).
19. K. G. Denbigh, *Trans. Faraday Soc.*, **36**, 936 (1940).
20. A. J. De Rossett, *J. Chem. Phys.*, **9**, 766 (1941).
21. F. D. Dexter, J. C. Miller, and W. Philippoff, *Trans. Soc. Rheol.*, **5**, 193 (1961).
22. J. T. Edsall in *Advances in Colloid Science*, Vol. 1, E. O. Kramer, Ed., 1942, p. 269.
23. J. T. Edsall, C. G. Gordon, J. W. Mehl, H. Scheinberg, and D. W. Mann, *Rev. Sci. Inst.*, **13**, 243 (1944).
24. J. Furukawa, T. Kotani, and S. Yamashita, *J. Appl. Polym. Sci.*, **13**, 2541 (1969).
25. F. H. Gortemaker, M. G. Hansen, B. DeCinoio, H. M. Laun, and H. Janeschitz-Kriegl, *Rheol. Acta*, **15**, 256 (1976).
26. E. F. Gurnee, *J. Appl. Phys.*, **50**, 1173 (1954).
27. C. D. Han, *Rheol. Acta*, **14**, 173 (1975).
28. C. D. Han, *J. Appl. Polym. Sci.*, **19**, 2403 (1975).
29. C. D. Han and L. H. Drexler, *J. Appl. Polym. Sci.*, **17**, 329 (1973).

30. C. D. Han and Y. J. Yoo, *J. Rheol.*, **24**, 55 (1980).

31. M. Hashiyama, R. G. Gaylord, and R. S. Stein, *Makromol. Chem. Suppl.*, **1**, 579 (1975).

32. J. J. Hermans, P. H. Hermans, D. Vermaas, and A. Weidinger, *Rec. Trav. Chem.*, **65**, 427 (1946).

33. P. H. Hermans, J. J. Hermans, D. Vermaas, and A. Weidinger, *J. Polym. Sci.*, **3**, 1 (1948).

34. P. H. Hermans and P. Platzek, *Kolloid Z.*, **88**, 68 (1939).

35. H. Hertz, *Wiedemanns Ann.*, **34**, 155 (1888).

36. H. Hertz, *Wiedemanns Ann.*, **40**, 577 (1890).

37. S. D. Hong, C. Chang, and R. S. Stein, *J. Polym. Sci.*, **58**, 185 (1962).

38. S. Hoshino, J. Powers, D. G. LeGrand, H. Kawai, and R. S. Stein, *J. Polym. Sci.*, **58**, 185 (1962).

39. R. H. Humphrey, *Proc. Phys. Soc.*, **35**, 217 (1923).

40. C. Huyghens, *Treatise on Light* (1690), Dover reprint, 1962.

41. A. I. Isayev, *Polym. Eng. Sci.*, **23**, 271 (1983).

42. H. G. Jerrard, *J. Opt. Soc. Am.*, **38**, 35 (1945).

43. H. J. Kang and J. L. White, *Int. Polym. Proc.*, **1**, 12 (1986).

44. W. Kuhn and F. Grun, *Kolloid Z.*, **101**, 248 (1943).

45. E. H. Land, *J. Opt. Soc. Am.*, **41**, 957 (1951).

46. A. S. Lodge, *Trans. Faraday Soc.*, **52**, 120 (1956).

47. H. A. Lorentz, *Ann. Phys. Chem.*, **9**, 641 (1880).

48. H. A. Lorentz, *Theory of Electrons*, 2nd ed. (1915), Dover reprint.

49. L. Lorenz, *Ann. Phys. Chem.*, **11**, 70 (1880).

50. T. Matsumoto and D. C. Bogue, *J. Polym. Sci., Polym. Phys. Ed.*, **15**, 1663 (1977).

51. J. C. Maxwell, *Trans. Roy Soc. Edinburgh*, **20**, 87 (1853).

52. J. C. Maxwell, *Phil. Trans. Roy. Soc.*, **155**, 459 (1865).

53. J. C. Maxwell, *Proc. Roy Soc.*, **22**, 46 (1873).

54. J. C. Maxwell, *A Treatise on Electricity and Magnetism*, 3rd ed., Constable, London, 1891.

55. F. H. Muller, *Kolloid Z.*, **95**, 138 (1941).

56. H. P. Nadella, H. M. Henson, J. E. Spruiell, and J. L. White, *J. Appl. Polym. Sci.*, **21**, 3003 (1977).

57. W. Nicol, *Edinburgh New Phil. J.*, **6**, 83 (1829).

58. S. Nomura and H. Kawai, *J. Polym. Sci.*, **A-2**(4), 797 (1966).

59. S. Nomura, H. Kawai, I. Kimura, and M. Kagiyama, *J. Polym. Sci.*, **A-2**(5), 479 (1967).

60. S. Nomura, H. Kawai, I. Kimura, and M. Kagiyama, *J. Polym. Sci.*, **A-2**(8), 383 (1970).

61. K. Oda, J. L. White, and E. S. Clark, *Polym. Eng. Sci.*, **18**, 53 (1978).

62. F. N. Peebles, H. J. Garber, and S. H. Jury, Proceedings of the 3rd Midwestern Conference on Fluid Mechanics, University of Minnesota, 1953.

63. W. Philippoff, *J. Appl. Phys.*, **32**, 984 (1956).

64. W. Philippoff, *Trans. Soc. Rheol.*, **4**, 159 (1960).

65. W. Philippoff, *Trans. Soc. Rheol.*, **4**, 169 (1960).

66. W. Philippoff, *Trans. Soc. Rheol.*, **5**, 149 (1961).

67. J. W. Prados and F. N. Peebles, *AIChE J.*, **5**, 225 (1959).

68. J. M. Preston, *Trans. Faraday Soc.*, **29**, 65 (1933).

69. R. J. Roe, *J. Appl. Phys.*, **36**, 22024 (1965).

70. R. L. Rowell and R. S. Stein, *J. Chem. Phys.*, **47**, 2985 (1967).

71. R. J. Samuels, *J. Polym. Sci.*, **A3**, 1741 (1965).

72. R. J. Samuels, *J. Appl. Polym. Sci.*, **26**, 1383 (1981).

73. D. W. Saunders, *Trans. Faraday Soc.*, **52**, 1414 (1956).

74. G. N. Schael, *J. Appl. Polym. Sci.*, **8**, 1717 (1964).

75. G. N. Schael, *J. Appl. Polym. Sci.*, **12**, 903 (1968).

76. Y. Shimomura, J. L. White, and J. E. Spruiell, *J. Appl. Polym. Sci.*, **27**, 3553 (1982).

77. A. Sommerfeld, *Electrodynamics*, Academic, New York, 1952.

78. A. Sommerfeld, *Optics*, Academic, New York, 1952.

79. R. S. Stein, *J. Chem. Phys.*, **21**, 1193 (1953).

80. R. S. Stein, *J. Polym. Sci.*, **24**, 383 (1957).

81. R. S. Stein, *J. Polym. Sci.*, **31**, 327 (1958).

82. R. S. Stein, *J. Polym. Sci.*, **31**, 355 (1958).

83. R. S. Stein, *J. Appl. Phys.*, **32**, 1280 (1961).

84. R. S. Stein, *J. Polym. Sci.*, **A-2**(7), 1021 (1969).

85. R. S. Stein, F. H. Holmes, and A. V. Tobolsky, *J. Polym. Sci.*, **14**, 443 (1954).

86. R. S. Stein and A. V. Tobolsky, *Text. Res. J.*, **19**, 302 (1949).

87. L. R. G. Treloar, *Trans. Faraday Soc.*, **43**, 284 (1947).

88. J. A. Van Aken and H. Janeschitz-Kriegl, *Rheol. Acta*, **20**, 419 (1981).

89. G. V. Vinogradov, A. I. Isayev, D. A. Mastataev, and Y. Y. Podolsky, *J. Appl. Polym. Sci.*, **22**, 665 (1978).

90. J. L. S. Wales, *Rheol. Acta*, **8**, 38 (1969).

91. J. L. S. Wales and W. Philippoff, *Rheol. Acta*, **12**, 25 (1973).

92. J. L. White, *J. Polym. Eng.*, **5**, 275 (1986).

93. J. L. White and J. E. Spruiell, *Polym. Eng. Sci.*, **21**, 859 (1981).

94. R. W. Wood, *Physical Optics*, 3rd ed., Macmillan, New York, 1934.

95. H. J. Yoo and C. D. Han, *J. Rheol.*, **25**, 115 (1981).

RHEOLOGICAL MODELING OF FLOW BEHAVIOR

VII

CONSTITUTIVE EQUATIONS FOR NON-NEWTONIAN FLUIDS

A. INTRODUCTION

In this chapter we turn our attention to the development of relationships between stress and deformation rate and history in polymer fluid systems. We begin with an historical discussion of the development of this area. We then turn to one-dimensional rheological theories of fluids with memory, that is, viscoelastic fluids, and of fluids with yield values. Differential and integral formulations of viscoelastic fluid behavior are distinguished and compared. The principles of formulation of three-dimensional constitutive equations for stress are outlined, emphasizing Oldroyd's approach using a convected coordinate system. This is successively applied to differential and integral viscoelastic fluid theories representing pure polymer melts. The formulation of theories of flow of suspensions and liquid crystalline materials is also considered. Three-dimensional rheological models of the flow of materials with yield values are described. The predictions of these theories are contrasted with experimental data on pure and filled polymer melts.

B. HISTORICAL PERSPECTIVE

The theory of the flow behavior of fluids with non-Newtonian flow characteristics would seem to date to the research of James Clerk Maxwell (43) during the late 1860s. Clerk Maxwell pointed out that in an elastic material the rate of increase of applied stress is proportional to the rate of increase of applied strain. In a viscoelastic material, the stress increases at a lesser rate than in the elastic case, and the difference might be conjectured to be proportional to the stress. Clerk Maxwell thus essentially wrote for the shear stress σ

$$\frac{d\sigma}{dt} = G\dot{\gamma} - C'\sigma = G\gamma - \frac{1}{\tau}\sigma \qquad (7\text{-B-1})$$

where G is a shear modulus, C' a proportionality constant, and τ the inverse of C', which is called a relaxation time. This one-dimensional expression is capable of explaining many well-known viscoelastic phenomena, such as stress relaxation following a step-change deformation and elastic recovery following a sudden release of a stressed object. Equation (7-B-1) predicts for steady shear flow a viscosity having a value τG, which is independent of rate of shear.

In 1874 Ludwig Boltzmann (10) gave a three-dimensional formulation for the stresses developed in a solid with incomplete memory which are subjected to a multistep strain history. Boltzmann's approach was to assume linearity and additivity of stress responses to the small imposed strains. He allowed for compressibility. In modern notation, Boltzmann wrote

$$\sigma = \lambda(\text{tr }\gamma)\mathbf{I} + 2G\gamma - 2\int_0^\infty [\Psi(t-s)\gamma(s) - Q(t-s)(\text{tr }\gamma(s))\mathbf{I}]\,ds \qquad (7\text{-B-2})$$

where $\gamma(s)$ is infinitesimal strain tensor, γ λ and G are Lame moduli, and $\Psi(t)$ and $Q(t)$ are relaxation functions. In a review published shortly afterwards, Clerk

Maxwell (*44*) noted the superiority of Boltzmann's formulation and relegated his own efforts to an historical introduction.

Clerk Maxwell's 1867 paper, however, stirred the imagination of the next generation of investigators concerned with the flow characteristics of materials with incomplete memory. These were largely eastern Europeans. The first of these investigators was the Russian Theodor Schwedoff (*65*), who was perhaps the first modern rheologist. In 1889–1900, he reported the steady shear flow and creep behavior of gelatin solution. Schwedoff recognized that these systems had both memory and a viscosity which decreases as a function of shear rate. Schwedoff expressed the response of these materials through a generalization of Maxwell's Eq. (7-B-1)

$$\frac{d\sigma}{dt} = G\dot{\gamma} - \frac{1}{\tau}[\sigma - Y]$$ (7-B-3)

where Y is a yield value. Schwedoff was able to represent both his creep and steady shear-flow data through Eq. (7-B-3). This he noted led to a shear stress–shear rate rleationship

$$\sigma = \left[\frac{Y}{\dot{\gamma}} + \tau G\right]\dot{\gamma} = Y + \tau G\dot{\gamma}$$ (7-B-4)

which indicated deviations from Newtonian flow due to the presence of the yield value Y.

During the opening years of the present century, attention turned to the formulation of three-dimensional large-strain formulations for the Maxwell model, Eq. (7-B-4). In a 1901 paper, L. Natanson (*47*) of the University of Cracow proposed a generalization in the form

$$\frac{D\boldsymbol{\sigma}}{Dt} = \lambda(\text{tr }\mathbf{d})\mathbf{I} + 2G\mathbf{d} - \frac{1}{\tau}[\boldsymbol{\sigma} + p\mathbf{I}]$$ (7-B-5)

where

$$\frac{D\boldsymbol{\sigma}}{Dt} = \frac{\partial\boldsymbol{\sigma}}{\partial t} + (\mathbf{v}\cdot\nabla)\boldsymbol{\sigma}$$ (7-B-6)

and λ and G are equivalent to Lame's constants of the theory of elasticity.

In a paper published shortly afterwards in the *Bulletin of the Academy of Cracow*, S. Zaremba (*95*) took exception to the formulation of Natanson largely because it did not represent the constitutive equation in a coordinate frame natural to the deforming medium. This, he argued, was a rigid coordinate system which not only translated but rotated with the fluid element. The stress components in the laboratory frame are the familiar σ_{ij}, but those that form the basic constitutive law are $\pi_{\alpha\beta}$, which may be related to σ_{ij} by

$$\pi_{\alpha\beta}(t) = \sum_i \sum_j l_{\alpha i}(t)l_{\beta j}(t)\sigma_{ij}(t)$$ (7-B-7)

The $l_{\alpha i}$ represent time-varying directional cosines relating the laboratory components to the translating and rotating coordinate system. If a simple partial time derivative of stress exists in the basic rheological model, it follows that

$$\frac{\partial \pi_{\alpha\beta}}{\partial t} = \sum_i \sum_j \left[l_{\alpha i} l_{\beta j} \frac{d\sigma_{ij}}{dt} + \left(\frac{d}{dt} l_{\alpha i} \right) l_{\beta j} \sigma_{ij} + l_{\alpha i} \left(\frac{d}{dt} l_{\beta j} \right) \right] \sigma_{ij} \qquad (7\text{-B-8})$$

and

$$\frac{\partial \pi_{\alpha\beta}}{\partial t} = \sum_i \sum_j l_{\alpha i} l_{\beta j} \frac{\mathscr{D}\sigma_{ij}}{\mathscr{D}t}$$

with

$$\frac{\mathscr{D}\sigma_{ij}}{\mathscr{D}t} = \frac{D\sigma_{ij}}{Dt} + \sum_m \sigma_{im}\omega_{mj} + \sum_m \sigma_{mj}\omega_{im}$$

where ω_{ij} is the vorticity tensor. A more complex stress derivative is thus necessary in the constitutive expression in a laboratory frame of reference. Zaremba also distinguished between distortions and volumetric relaxation writing finally

$$\frac{\mathscr{D}\sigma}{\mathscr{D}t} = \lambda(\operatorname{tr} \mathbf{d})\mathbf{I} + 2G\mathbf{d} - \left[\frac{1}{\tau}\sigma + \frac{1}{\tau_V}(\operatorname{tr} \sigma)\mathbf{I} \right] \qquad (7\text{-B-9})$$

where τ is a distortional relaxation time and τ_V a volumetric relaxation time. He applied this formulation to flow between coaxial cylinders, where he finds normal stresses arise.

Little attention was given to either Schwedoff or Zaremba by their contemporaries. A quarter century after each of their papers, their work was independently reproduced, and even a longer period passed before their views, then credited to others, were accepted. In 1913–1916 variable viscosity behavior was rediscovered (4). Bingham (4) proposed models involving yield values. From this period, non-Newtonian viscosity was widely accepted and used to represent data. Elastic behavior was recognized only in a qualitative sense.

In 1929 Hencky (30), noting Natanson's paper, recognized its error and rederived an incompressible form of Eq. (7-B-9). Four years later, Eisenschitz (21) recognized the validity of Hencky's theory and used it to interpret shear viscosity data on a cellulose nitrate solution. In his analysis, Eisenschitz derives expressions for normal stresses, but does not recognize their significance.

Hohenemser and Prager (32) in 1932 presented the first three-dimensional theory of plastic materials with differentially viscous behavior above the yield value. In so doing, they introduced the use of invariants into the theory of non-Newtonian fluids. Their paper builds on von Mises theory of plastic yielding. They write

$$\sigma = \frac{1}{3}(\operatorname{tr} \sigma)\mathbf{I} + T, \qquad \mathbf{T}\left(1 - \frac{Y}{\sqrt{\frac{1}{2}\operatorname{tr} \mathbf{T}^2}} \right) = 2\eta_B\mathbf{d} \qquad (7\text{-B-10a, b})$$

where $\mathbf{T}$ is the deviatoric stress tensor and $\operatorname{tr}\mathbf{T}^2$ its second invariant.

The rheologists of the 1930s and early 1940s gave little attention to the papers by Hencky, Eisenschitz, and Hohenemser and Prager and used one-dimensional flow models. It is only with the papers of Reiner (58), Oldroyd (49–51), and Rivlin (60–62) in the period 1945–1950 that three-dimensional rheological theories began to be accepted. Reiner (58), seeking to explain the observation of dilation during a shear flow, used the Cayley–Hamilton theorem to derive the expression

$$\sigma_0 = a_0\mathbf{I} + 2\eta\mathbf{d} + 4\eta_C\mathbf{d}^2 \tag{7-B-11}$$

where a_0, η, and η_C depend upon $\operatorname{tr}\mathbf{d}$, $\operatorname{tr}\mathbf{d}^2$, $\operatorname{tr}\mathbf{d}^3$. Here η is the shear viscosity and η_C came to be called the cross viscosity. Reiner showed that η_C leads to normal stresses in shear flow. Rivlin (60) argued that the existence of a cross viscosity explained Weissenberg's (80) observations of normal stress effects.

Oldroyd's development followed a different route. He first independently derived (49) and generalized (50) Hohenemser and Prager's theory of Bingham plastics. In a later paper, Oldroyd (51), unaware of Zaremba's or Hencky's papers, proposed a formulation of the nonlinear viscoelastic fluid based upon a coordinate system embedded in and deforming with the medium. This he called a convected coordinate system. Oldroyd's paper was based on his intuition that the normal stresses observed by Weissenberg were due to the viscoelastic properties of the fluids involved. These he saw must come from the stress derivatives. This necessitates careful consideration of the basic formulation of three-dimensional viscoelastic fluid models and leads to the idea of special embedded coordinate systems. In any case, Oldroyd's paper inspired the development of the theory of nonlinear viscoelasticity by various researchers during the next 10–20 years. Oldroyd (51) himself developed expressions based on generalizing differential models similar to Eq. (7-B-1). He begins with models of the form

$$\sigma + \tau_1\frac{d\sigma}{dt} = 2\eta\left[\frac{d\gamma}{dt} + \tau_2\frac{d^2\gamma}{dt^2}\right] \tag{7-B-12}$$

which generalize to

$$\sigma = -p\mathbf{I} + \mathbf{P}$$

$$\mathbf{P} + \tau_1\frac{\delta\mathbf{P}}{\delta t} = 2\eta\left[\mathbf{d} + \tau_2\frac{\delta\mathbf{d}}{\delta t}\right] \tag{7-B-13}$$

where τ_1 and τ_2 have units of time and η has units viscosity. The form of the derivatives was found not to be unique, but depended upon whether a covariant or contravariant formulation was used. The former was called an A fluid and the latter a B fluid. Oldroyd (52) noted that as the form of Eq. (7-B-12) is linear, any combination of second- and higher-order products of $\mathbf{P}$ and $\mathbf{d}$ could be introduced.

More general perspectives of viscoelastic fluid behavior were developed in succeeding years under the inspiration of the papers of Oldroyd. This may be seen in the papers of Rivlin (28, 63, 64), Criminale, Ericksen, and Filbey (17), Coleman and Noll (15, 16) and others. The stress tensor was expressed as a general function of the rates and accelerations of deformation or as a functional of deformation history. The results obtained were highly complex. This led other investigators (3, 6–9, 11, 35, 50, 67, 72, 78, 87, 92) during the next two decades to develop a range of constitutive equations for viscoelastic fluids which were simpler in character and could be used to correlate experimental data for polymer solutions and melts. While Zapas (93, 94) made early quantitative studies on elastomers, extensive quantitative comparisons of experiments on polymer melts and theory are only found in the 1970s (13, 14, 70, 78, 79).

Mechanistic constitutive theories have been derived for networks of polymer chains. Such theories date to Green and Tobolsky in 1946 (29) and Yamamoto from 1953 (91), and were developed in later years by Lodge (40) and others (5, 19, 55, 90).

Constitutive equations have been developed for a wide range of other materials in the past generation. Substances with structure at a state of rest, including liquid crystals (24, 38), have been characterized. Largely owing to the efforts of Batchelor (2) and Hinch and Leal (31), three-dimensional constitutive equations have been developed for dilute suspensions of rigid particles. Phenomenological constitutive equations have also been applied to polymer systems filled with small particles. In 1975, Hutton (33) proposed representing the stress field in such systems in terms of a yield surface plus a functional of the deformation history. More recently, White, Lobe, Tanaka, and Suetsugu (39, 69, 83, 84, 88) have developed more specific formulations along these lines and contrasted them to experiments.

C. ONE-DIMENSIONAL RHEOLOGICAL THEORY

1. Viscoelasticity

Some discussion of the characteristics of the one-dimensional formulation of viscoelasticity is really needed before we begin consideration of the three-dimensional nonlinear theory. The general one-dimensional linear theory may be derived by considering the stress to be equal to the sum of responses to a series of consecutive small strains. We presume that a one-dimensional step strain γ_0 given rise to a step rise in stress σ followed by relaxation through

$$\sigma(t) = G(t)\gamma_0 \tag{7-C-1}$$

where $G(t)$ is a relaxation modulus. A sequence of small steps at times $t_1, t_2, t_3, \ldots$ gives rise to the stress

$$\sigma(t) = \sum_i G(t - t_i)\gamma(t_i) = \int_{-\infty}^{t} G(t - s)\, d\gamma(s) = \int_{-\infty}^{t} G(t - s)\frac{d\gamma}{ds}\, ds \tag{7-C-2}$$

where, we have noted, the asymptote as the size of the strain steps becomes infintesimally small. Equation (7-C-2) is called Boltzmann's superposition integral theory, Boltzmann (*10*) actually began his paper with the more complex Eq. (7-B-2).

There is a second formulation of viscoelasticity which traces to Clerk Maxwell (*43*). This relates rates of change of stress to rates of change of deformation. Clerk Maxwell's basic model is expressed in Eq. (7-B-1), which may be rewritten as

$$\frac{d\sigma}{dt} + \frac{1}{\tau}\sigma = G\frac{d\gamma}{dt} \qquad (7\text{-}C\text{-}3)$$

This may be recognized as a first-order ordinary nonhomogeneous differential equation which may be solved using an integrating factor $e^{t/\tau}$. This leads to

$$\sigma = \int_{-\infty}^{t} G\,e^{-(t-s)/\tau}\frac{d\gamma}{ds}\,ds \qquad (7\text{-}C\text{-}4)$$

Comparing Eqs. (7-C-3) and (7-C-4), it may be seen that the Maxwell model is the special case of Eq. (7-C-2) for which

$$G(t) = G\,e^{-t/\tau} \qquad (7\text{-}C\text{-}5)$$

It is traditional to express the relaxation modulus through a series of exponentials

$$G(t) = \sum_i G_i\,e^{-t/\tau_i} \qquad (7\text{-}C\text{-}6)$$

or through a continuous spectrum $H(\tau)$ as

$$G(t) = \int_0^{\infty} H(\tau)e^{-t/\tau}\frac{d\tau}{\tau} \qquad (7\text{-}C\text{-}7)$$

Generally, the rheological behavior of polymers in the linear viscoelastic range is specified through the spectrum of relaxation times $H(\tau)$.

It is common following Boltzmann to express Eq. (7-C-2) in terms of an integral over strain history rather than rate of deformation. This may be obtained from Eq. (7-C-2) by integrating by parts to yield

$$\sigma(t) = G(0)\gamma(t) - G(\infty)\gamma(-\infty) - \int_{-\infty}^{t} \gamma(s)\,dG(t-s) \qquad (7\text{-}C\text{-}8)$$

Noting that $G(\infty)$ is zero for fluids, Eq. (7-C-8) may be expressed in the equivalent form

$$\sigma(t) = G(0)\gamma(t) + \int_0^\infty \phi(z)\gamma(t-z)\,dz \tag{7-C-9}$$

where z is $t - s$ and

$$\phi(t) = -\frac{dG(t)}{dt} \tag{7-C-10}$$

For viscoelastic fluids, there is no preferred state and strains are usually measured from the instantaneous configuration. Thus

$$\gamma(t) = 0$$
$$\sigma(t) = \int_0^\infty \phi(z)\gamma(z)\,dz \tag{7-C-11}$$

where z is measured background from time zero.

For smooth deformations, the strain $\gamma(z)$ may be written as a Taylor series in time z

$$\gamma(z) = \sum \frac{z^n}{n!}\left(\frac{d^n\gamma}{dz^n}\right)_{z=0} \tag{7-C-12}$$

where $(d^n\gamma/dz^n)_0$ is the nth acceleration at time t. Substitution of Eq. (7-C-10) into Eq. (7-C-12) yields the linear differential equation, expansion

$$\sigma(t) = \sum_{n=1} a_n\left(\frac{d^n\gamma}{dz^n}\right)_0 (t) \tag{7-C-13}$$

where

$$a_n = \frac{1}{n!}\int_0^\infty z\phi(z)\,dz \tag{7-C-14}$$

2. Materials with Yield Values

Another class of rheological models of interest to industrial polymer fluids are systems with yield values. The simplest model would be one in which (1) there is no deformation or a linear elastic response when the stresses are below a yield stress Y and (2) the stresses are a constant Y above it, that is,

$$\sigma < Y, \quad \left\{ \quad \text{or} \quad \begin{array}{l} \gamma = 0 \\ \sigma = G_1\gamma \end{array} \right. \qquad \begin{array}{l} (7\text{-}C\text{-}15a) \\ (7\text{-}C\text{-}15b) \end{array}$$

with γ in Eq. (15b) determined from the rest-state configuration, which must be redefined after each period of flow. When σ reaches Y,

$$\sigma = Y, \qquad \gamma = \text{indeterminate} \qquad\qquad (7\text{-C-}16)$$

Equation (7-C-15) is known as a *perfect plastic* or *rigid plastic* if Eq. (7-C-15a) is also valid. It is described as an *elastic plastic* if Eq. (7-C-15b) represents the behavior below the yield value.

In Bingham's theory of plastic flow [compare Eq. (7-B-4) of Schwedoff (65)], the material exhibits a viscous resistance to deformation above the yield value. This is expressed as

$$\sigma = Y + \eta_B \dot{\gamma} \qquad\qquad (7\text{-C-}17)$$

This defines the Bingham plastic. We may refer to this behavior as that of a *viscous plastic*. This may be further classified as a *rigid viscous plastic* or *elastic viscous plastic*, depending upon whether Eq. (7-C-15a) or (7-C-15b) is valid below the yield value.

It is possible to generalize this formulation to include viscoelastic response above the yield value. This was first devised by Schwedoff (65), as stated in our Eq. (7-B-3). This may be considered a differential equation in which σ may be solved to give

$$\frac{d\sigma}{dt} = G\dot{\gamma} - \frac{1}{\tau}(\sigma - Y), \qquad \sigma = Y + \int_{-\infty}^{t} G e^{-(t-s)/\tau} \gamma(s)\, ds \qquad (7\text{-C-}18a, b)$$

This of course may be considered a special case of a more general expression

$$\sigma = Y + \int_{-\infty}^{t} \phi(t - s)\gamma(s)\, ds \qquad\qquad (7\text{-C-}19)$$

We may classify it as *rigid viscoelastic plastic* and *elastic viscoelastic plastic* according to whether Eq. (7-C-15a) or Eq. (7-C-15b) represents the behavior below the yield value.

3. Mechanical Models

We may represent the various rheological theories described earlier in this section using mechanical analogs. Three elements are included in these models: (1) a spring, representing elastic behavior; (2) a dashpot, representing viscous behavior; and (3) a slider, expressing plastic yielding. These elements are shown in Figure 7-C-1a–c. The displacements in these models represent strain, and the tension deforming them, stress.

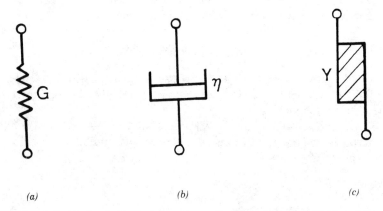

FIGURE 7-C-1. Mechanical models representing rheological behavior: (*a*) spring—linear elastic solid; (*b*) dashpot—linearly viscous fluid; (*c*) slider—rigid perfect plastic.

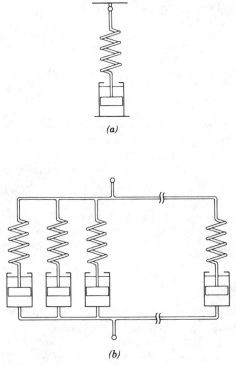

FIGURE 7-C-2. Mechanical models representing viscoelastic behavior: (*a*) Maxwell model; (*b*) generalized Maxwell or Wiechert model.

The strain response of a spring and dashpot in series (Fig. 7-C-2a) is

$$\gamma = \frac{\sigma}{G} + \int \frac{\sigma}{\eta} \, dt \qquad \text{(7-C-20)}$$

$$\frac{d\gamma}{dt} = \frac{1}{G} \frac{d\sigma}{dt} + \frac{\sigma}{\eta} \qquad \text{(7-C-21)}$$

This is equivalent to Maxwell's equation. A series of parallel spring–dashpot models with coefficients G_i, η_i (Fig. 7-C-2b) is equivalent to

$$\sigma = \sum \sigma_i$$

$$\frac{d\gamma}{dt} = \frac{1}{G_i} \frac{d\sigma_i}{dt} + \frac{\sigma_i}{\eta_i} \qquad \text{(7-C-22)}$$

Solving these equations leads to

$$\sigma = \int_{-\infty}^{t} \sum_i G_i e^{-(t-s)/\tau} \gamma \, ds \qquad \text{(7-C-23)}$$

which is essentially Boltzmann's superposition principle.

A slider and a dashpot in parallel leads to (Fig. 7-C-3a)

$$\gamma = 0, \qquad \sigma < Y \qquad \text{(7-C-24)}$$

$$\sigma = Y + \eta \frac{d\gamma}{dt}, \qquad \sigma > Y \qquad \text{(7-C-25)}$$

which is the Bingham plastic model.

A viscoelastic–plastic materials may be represented by a slider in parallel with one or more spring–dashpot series combinations (Fig. 7-C-3b, c). Figure (7-C-3b) is equivalent of Schwedoff's model. We may write for $\sigma > Y$,

$$\sigma = Y + \sum \sigma_i \qquad \text{(7-C-26)}$$

where the σ_i are given by Eq. (7-C-22). This leads to

$$\sigma = Y + \int_{-\infty}^{t} \sum_i G_i e^{-(t-s)/\tau_i} \gamma \, ds \qquad \text{(7-C-27)}$$

The Bingham and viscoelastic plastic models of Figure 7-C-3 behave rigidly below their yield values, that is, γ will be zero for σ less than Y, as with Eq. (7-C-15a). A more realistic model would be to include elastic behavior for stresses less than Y. This can be accomplished by placing a spring of modulus G_0 in series with

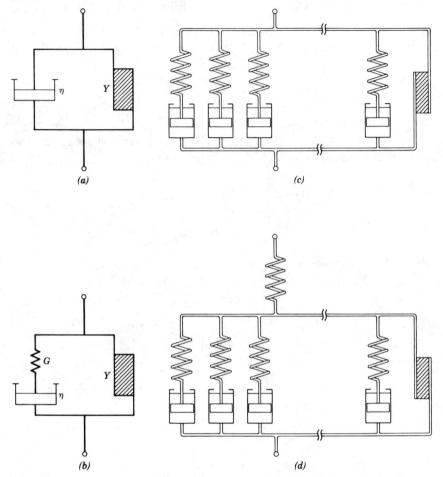

FIGURE 7-C-3. Mechanical model representing differentially viscous viscoelastic materials with a yield value: (a) Bingham plastic; (b) Schwedoff model; (c) generalized viscoelastic plastic; (d) viscoelastic plastic model allowing for elastic behavior below yield value.

the models of Figure 7-C-3. This leads to a response equivalent to Eq. (7-C-15b) at low stresses (see Fig. 7-C-3d).

D. FORMULATION OF THREE-DIMENSIONAL LARGE-STRAIN CONSTITUTIVE EQUATIONS

The principles of formulation of constitutive equations for the mechanical behavior of continuous media were a subject of concern as far back as the efforts of Cauchy (*12*) and Stokes (*68*) dealing with elastic solids and linear viscous fluids. Zaremba (*95*) presented a more complete perspective in his 1903 paper. It was, however, only in the 1950s that a general understanding of the concepts involved came to be widely accepted and used. Several different but equivalent formulations were developed in these years.

Oldroyd (*51, 53*) approached the problem of what he called "rheological equations of state" as follows (*51*):

> Equations of state must be considered as equations defining the properties of an arbitrary material element moving as part of a continuum. The quantities which may be involved in the equations are then quantities associated with that particular element over a period of time during which the element has moved and has been deformed continuously in an arbitrary manner. Moreover, only those tensor quantities need be considered which have a significance for the material element independent of its motion as a whole in space. In order to specify these quantities in the general case, it is not at first convenient to use a frame of reference fixed in space, but a convected coordinate system. The coordinate surfaces $\xi_j = $ constant are chosen as surfaces drawn in the material and deforming continuously with it, and the ξ_j and the time t are taken as independent variables. The device of representing all tensor quantities associated with the material by their convected components (i.e. the components referred to a convected frame of reference), allows similar quantities associated with the same material at different times to be added, component by component to give a similar tensor quantity as the sum.

Noll, in conjunction with Truesdell (*48, 74*), has proposed an alternate formulation. They express a "principle of isotropy of space" or a "principle of material indifference," which is expressed simply as (*74*): "Constitutive equations must be invariant under changes of frame of reference."

These are in fact, as noted by Truesdell and Noll (*74*), two different principles—that of Cauchy, Stokes, and Noll, in which constitutive equations are required to be independent of superimposed rigid rotations, and that of Zaremba and Oldroyd, which used special coordinate systems. A constitutive equation of the Zaremba–Oldroyd form satisfies the Cauchy–Stokes–Noll form. The converse is not as clear, but in practice seems to be the case.

In this chapter, we follow the Zaremba–Oldroyd approach and in particular that of the latter author. The mathematical level required to present this is advanced and requires the use of tensor analysis to develop it. This application of tensors seems necessary and can only be avoided by introducing great oversimplifications. The only help we can offer the reader is to reference the literature and suggest alternate reading. The volumes of Aris (*1*), Bird et al. (*6*) Frederickson (*24*), and Truesdell and Noll (*74*) all represent useful reading in this area from different perspectives.

We consider the convected frame of reference to be the basic one. The convected coordinate system is defined in terms of a metric tensor $\gamma_{\alpha\beta}$ through

$$(ds)^2 = \gamma_{\alpha\beta} \, d\xi^\alpha \, d\xi^\beta \tag{7-D-1}$$

where we distinguish between covariant (subscript) and contravariant (super-script) components in this generally curvilinear frame. Summation notation is used, that is, a repeated subscript and superscript (or combination of subscripts) implies summations. For a pedagogical treatment which generally follows our formulation, see Aris (*1*). In terms of a fixed space cartesian frame x^i,

$$\gamma_{\alpha\beta} = \frac{\partial x^i}{\partial \xi^\alpha} \frac{\partial x^j}{\partial \xi^\beta} \delta_{ij} \tag{7-D-2}$$

The ξ^α coordinate system is so defined that it is not explicitly dependent upon time. Thus

$$\frac{d\xi^\alpha}{dt} = 0, \qquad \frac{d}{dt}(ds)^2 = \frac{d\gamma_{\alpha\beta}}{dt} d\xi^\alpha d\xi^\beta \tag{7-D-3a,b}$$

It should also be seen that $d\gamma_{\alpha\beta}/dt$ is equivalent to the Stokes rate of deformation tensor [Eq. (2-D-3)] in a convected frame of reference. Referring to Eq. (7-D-2) we have

$$\frac{d\gamma_{\alpha\beta}}{dt} = \frac{\partial x^i}{\partial \xi^\alpha} \frac{\partial x^i}{\partial \xi^\beta} 2d_{ij} \tag{7-D-4}$$

We may also define a conjugate metric tensor $\gamma^{\alpha\beta}$ through

$$\gamma^{\alpha\beta}\gamma_{\beta\lambda} = \delta^\alpha_\lambda, \qquad \gamma^{\alpha\beta} = \frac{\partial \xi^\alpha}{\partial x^i} \frac{\partial \xi^\beta}{\partial x^j} \delta^{ij} \tag{7-D-5}$$

This also represents a deformation tensor. Differentiating Eq. (7-D-5) with time yields

$$\frac{d\gamma^{\alpha\beta}}{dt} \gamma_{\beta\lambda} + \gamma^{\alpha\beta} \frac{d\gamma_{\beta\lambda}}{dt} = 0$$

$$\frac{d\gamma^{\alpha\delta}}{dt} = -\gamma^{\alpha\beta}\gamma^{\lambda\delta} \frac{d\gamma_{\beta\lambda}}{dt} \tag{7-D-6}$$

The rate of change of $\gamma^{\alpha\delta}$ with time corresponds to a negative rate of deformation tensor.

Constitutive equations in convected coordinate systems can be expressed in terms of the kinematic tensors $\gamma_{\alpha\beta}$, $d\gamma_{\alpha\beta}/dt,\ldots$ and their relationship with *covariant* stress components $\pi_{\alpha\beta}$.

$$\pi_{\alpha\beta} = \frac{\partial x^i}{\partial \xi^\alpha} \frac{\partial x^j}{\partial \xi^\beta} \sigma_{ij} \tag{7-D-7}$$

The simplest example is to take a linear relationship between $\pi_{\alpha\beta}$ and $d\gamma_{\alpha\beta}/dt$

of form

$$\pi_{\alpha\beta} = -p\gamma_{\alpha\beta} + \eta\frac{d\gamma_{\alpha\beta}}{dt} \qquad (7\text{-D-}8)$$

Using Eq. (7-D-2), Eq. (7-D-4), and Eq. (7-D-7) leads to, in a fixed-space coordinate frame:

$$\frac{\partial x^i}{\partial \xi^\alpha}\frac{\partial x^j}{\partial \xi^\beta}\sigma_{ij} = \frac{\partial x^i}{\partial \xi^\alpha}\frac{\partial x^j}{\partial \xi^\beta}[-p\delta_{ij} + 2\eta d_{ij}] \qquad (7\text{-D-}9)$$

This is, of course, equivalent to the Newtonian fluid.

Equivalent formulations are possible using contravariant components. This involves $\gamma^{\alpha\beta}, d\gamma^{\alpha\beta}/dt$, and the contravariant components $\pi^{\alpha\beta}$ of the stress tensor. We may write

$$\pi^{\alpha\beta} = -p\gamma^{\alpha\beta} + \eta\left(-\frac{d\gamma^{\alpha\beta}}{dt}\right) \qquad (7\text{-D-}10)$$

First we evaluate $d\gamma^{\alpha\beta}/dt$:

$$\frac{d}{dt}\gamma^{\alpha\beta} = \frac{d}{dt}\left(\frac{\partial \xi^\alpha}{\partial x^i}\frac{\partial \xi^\beta}{\partial x^j}\delta^{ij}\right) = \frac{d}{dt}\left(\frac{\partial \xi^\alpha}{\partial x^i}\right)\frac{\partial \xi^\beta}{\partial x^j}\delta^{ij} + \frac{\partial \xi^\alpha}{\partial x^i}\frac{d}{dt}\left(\frac{\partial \xi^\beta}{\partial x^j}\right)\delta^{ij} \qquad (7\text{-D-}11a)$$

and noting that

$$\frac{d}{dt}\left[\frac{\partial \xi^\alpha}{\partial x^i}\frac{\partial x^i}{\partial \xi^\beta}\right] = \frac{d}{dt}\left[\frac{\partial \xi^\alpha}{\partial x^i}\right]\frac{\partial x^i}{\partial \xi^\beta} + \frac{\partial \xi^\alpha}{\partial x^m}\frac{\partial v^m}{\partial \xi^\beta} = 0 \qquad (7\text{-D-}11b)$$

$$\frac{d}{dt}\left(\frac{\partial \xi^\alpha}{\partial x^i}\right) = -\frac{\partial \xi^\alpha}{\partial x^m}\frac{\partial v^m}{\partial x^i}$$

we obtain

$$\frac{d\gamma^{\alpha\beta}}{dt} = -\frac{\partial \xi^\alpha}{\partial x^m}\frac{\partial \xi^\beta}{\partial x}\left(\frac{\partial v^m}{\partial x^j} + \frac{\partial v^j}{\partial x^m}\right) \qquad (7\text{-D-}11c)$$

If we introduce a contravariant transformation for $\pi^{\alpha\beta}$ to σ^{ij} in Eq. (7-D-10) we obtain

$$\pi^{\alpha\beta} = \frac{\partial \xi^\alpha}{\partial x^i}\frac{\partial \xi^\beta}{\partial x^j}\sigma^{ij} \qquad (7\text{-D-}12)$$

$$\frac{\partial \xi^\alpha}{\partial x^i}\frac{\partial \xi^\beta}{\partial x^j}\sigma^{ij} = \frac{\partial \xi^\alpha}{\partial x^i}\frac{\partial \xi^\beta}{\partial x^j}[-p\delta^{ij} + 2\eta d^{ij}] \qquad (7\text{-D-}13)$$

This constitutive equation is equivalent to (7-D-9).

E. DEFORMATION TENSORS AND THE THEORY OF DEFORMATION OF AN ELASTIC SOLID

The characterization of Eq. (7-D-1) and (7-D-5) as deformation tensors requires further development. Consider an elastic solid which may initially be represented in terms of coordinate position X^A. In the deformed state it may be expressed in terms of coordinate position x^i. We may rewrite Eq. (7-D-1) in the form, where we use cartesian spatial coordinates,

$$(dS)^2 = \gamma^0_{\alpha B} d\xi^\alpha d\xi^\beta = dX^A dX^A = c_{ij} dx^i dx^j \tag{7-E-1}$$

$$(ds)^2 = \gamma_{\alpha\beta} d\xi^\alpha d\xi^\beta = dx^i dx^i = C_{AB} dX^A dX^B \tag{7-E-2}$$

Here dS, dX^A, and $\gamma^0_{\alpha\beta}$ refer to the initial state and ds, dx^i, and $\gamma_{\alpha\beta}$ refer to the deformed state. The term c_{ij} is known as the Cauchy deformation tensor and C_{AB} as the Green deformation tensor (75). They are given by

$$c_{ij} = \gamma^0_{\alpha\beta} \frac{\partial \xi^\alpha}{\partial x^i} \frac{\partial \xi^\beta}{\partial x^j} = \frac{\partial X^A}{\partial x^i} \frac{\partial X^A}{\partial x^j} \tag{7-E-3}$$

$$C_{AB} = \gamma_{\alpha\beta} \frac{\partial \xi^\alpha}{\partial X^A} \frac{\partial \xi^\beta}{\partial X^B} = \frac{\partial x^i}{\partial A^A} \frac{\partial x^j}{\partial X^A} \tag{7-E-4}$$

The Cauchy deformation tensor is expressed in terms of the deformed state and the Green tensor in terms of the undeformed state.

The Cauchy measure for deformation is not unique. Many other equivalent deformation tensors are discussed in the literature. The most important of these deformation tensors is that of Finger, which is defined through (75)

$$\mathbf{c} \cdot \mathbf{c}^{-1} = \mathbf{I}$$

or $\tag{7-E-5}$

$$c_{ij}(c^{-1})^{jk} = \delta_{ik}$$

The Finger tensor $\mathbf{c}^{-1}$ has the components

$$(c^{-1})^{ij} = \frac{\partial x^i}{\partial X^A} \frac{\partial x^j}{\partial X^A} \tag{7-E-6}$$

It is a contravariant second-order tensor.

The Finger tensor may be interpreted in terms of a differential area element. Consider the invariance of a differential material element of mass

$$dm = \rho^0 dX^A dA_A = \rho dx^i da_i \tag{7-E-7}$$

where dA_A and da_i represent differential area elements in the initial and deformed state and ρ^0 and ρ are the corresponding densities from Eq. (7-E-7). It follows that (75)

$$dA_A = J\frac{\partial x^i}{\partial X^A}da_i \tag{7-E-8}$$

$$(dA)^2 = dA_A(t')dA_A(t') = J^2(c^{-1})^{ij}da_ida_j \tag{7-E-9}$$

We now turn to the formulation of constitutive equations for an elastic solid in which the stress tensor varies linearly with the deformation measure. Covariant and contravariant formulations have the form

$$\pi_{\alpha\beta}(t) = -p\gamma_{\alpha\beta}(t) + G[\gamma_{\alpha\beta}(t) - \gamma_{\alpha\beta}(s)] \tag{7-E-10}$$

$$\pi^{\alpha\beta}(t) = -p\gamma^{\alpha\beta}(t) + G[\gamma^{\alpha\beta}(s) - \gamma^{\alpha\beta}(t)] \tag{7-E-11}$$

As we have noted, $\gamma_{\alpha\beta}$ is equivalent to the Cauchy deformation measure and $\gamma^{\alpha\beta}$ to the Finger deformation measure. Applying Eqs. (7-D-7) and (7-D-12) to Eqs. (7-E-10) and (7-E-11) yields

$$\frac{\partial x^i}{\partial \xi^\alpha}\frac{\partial \xi^j}{\partial \xi^\beta}\sigma_{ij} = \frac{\partial x^i}{\partial \xi^\alpha}\frac{\partial x^j}{\partial \xi^\beta}[-p\delta_{ij} + G(\delta_{ij} - c_{ij})] \tag{7-E-12}$$

$$\frac{\partial \xi^\alpha}{\partial x^i}\frac{\partial \xi^\beta}{\partial x^j}\sigma_{ij} = \frac{\partial \xi^\alpha}{\partial x^i}\frac{\partial \xi^\beta}{\partial x^j}[-p\delta_{ij} + G(c_{ij}^{-1} - \delta_{ij})] \tag{7-E-13}$$

These constitutive expressions for the stress are quite different because the forms of the c_{ij} and c_{ij}^{-1} measures are not the same. For specific applied deformations, the forms of c_{ij} and c_{ij}^{-1} yield differing responses. In the case of uniaxial extension,

$$x_1(t) = \lambda X_1, \qquad x_2(t) = \frac{1}{\sqrt{\lambda}}X_2, \qquad x_3(t) = \frac{1}{\sqrt{\lambda}}X_3 \tag{7-E-14}$$

we obtain

$$\mathbf{c} = \begin{bmatrix} 1/\lambda^2 & 0 & 0 \\ 0 & \lambda & 0 \\ 0 & 0 & \lambda \end{bmatrix}, \qquad \mathbf{c}^{-1} = \begin{bmatrix} \lambda^2 & 0 & 0 \\ 0 & 1/\lambda & 0 \\ 0 & 0 & 1/\lambda \end{bmatrix} \tag{7-E-15}$$

This makes the stresses predicted by Eqs. (7-E-12) and (7-E-13) different for the same deformations. They reduce, however, to the same form (linear elasticity) when the extension ratio λ is decreased toward unity.

Reiner (59) has sought to develop a more general formulation of nonlinear elasticity theory which is not dependent upon the type of deformation measure.

Writing

$$\boldsymbol{\sigma} = \chi_0 \mathbf{I} + \chi_1 \mathbf{c} + \chi_2 \mathbf{c}^2 + \chi_3 \mathbf{c}^3 + \cdots \tag{7-E-16}$$

and using the Cayley–Hamilton theorem, Eq. (2-J-14), the stress tensor may be expressed as

$$\boldsymbol{\sigma} = v_0 \mathbf{I} + v_1 \mathbf{c} + v_2 \mathbf{c}^2 \tag{7-E-17a}$$

and using Eq. (7-E-5) as

$$\boldsymbol{\sigma} = \mu_0 \mathbf{I} + \mu_1 \mathbf{c}^{-1} + \mu_2 \mathbf{c} \tag{7-E-17b}$$

where the μ_j depends upon the invariants of $\mathbf{c}$ or $\mathbf{c}^{-1}$. Finger and Rivlin (61, 63, 64) have developed an alternate formulation of nonlinear elasticity in terms of the strain energy function W and the invariants of $\mathbf{c}^{-1}$. For incompressible materials, they obtain the same form as Eq. (7-E-17).

$$\boldsymbol{\sigma} = -p\mathbf{I} + 2\frac{\partial W}{\partial I_1}\mathbf{c}^{-1} - 2\frac{\partial W}{\partial I_2}\mathbf{c} \tag{7-E-18}$$

F. VISCOELASTIC FLUIDS

1. General Remarks

In this section we will seek to extend the observations of the previous section to fluids with memory. There are alternate approaches to this problem. Following Maxwell, Natanson, and Zaremba, we may express the constitutive equation for a viscoelastic fluid as a differential equation for the stress. Alternatively, following Boltzmann, the stress may be expressed in terms of integrals over the deformation history.

In either case, we must begin with expressions valid for small strains, place them in a convected frame, and transform them back to a fixed space frame. As indicated in the previous section, this cannot be done uniquely for elastic solids. The covariant and contravariant formulations lead to different deformation tensors. A similar problem arises for viscoelastic media. This problem was realized by Oldroyd (51) in his original paper. His eventual resolution was that nonlinear generalizations of linear theories are inherently nonunique and that one must always add arbitrary quadratic terms to account for this (40).

An alternate approach is suggested by the work of Lodge (40). The theory of a network of flexible polymer chains, that is, the kinetic theory of rubber elasticity (73), has a constitutive equation equivalent to

$$\sigma_{ij} = -p\delta_{ij} + vkTc_{ij}^{-1} \tag{7-F-1}$$

where v is the cross-link density, k is Boltzmann's constant, and T is the absolute temperature. The realization of the form of Eq. (7-F-1) is inherent in the work of Treloar (73) and Rivlin (61), who termed it a neo-Hookean solid. Equation (7-F-1), together with Eqs. (7-C-3) and (7-C-4), suggests that the starting point for nonlinear viscoelastic models should be a generalization of the Maxwell model, as suggested by White and Metzner (87) [essentially by Oldroyd (51)] to be

$$\pi^{\alpha\beta} = -p\gamma^{\alpha\beta} + \pi'^{\alpha\beta}, \qquad \frac{d\pi'_{\alpha\beta}}{dt} = -G\frac{d\gamma^{\alpha\beta}}{dt} - \frac{1}{\tau}\pi'^{\alpha\beta} \qquad (7\text{-F-2a,b})$$

and the Boltzmann superposition model, as suggested by Lodge (40) to be

$$\pi^{\alpha\beta} = -p\gamma^{\alpha\beta} + \int_{-\infty}^{t} \phi(t-s)\gamma^{\alpha\beta}(s)ds \qquad (7\text{-F-3})$$

Equations (7-F-2) and (7-F-3) form the starting point for most useful simple nonlinear viscoelastic tensor differential and integral models.

A more general approach has been taken by Noll and Coleman (15, 16, 74), Green and Rivlin (28), and others which involves considering the stress tensor to be an arbitrary *hereditary* functional of the deformation history. In an hereditary functional, events in the recent past are more important than those in the distant past. They express this in terms of the Cauchy deformation tensor, which in our notation is equivalent to a covariant formulation in the convected coordinate system, that is,

$$\pi_{\alpha\beta} = -p\gamma_{\alpha\beta} + F_{\alpha\beta}[\gamma_{\alpha\beta}(z)]_{0}^{\infty} \qquad (7\text{-F-4})$$

The alternate formulation of the author (81, 82) uses a Finger deformation measure that is equivalent to

$$\pi^{\alpha\beta} = -p\gamma^{\alpha\beta} + G^{\alpha\beta}[\gamma^{\alpha\beta}(z)]_{0}^{\infty} \qquad (7\text{-F-5})$$

This is expected to lead to a less cumbersome formulation because of the relationship of this form to Eqs. (7-F-1) and (7-F-3). In the succeeding sections, we will consider both the differential and integral approaches in turn.

2. Differential Models

We may readily convert the differential Maxwell model of Eq. (7-F-2) to a fixed-space coordinate system. Representing the extra stress with the symbol P_{ij}, Eq. (7-B-11) allows us to write

$$\frac{d}{dt}\left[\frac{\partial\xi^{\alpha}}{\partial x^{i}}\frac{\partial\xi^{\beta}}{\partial x^{j}}P_{ij}\right] = -2G\frac{d}{dt}\left[\frac{\partial\xi^{\alpha}}{\partial x^{m}}\frac{\partial\xi^{\beta}}{\partial x^{m}}\right] - \frac{1}{\tau}\frac{\partial\xi^{\alpha}}{\partial x^{i}}\frac{\partial\xi^{\beta}}{\partial x^{j}}P_{ij} \qquad (7\text{-F-6})$$

Differentiating through Eqs. (7-F-6), as in Eq. (7-D-6) and (7-D-11a), and specifically using Eq. (7-D-11b), allows us to reduce Eq. (7-F-6) to

$$\frac{\partial \xi^\alpha}{\partial x^i}\frac{\partial \xi^\beta}{\partial x^j}\left[\frac{DP_{ij}}{Dt} - \frac{\partial v_i}{\partial x_m}P_{mj} - \frac{\partial v_j}{\partial x_m}P_{im}\right]$$
$$= \frac{\partial \xi^\alpha}{\partial x^i}\frac{\partial \xi^\beta}{\partial x^j}\left[G\left(\frac{\partial v_j}{\partial x_j} + \frac{\partial v_j}{\partial x_i}\right) - \frac{1}{\tau}P_{ij}\right] \qquad (7\text{-}F\text{-}7)$$

The cartesian fixed-space coordinates may be written

$$\frac{\delta P_{ij}}{\delta t} = 2Gd_{ij} - \frac{1}{\tau}P_{ij}$$

or   (7-F-8a)

$$P_{ij} = 2\tau Gd_{ij} - \tau\frac{\delta P_{ij}}{\delta t}$$

with

$$\frac{\delta P_{ij}}{\delta t} = \frac{DP_{ij}}{Dt} - \frac{\partial v_i}{\partial x_m}P_{mj} - \frac{\partial v_j}{\partial x_m}P_{im} \qquad (7\text{-}F\text{-}8b)$$

This is the form given by White and Metzner (87). Here, $\delta P_{ij}/\delta t$ represents a convected derivative of the stress P_{ij}. It is equal to the term in brackets on the left-hand side of Eq. (7-F-7).

If we apply Eq. (7-F-8) to shear flow,

$$v_1 = \dot\gamma x_2, \qquad v_2 = v_3 = 0 \qquad (7\text{-}F\text{-}9)$$

This leads to

$$P_{12} = \tau G\dot\gamma - 0, \qquad P_{11} = 0 - (-2\tau P_{12}\dot\gamma), \qquad P_{22} = 0 - 0 \quad (7\text{-}F\text{-}10a\text{--}c)$$

or

$$\sigma_{12} = P_{12} = \tau G\dot\gamma, \qquad \eta = \tau G \qquad (7\text{-}F\text{-}11a)$$

$$N_2 = 0, \qquad N_1 = 2\tau\eta\dot\gamma^2 = 2\sigma_{12}^2/G \qquad (7\text{-}F\text{-}11b,c)$$

This predicts a constant shear viscosity and a principal normal stress difference N_1 that increases with the square of the shear rate of shear stress.

If the covariant formulation using $\pi_{\alpha\beta}$ and $\gamma_{\alpha\beta}$ is used in place of Eq. (7-F-1) we obtain

$$\frac{\delta_{\text{cov}}P_{ij}}{\delta t} = 2Gd_{ij} - \frac{1}{\tau}P_{ij} \qquad (7\text{-}F\text{-}12a)$$

where

$$\frac{\delta_{\text{cov}}P_{ij}}{\delta t} = \frac{DP_{ij}}{Dt} + \frac{\partial v_m}{\partial x_i}P_{mj} + \frac{\partial v_m}{\partial x_j}P_{im} \qquad (7\text{-}F\text{-}12b)$$

In shear flow this leads to Eq. (7-F-11a) for σ_{12} and η but

$$N_2 = -N_1, \qquad N_1 = 2\tau\eta\dot{\gamma}^2 = \frac{2\sigma_{12}^2}{G} \qquad (7\text{-F-}13)$$

This differs considerably from experiment and the formulation of Eq. (7-F-8) is thus preferred.

In order to represent experimental data on polymer melts, it is necessary that the viscosity vary with shear rate. This may be done by making τ a function of $\dot{\gamma}$ for shear flows. We must be careful to note that any relationship between τ and rates of deformation must be framed in a manner that is independent of coordinate system. This can be accomplished by using the invariants of the rate of deformation tensor **d**. These may be taken as (compare Section 2-J).

$$I_d = \text{tr } \mathbf{d}, \qquad II_d = 2\,\text{tr } \mathbf{d}^2, \qquad III_d = 4\,\text{tr } \mathbf{d}^3 \qquad (7\text{-F-}14a\text{–}c)$$

It may be shown that in shear flow, the three invariants have the values

$$I_d = 0, \qquad II_d = \dot{\gamma}^2, \qquad III_d = 0 \qquad (7\text{-F-}15)$$

A shear-rate dependece of τ may be defined through the functionality

$$\tau = \tau(II_d) \qquad (7\text{-F-}16)$$

Different forms of $\tau(II_d)$ have been proposed in the literature. These include a power-law form

$$\tau = \frac{K}{G} II_d^{(n-1)/2} \rightarrow \left[\frac{K}{G}\dot{\gamma}^{n-1} \text{ (in shear flow)} \right] \qquad (7\text{-F-}17a)$$

This form has the difficulty that it does not go to a constant relaxation time τ_0 at low deformation rates. A form that exhibits this behavior, and has been used more successful is

$$\tau = \frac{\tau}{1 + a\tau_0 II_d^{1/2}} \rightarrow \left[\frac{\tau_0}{1 + a\tau_0\dot{\gamma}} \text{ in shear flow} \right] \qquad (7\text{-F-}17b)$$

The predicted shear viscosity behavior of Eq. (7-F-8) with Eq. (7-F-17b) for its relaxation time is shown in Figure 7-F-1.

We now consider uniaxial extension

$$v_1 = \dot{\gamma}_E x_1, \qquad v_2 = -\left(\frac{\dot{\gamma}_E}{2}\right)x_2, \qquad v_3 = -\left(\frac{\dot{\gamma}_E}{2}\right)x_3 \qquad (7\text{-F-}18)$$

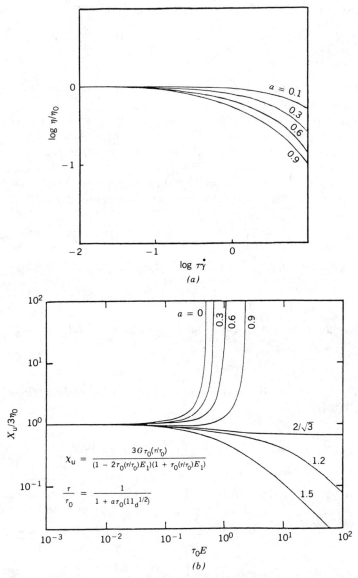

FIGURE 7-F-1. Predictions of convected Maxwell model with relaxation time given by Eq. (7-F-16b): (a) viscosity–shear rate; (b) elongational viscosity–stretch rate.

For long-duration flow, the tensile stress and elongational viscosity are

$$\sigma_{11} = \frac{3G\tau\dot{\gamma}_E}{(1 + \tau\dot{\gamma}_E)(1 - 2\tau\dot{\gamma}_E)} \tag{7-E-19a}$$

$$\chi = \frac{3G\tau}{(1 + \tau\dot{\gamma}_E)(1 - 2\tau\dot{\gamma}_E)} \tag{7-F-19b}$$

Equation (7-F-19b) predicts χ is equal to $3G\tau$ at low stretch rates and increases with $\dot{\gamma}_E$ at higher stretch rates. Note that χ goes to infinity when $\dot{\gamma}_E$ increases to a value of $1/2\tau$. This does not account for the dependence upon deformation rate. If we use Eq. (7-F-17b) for the relaxation time, we have

$$\tau = \frac{\tau_0}{1 + \sqrt{3}\, a\tau_0 \dot{\gamma}_e} \qquad \text{(7-F-20a)}$$

This leads to

$$\chi = \frac{3G\tau_0}{[1 - (2 - \sqrt{3}a)\tau_0\dot{\gamma}_E][1 + (1 + \sqrt{3}a)\tau_0\dot{\gamma}_E]} \qquad \text{(7-F-20b)}$$

The effect of a nonzero a is to make χ increase less rapidly with $\dot{\gamma}_E$. Indeed, when a reaches a value of $2/\sqrt{3}$, χ no longer increases in an unbounded manner. This is shown in Figure 7-F-1b.

Other differential formulations have appeared in the literature. Indeed, Oldroyd's formulation (51, 52) is based on Eq. (7-B-13)

$$\mathbf{P} + \tau_1 \frac{\delta \mathbf{P}}{\delta t} = 2\eta \left[\mathbf{d} + \tau_2 \frac{\delta \mathbf{d}}{\delta t} \right] \qquad \text{(7-F-21)}$$

where $\delta/\delta t$ is defined as in Eq. (7-F-8b). Oldroyd found that this form gave too simple viscous and normal stress response as a function of shear rate, namely, a constant shear viscosity and normal stress coefficient. He noted that when a linear theory is generalized as described above, and includes terms of second order in $\mathbf{d}$, as does $\delta\mathbf{d}/\delta t$, one could in an equivalent approximation add in similar arbitrary second-order terms, perhaps arising from other mechanisms. Thus we would have in place of Eq. (7-F-21) (52)

$$\mathbf{P} + \tau_1 \frac{\delta \mathbf{P}}{\delta t} + \mu_0 (\text{tr } \mathbf{P} \cdot \mathbf{d})\mathbf{I} + \mu_1 (\text{tr } \mathbf{P})\mathbf{d} + \mu_2 (\mathbf{P} \cdot \mathbf{d} + \mathbf{d} \cdot \mathbf{P})$$

$$= 2\eta \left[\mathbf{d} + \tau_2 \frac{\delta \mathbf{d}}{\delta t} \right] + \mu_3 (\text{tr } \mathbf{d}^2)\mathbf{I} + \mu_4 \mathbf{d}^2 \qquad \text{(7-F-22)}$$

These additional parameters, $\mu_0 - \mu_4$, serve a similar role to White and Metzner's $\tau(\text{II}_d)$.

An alternate interpretation of Oldroyd's Eq. (7-F-22) is to argue that the problem is in his choice of convected derivatives. Thus, as suggested by Giesekus (25), we might write (compare Keunings and Crochet (36))

$$\mathbf{P} + \tau_1 \frac{\Delta}{\Delta t}\mathbf{P} = 2\eta \left[\mathbf{d} + \tau_2 \frac{\Delta}{\Delta t}\mathbf{d} \right] \qquad \text{(7-F-23)}$$

where

$$\frac{\Delta \mathbf{P}}{\Delta t} = \frac{D\mathbf{P}}{Dt} - \nabla \mathbf{v} \cdot \mathbf{P} - \mathbf{P} \cdot \nabla \mathbf{v} + \lambda_1 (\text{tr } \mathbf{P} \cdot \mathbf{d}) \mathbf{I}$$

$$+ \lambda_2 (\text{tr } \mathbf{P}) \mathbf{d} + \lambda_3 (\mathbf{P} \cdot \mathbf{d} + \mathbf{d} \cdot \mathbf{P}) \qquad (7\text{-}F\text{-}24)$$

If we neglect the terms in λ_1 and λ_2 that contain traces, this derivative may be expressed as the sum of the derivatives occurring in the contravariant and covariant formulations of Eq. (7-F-21), that is, as

$$\frac{\Delta P}{\Delta t} = \kappa \frac{\delta \mathbf{P}}{\delta t} + (1 - \kappa) \frac{\delta_{\text{cov}} \mathbf{P}}{\delta t} \qquad (7\text{-}F\text{-}25)$$

where $\delta_{\text{cov}} P_{ij}/\delta t$ is given by Eq. (7-F-12b).

Various theories have been devised in recent years which may be expressed in this form. Many of these follow arguments along the line of rewriting Eq. (7-F-8) as

$$\tau \left[\frac{D\mathbf{P}}{Dt} - \mathbf{L} \cdot \mathbf{P} - \mathbf{P} \cdot \mathbf{L} \right] + \mathbf{P} = 2\eta \mathbf{d} \qquad (7\text{-}F\text{-}26)$$

Where $\mathbf{L}$ is an effective velocity gradient associated with nonaffine effects. The theories of Johnson and Segalman (35) and Phan-Tien and Tanner (54, 55) may be represented in this manner. In one form of the Johnson–Segalman theory, Eq. (7-F-26) reduces to

$$\tau \frac{\Delta P}{\Delta t} + \mathbf{P} = 2\eta \mathbf{d} \qquad (7\text{-}F\text{-}27)$$

where $\Delta P/\Delta t$ is given by Eq. (7-F-25) and τ and η are constants. In the Phan-Tien–Tanner theory (54), the authors proposed:

$$\tau \frac{\Delta P}{\Delta t} + (1 + \alpha \text{ tr } \mathbf{P})\mathbf{P} = 2\eta \mathbf{d} \qquad (7\text{-}F\text{-}28a)$$

In the more recent modified theory of Phan-Tien (55), the constant viscosity of Eq. (7-F-28a) is replaced with

$$\tau \frac{\Delta P}{\Delta t} + (1 + \alpha \text{ tr } \mathbf{P})\mathbf{P} = 2\eta_0 \left[\frac{1 + \xi(2 - \xi)\tau^2 \text{II}_{\text{d}}}{(1 + \lambda^2 \text{II}_{\text{d}})^m} \right] \mathbf{d} \qquad (7\text{-}F\text{-}28b)$$

Another formulation is the model of Leonov (37) [see also Eq. (7-F-69). Luo and Tanner (41) express this as

$$\tau \frac{\Delta \mathbf{P}'}{\Delta t} + \frac{1}{2} \left[\frac{1}{2\mu} \mathbf{P}' \cdot \mathbf{P}' - 2\mu \mathbf{P}' + \frac{1}{3} \left\{ 2\mu \text{ tr } \mathbf{P}'^{-1} - \frac{1}{2\mu} \text{ tr } \mathbf{P}' \right\} \mathbf{P}' \right] = 0 \quad (7\text{-}F\text{-}29a)$$

where

$$\sigma = -p\mathbf{I} + 2\eta\mathbf{d} + \mathbf{P}' \qquad (7\text{-}F\text{-}29b)$$

These models, Eqs. (7-F-28) and (7-F-29), are capable of predicting a non-Newtonian shear viscosity, a principal normal stress difference, a negative second normal stress difference (associated with κ), and an elongational viscosity function exhibiting a maximum.

3. Integral Models—Large Deformation

Lodge's integral model, Eq. (7-F-3), has served as a basis for a large literature on nonlinear viscoelastic integral models. This expression may be converted to a fixed-space coordinate system in the same manner as the differential models. We may write

$$\frac{\partial \xi^\alpha}{\partial x^i}\frac{\partial \xi^\beta}{\partial x^j}\sigma_{ij} = -p\frac{\partial \xi^\alpha}{\partial x^i}\frac{\partial \xi^\beta}{\partial x^j}\delta_{ij} + \int_{-\infty}^{t}\phi(t-s)\frac{\partial \xi^\alpha}{\partial \bar{x}^m}\frac{\partial \xi^\beta}{\partial \bar{x}^m}ds \qquad (7\text{-}F\text{-}30)$$

which is equivalent to

$$\frac{\partial \xi^\alpha}{\partial x^i}\frac{\partial \xi^\beta}{\partial x^j}\sigma_{ij} = \frac{\partial \xi^\alpha}{\partial x^i}\frac{\partial \xi^\beta}{\partial x^j}\left[-p\delta_{ij} + \int_{-\infty}^{t}\phi(t-s)c_{ij}^{-1}(s)\,ds\right] \qquad (7\text{-}F\text{-}31a)$$

where

$$c_{ij}^{-1} = \frac{\partial x^i(t)}{\partial \bar{x}^m(s)}\frac{\partial x^j(t)}{\partial \bar{x}^m(s)} \qquad (7\text{-}F\text{-}31b)$$

Equation (7-F-31) us equivalent to

$$\sigma = -p\mathbf{I} + \int_{-\infty}^{t}\phi(t-s)\mathbf{c}^{-1}\,ds \qquad (7\text{-}F\text{-}32)$$

If we consider simple shear flow, as defined by Eq. (7-F-9), the stress field will be

$$\sigma_{12} = \left[\int_0^\infty z\phi(z)\,dz\right]\dot{\gamma}$$

$$\sigma_{11} = -p + \left[\int_0^\infty z^2\phi(z)\,dz\right]\dot{\gamma}^2 \qquad (7\text{-}F\text{-}33)$$

$$\sigma_{22} = \sigma_{33} = -p$$

The two normal stress differences are

$$N_1 = \int_0^\infty z^2 \phi(z)\, dz, \qquad N_2 = 0 \tag{7-F-34}$$

If we represent $\phi(t)$ by Eq. (7-C-10) with Eq. (7-C-6), that is,

$$\phi(t) = \sum_i \frac{G_i}{\tau_i} e^{-t/\tau_i} \tag{7-F-35}$$

The shear stress and normal stress differences are

$$\sigma_{12} = \left[\sum_i G_i \tau_i \right] \dot{\gamma} = \eta \dot{\gamma}$$

$$N_1 = 2\left[\sum_i G_i \tau_i^2 \right] \dot{\gamma}^2 = 2\bar{\tau}\eta\dot{\gamma}^2 \tag{7-F-36}$$

$$N_2 = 0$$

where

$$\bar{\tau} = \sum_i G_i \tau_i^2 \bigg/ \sum_i G_i \tau_i \tag{7-F-37}$$

These expression indicate a constant viscosity and a principal normal stress difference which varies with the square of the shear rate. The second normal stress difference is zero. The results are very similar to Eq. (7-F-10) for the convected Maxwell model. It indeed becomes identical when $\phi(t)$ has only a single relaxation time.

Various procedures have been proposed to generalize Eq. (7-F-32) so that it can predict a broader range of response. Spriggs et al. (67) have suggested that it could be expanded to include a nonzero second normal stress difference by writing

$$\boldsymbol{\sigma} = -p\mathbf{I} + \int_{\infty}^t \phi(t-s)\left[\left(1 + \frac{\varepsilon}{2}\right)\mathbf{c}^{-1} - \frac{\varepsilon}{2}\mathbf{c} \right] ds \tag{7-F-38}$$

Equation (7-F-35) still satisfies the principle of material indifference and in shear flows yields

$$\sigma_{12} = \left[\int_0^\infty z\phi(z)\, ds \right] \dot{\gamma}$$

$$N_1 = \left[\int_0^\infty z^2 \phi(z)\, dz \right] \dot{\gamma}$$

$$N_2 = -\frac{\varepsilon}{2}\left[\int_0^\infty z^2 \phi(z)\, dz \right] \dot{\gamma}^2 \tag{7-F-39}$$

We may introduce a nonlinear viscosity into the integral model in various manner. It may be accomplished through introducing functions of II_d into $\phi(t)$ to produce a function $m(t, II_d)$. This has been applied by Spriggs et al. (67), Bogue (8), Bird and Carreau (7), Bogue and White (9), Carreau (11), and others:

$$\boldsymbol{\sigma} = -p\mathbf{I} + \int_{-\infty}^{t} m(t - s, II_d)\left[\left(1 + \frac{\varepsilon}{2}\right)\mathbf{c}^{-1} - \frac{\varepsilon}{2}\mathbf{c}\right]ds \qquad (7\text{-}F\text{-}40)$$

The forms of $m(t, II_d)$ which were proposed in place of $\phi(t)$ are
Spriggs et al. (67), Bird and Carreau (7)

$$m(t, II_d) = \sum_i \frac{G}{\tau_i} \frac{1}{1 + 2c_i^2 \tau_i^2 II_d} e^{-t/\tau_i} \qquad (7\text{-}F\text{-}41)$$

Bogue (8), Bogue and White (9), Chen et al. (13, 14)

$$m(t, II_d) = \sum_i \frac{G_i}{\tau_{i\,\text{eff}}} e^{-t/\tau_{i\,\text{eff}}} \qquad (7\text{-}F\text{-}42a)$$

where

$$\tau_{i\,\text{eff}} = \frac{\tau_{i0}}{1 + a\tau_{i0}\overline{II_d^{1/2}}} \qquad (7\text{-}F\text{-}42b)$$

with

$$\overline{II_d^{1/2}} = \tfrac{1}{2}\int_0^t II_d^{1/2}(s)\,ds \qquad (7\text{-}F\text{-}42c)$$

Carreau (11)

$$m(t, II_d) = \sum \frac{G_i}{\tau_i f_i(II_d)} e^{-\int_{t'}^{t} \frac{dt''}{\tau_i g_i(II_d)}} \qquad (7\text{-}F\text{-}43)$$

A second approach to introducing nonlinearities into $\phi(t)$ is through the use of strain invariants. Such an approach was first used by Bernstein et al. (3) and later expanded upon by Zapas (93, 94), White and Tokita (89), and Wagner (78, 79). Both White and Tokita and Wagner argue that $m(t)$ should be separable into a function of time and strain invariants, that is,

$$m(t, I_{c-1}, II_{c-1}) = \phi(t)h(I_{c-1}, II_{c-1}) \qquad (7\text{-}F\text{-}44a)$$

where I_{c-1} and II_{c-1} are invariants of the Finger strain measure. Wagner chooses the specific form

$$h = e^{-N[\frac{1}{2}(I_{c-1}^2 - II_{c-1})]} \qquad (7\text{-}F\text{-}44b)$$

with ε taken to be zero. Bernstein et al. (3, 93, 94) use a formalism suggested by the theory of nonlinear elasticity (61, 62, 73). They write

$$\boldsymbol{\sigma} = -p\mathbf{I} + \int_{-\infty}^{t} \left[\frac{\partial U}{\partial I_1} \mathbf{c}^{-1} - \frac{\partial U}{\partial I_2} \mathbf{c} \right] ds \tag{7-F-45a}$$

where I_1 and I_2 are invariants defined by

$$I_1 = I_{c-1}, \qquad I_2 = \tfrac{1}{2}[I_{c-1}^2 - II_{c-1}] \tag{7-F-45b, c}$$

and U is analogous to the strain energy function of nonlinear elasticity. Various forms of U have been proposed. Zapas (93) takes

$$U = \frac{\alpha(t)}{2}(I_1 - 3)^2 + 4.5\beta(t) \ln \left[\frac{I_1 + I_2 + 3}{9} \right]$$

$$+ 24[\beta(t) - c(t)] \ln \left[\frac{I_1 + 15}{I_2 + 15} \right] + c(t) \tag{7-F-46}$$

Here $\alpha(t)$, $\beta(t)$, and $c(t)$ are functions of time.

Other formalisms are possible. Takahashi et al. (72) use the general formalism used here, but take $m(t)$ to be dependent upon the invariants of the extra stress. They take essentially

$$m(t) = m(t, I_p, II_p, III_p) \tag{7-F-47}$$

Takahashi et al. choose a dependence upon I_p. Goddard and Miller (27) and Bird et al. (6) describe formulations of integral models based on rigidly rotating frames, in the sense of Zaremba and Hencky, rather than on convected coordinate frames.

4. Comparison of Empirical Models and Experiments

Differential viscoelastic models tend to qualitatively rather than quantitaively represent the behavior of polymer melts. A distribution of relaxation times is required. The various integral theories correlate small-strain viscoelastic data on polymer solutions and melts, and many of the integral theories also do a satisfactory job with representation of the shear viscosity and normal stress behavior. Critical comparisons of theories have generally used transient shear-flow experiments and transient and steady-state elongational flow experiments. Many studies of this general type have appeared for melts (13, 14, 78, 93, 94). Figure 7-F-2 shows predictions of an integral model with shear-flow experiments on a polymer melt. There seems to be a finite number of empirical integral models that can effectively correlate a wide range of linear and nonlinear experiments.

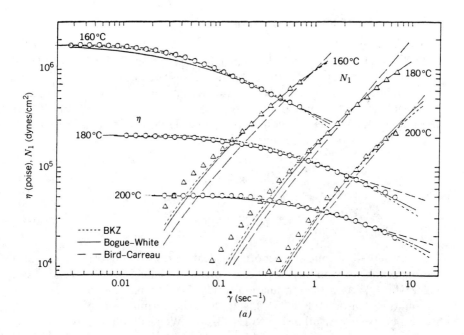

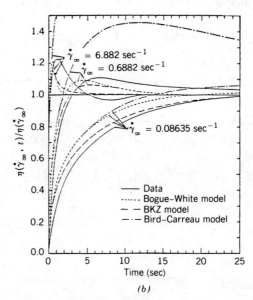

FIGURE 7-F-2. Comparison of theory and experiment for integral viscoelastic fluid models with data on polymer melts. (*a*) Comparison of the primary models with the steady shear data for polystyrene. (*b*) Comparison of the primary models with shear stress growth data for low-density polyethylene at 160°C.

5. General Constitutive Formulations

A much more general formulation of the theory of viscoelastic fluids subjected to large deformations has been developed by Green and Rivlin (28), Noll and Coleman (15, 16, 74), and others, as noted in Section 7-B and 7-F-1. We seek to develop this approach here. If we presume the stress to be a hereditary functional of the deformation history, we obtain Eqs. (7-F-4) or (7-F-5), depending upon the deformation measure used. Transforming these expressions to a fixed-space coordinate frame, we obtain

$$\sigma = -p\mathbf{I} + \mathbf{F}[\mathbf{c}(z)]_0^\infty \tag{7-F-48}$$

$$\sigma = -p\mathbf{I} + \mathbf{G}[\mathbf{c}^{-1}(z)]_0^\infty \tag{7-F-49}$$

There are actually a finite number of equivalent representations of the influence of deformation history on the stress field. Noll and Coleman (15, 16, 74) and Green and Rivlin (28) used the formulation of Eq. (7-F-38), based upon the Cauchy deformation tensor. White (81, 82) applied the formulation of Eq. (7-F-39) because of its close relationship to Eq. (7-F-1). We shall follow this here.

The method of representation of G in a more specific form is suggested by the theory of functionals as described by Volterra (77). A scalar functional G which obeys certain minimal smoothness requirements may be expressed:

$$G[y(t)] = k_0 + \int_a^b h_1(\xi)y(\xi)d\xi + \int_a^b \int_a^b k_2(\xi_1,\xi_2)y(\xi_1)y(\xi_2)d\xi_1 d\xi_2 + \cdots$$

$$+ \int_a^b \cdots \int_a^b k_n(\xi_1,\xi_2,\ldots,\xi_n)y(\xi_1)y(\xi_2),\ldots,y(\xi_n)d\xi_1 d\xi_2 \cdots d\xi_n \tag{7-F-50a}$$

For a scalar *hereditary* functional, that is, one depending upon history, so that happenings in the distant past are less important than those in the recent past,

$$G[y(t)] = h_0 + \int_a^b k_1(b-\xi)y(\xi)d\xi + \int_a^b \int_a^b h_2(b-\xi_1,b-\xi_2)y(\xi_1)y(\xi_2)d\xi_1 d\xi_2$$

$$+ \int_a^b \cdots \int_a^b k_n(b-\xi_1,b-\xi_2,\ldots,b-\xi_n)y(\xi_1)y(\xi_2)\cdots y(\xi_n)$$

$$\times d\xi_1 d\xi_2 \cdots d\xi_n + \cdots \tag{7-F-50b}$$

For an hereditary tensor functional of $\mathbf{c}^{-1}$, we may write

$$\mathbf{G}[\mathbf{c}^{-1}(z)] = \int_{-\infty}^t \phi(t-s)\mathbf{c}^{-1}(d)\,ds + \int_{-\infty}^t \int_{-\infty}^t [\psi_1(t-s_1,t-s_2)(\mathbf{c}^{-1})^2$$

$$+ \psi_2(t-s_1 \cdot t-s_2)(\operatorname{tr}\mathbf{c}^{-1})\mathbf{c}^{-1}]\,ds_1\,ds_2 + \cdots \tag{7-F-51}$$

where we have noted that products of c^{-1} of order two and higher are not unique and we must include all possibilities.

White and Tokita (89) have noted that if all the higher-order kernels in the integrals, such as $\psi_1(z_1, z_2)$ and $\psi_2(z_1, z_2)$, contain only a single memory function, Eq. (7-F-41) reduces to an expression of the form

$$\mathbf{G} = \int_{-\infty}^{t} M(t-s)[\mathbf{c}^{-1} + \alpha(\mathbf{c}^{-1})^2]\, dz$$

$$= \int_{-\infty}^{t} m(t-s)\left[\left(1 + \frac{\varepsilon}{2}\right)\mathbf{c}^{-1} - \frac{\varepsilon}{2}\mathbf{c}\right] dz \qquad (7\text{-}F\text{-}52)$$

where the memory function $m(z)$ depends upon I_{c-1} and II_{c-1}. The Cayley–Hamilton theorem is used in the reduction. The BKZ and Wagner constitutive equations are of this form

6. Acceleration Tensor Expansions

Equations (7-C-11) and (7-C-15) indicate how Boltzmann's superpositional integral may be simplified for smooth flows. A similar development is possible for three-dimensional nonlinear viscoelastic fluids. We may write the conjugate metric tensor $\gamma^{\alpha\beta}(z)$ as

$$\gamma^{\alpha\beta}(z) = \sum_{n=1}^{\infty} \frac{z^n}{n!}\left(\frac{d^n \gamma^{\alpha\beta}}{dz^n}\right)_{z=0} \qquad (7\text{-}F\text{-}53a)$$

This may be transformed to a fixed-space coordinate system

$$\mathbf{c}^{-1} = \mathbf{I} + \sum_{n=1}^{\infty} \frac{z^n}{n!}(-1)^{n+1}\mathbf{B}_n \qquad (7\text{-}F\text{-}53b)$$

where the $\mathbf{B}_n$ are acceleration tensors of the form

$$\mathbf{B}_1 = 2\mathbf{d} = \nabla\mathbf{v} + (\nabla\mathbf{v})^{\mathrm{T}}, \qquad \mathbf{B}_{n+1} = \frac{D\mathbf{B}_n}{Dt} - \nabla\mathbf{v}\cdot\mathbf{B}_n - \mathbf{B}_n\cdot\nabla\mathbf{v} \quad (7\text{-}F\text{-}54a, b)$$

If we introduce the acceleration tensor for $\mathbf{c}^{-1}$ in terms of $\mathbf{B}_n$ into Eq. (7-F-47), which is valid for smooth "differentiably" flow, Eqs. (7-F-45) and (7-F-47) may be expanded into

$$\sigma = -p\mathbf{I} + a_1\mathbf{B}_1 + a_2\mathbf{B}_2 + a_3\mathbf{B}_1^2 + a_4\mathbf{B}_3 + a_5(\mathrm{tr}\,\mathbf{B}_1^2)\mathbf{B}_1$$
$$+ a_6(\mathbf{B}_1\mathbf{B}_2 + \mathbf{B}_2\mathbf{B}_1) + \cdots \qquad (7\text{-}F\text{-}55)$$

where the a_j are constant coefficients. Expansions of this type were first specifically discussed by Rivlin and Ericksen (64). Coleman and Noll (15) [see also Green and Rivlin (28)] developed expressions similar to Eq. (7-F-55) in the manner described above for $\mathbf{c}$ in terms of acceleration tensors based on $\mathbf{c}$ rather than $\mathbf{c}^{-1}$.

If we limit the expansion of Eq. (7-F-51) to terms of order 1 in velocity, we have

$$\boldsymbol{\sigma} = -p\mathbf{I} + a_1\mathbf{B}_1 \tag{7-F-56}$$

This is known as a first-order fluid. It is equivalent to the Newtonian fluid. From Eq. (7-F-47) [compared Eq. (7-C-14)],

$$a_1 = \int_0^\infty z\phi(z)\,dz \tag{7-F-57}$$

and from Eq. (7-C-10), integration by parts yields

$$a_1 = \int_0^\infty G(z)\,dz \tag{7-F-58}$$

If we include terms of order 2 in the velocity gradient,

$$\boldsymbol{\sigma} = -p\mathbf{I} + a_1\mathbf{B}_1 + a_2\mathbf{B}_2 + a_3\mathbf{B}_1^2 \tag{7-F-59}$$

This is the second-order fluid of Coleman and Noll (15). The two coefficients a_2 and a_3 may also be written in terms of integrals of the memory functional expansion, Eq. (7-F-51), as

$$a_2 = -\tfrac{1}{2}\int_0^\infty z\phi(z)\,dz = -\int_0^\infty zG(z)\,dz \tag{7-F-60}$$

$$a_3 = \int_0^\infty \int_0^\infty z_1 z_2 \psi(z_1,z_2)\,dz_1\,dz_2 \tag{7-F-61}$$

This clearly indicates that the coefficient a_2 is negative. We may write

$$a_2 = -\tau a_1 \tag{7-F-62a}$$

where $\bar{\tau}$ is a mean relaxation time given by

$$\bar{\tau} = \int_0^\infty sG(s)\,ds \bigg/ \int_0^\infty G(s)\,ds \tag{7-F-62b}$$

Equivalently, from the theory of linear viscoelasticity (15),

$$a_2 = -J_e a_1^2 \tag{7-F-63a}$$

where

$$J_e = \int_0^\infty sG(s)\,ds \Big/ \left[\int_0^\infty G(s)\,ds\right]^2 \qquad \text{(7-F-63b)}$$

J_e is the steady-state compliance (recoil per unit stress).

The second-order fluid is the simplest of viscoelastic fluid models and it is tempting to apply it widely. This should, however, be done only with the greatest caution. The presumption of Eq. (7-F-53) is very severe. Viscoelastic materials respond to sudden impulses through propagation of shockwaves; such responses are not contained in Eq. (7-F-53). It is not possible to apply Eq. (7-F-59) or Eq. (7-F-55) to such problems.

We may similarly develop a third-order fluid. This has the form

$$\boldsymbol{\sigma} = -p\mathbf{I} + a_1\mathbf{B}_1 + a_2\mathbf{B}_2 + a_3\mathbf{B}_1^2 + a_4\mathbf{B}_3$$
$$+ a_5(\text{tr }\mathbf{B}_1^2)\mathbf{B}_1 + a_6(\mathbf{B}_1\mathbf{B}_2 + \mathbf{B}_2\mathbf{B}_1) \qquad \text{(7-F-64)}$$

The coefficients a_4, a_5, and a_6 may be similarly written in terms of the relaxation kernel functions with a_4 depending on $\phi(t)$ or $G(t)$. Equation (7-F-64) is, however, sufficiently complicated that it is of much less interest for studying flow problems.

Consider the behavior of these fluids in shear flow. If we introduce Eq. (7-F-9), the stress components for Eq. (7-F-64) in shear flow are

$$\sigma_{12} = [a_1 + 2(a_5 - a_6)\gamma^2]\dot\gamma$$
$$N_1 = \sigma_{11} - \sigma_{22} = -2a_2\dot\gamma^2 = J_e a_1^2\dot\gamma^2 \qquad \text{(7-F-65)}$$
$$N_2 = \sigma_{22} - \sigma_{33} = a_3\dot\gamma^2$$

A non-Newtonian shear viscosity is predicted. The principal normal stress difference N_1 must be positive.

7. Criminale–Ericksen–Filbey Theory (17)

If we limit ourselves to shear flows, we may also contract the acceleration tensor expansion. This follows from

$$\mathbf{B}_1 = \begin{bmatrix} 0 & \dot\gamma & 0 \\ \dot\gamma & 0 & 0 \\ 0 & 0 & 0 \end{bmatrix} \qquad \mathbf{B}_2 = \begin{bmatrix} -2\dot\gamma^2 & 0 & 0 \\ 0 & 0 & 0 \\ 0 & 0 & 0 \end{bmatrix} \qquad \text{(7-F-66)}$$

$$\mathbf{B}_n = 0, \qquad (n > 3)$$

Any acceleration tensor expansion may be shown to contract to

$$\boldsymbol{\sigma} = -p\mathbf{I} + f[\mathbf{B}_1, \mathbf{B}_2]$$
$$= -p\mathbf{I} + b_1\mathbf{B}_1 + b_2\mathbf{B}_2 + b_3\mathbf{B}_1^2 \qquad \text{(7-F-67)}$$

where the b_j depend upon $\dot{\gamma}^2$. It may be shown that the shear viscosity and normal stress coefficients are given by

$$\eta = b_1(\dot{\gamma}^2), \qquad \Psi_1 = 2b_2(\dot{\gamma}^2), \qquad \Psi_2 = b_3(\dot{\gamma}^2) \tag{7-F-68}$$

It is also possible to apply Eq. (7-F-67) to more complex shear flows such as those where

$$\mathbf{v} = v_1(x_2, x_3)\mathbf{e}_1 + O\mathbf{e}_2 + O\mathbf{e}_3 \tag{7-F-69a}$$

$$\mathbf{v} = v_1(x_2)\mathbf{e}_1 + O\mathbf{e}_2 + v_3(x_2)\mathbf{e}_3 \tag{7-F-69b}$$

8. Recoverable Strain Formulations

Various rheologists in the later 1940s and 1950s sought to express nonlinear viscoelastic behavior using formulations equivalent to that of nonlinear elasticity. This argument was successively used by Weissenberg (*80*), Reiner (*59*), Mooney (*46*), and Philippoff (*56*). They write

$$\boldsymbol{\sigma} = -p\mathbf{I} + \mu_1' \mathbf{c}_{\text{rec}}^{-1} + \mu_2' \mathbf{c}_{\text{rec}} \tag{7-F-70a}$$

or

$$\boldsymbol{\sigma} = -p\mathbf{I} + 2\frac{\partial W}{\partial I_1}\mathbf{c}_{\text{rec}}^{-1} - 2\frac{\partial W}{\partial I_2}\mathbf{c}_{\text{rec}} \tag{7-F-70b}$$

where $\mathbf{c}_{\text{rec}}^{-1}$ and $\mathbf{c}_{\text{rec}}$ represent recoverable strain measures.

The difficulty with these formulations is that they leave the tensors $\mathbf{c}_{\text{rec}}$ and $\mathbf{c}_{\text{rec}}^{-1}$ largely undefined. More recently, a formulation by Leonov (*37*) has sought to overcome this problem. The stress tensor is expressed in terms of an elastic deformation tensor whose time-dependent behavior is governed by a separate constitutive expression. The Leonov theory was expressed earlier in Eq. (7-F-29) in the Luo–Tanner form. The Leonov equation has been supported in more recent years in papers by Isayev and his coworkers (*34, 76*).

9. Molecular Theories of Flexible Polymer Chains

Our intention in this chapter and book is to take a phenomenological view of polymer flow behavior. However, there has been enormous activity for generations to develop molecular models of flow or systems of flexible polymer chains (*5, 19, 29, 40, 55, 90, 91*). Its influence is pervasive. At the beginning of Section 7-F-1 we used arguments drawn from this area to justify preference for Finger tensor formulations. Those formulations originally grew out of the molecular theory of rubber elasticity (*73*) based on the elastic deformation of cross-linked flexible chains in a Gaussian approximation of end-to-end distances. The earliest formulations were those of Green and Tobolsky (*29*) and Yamamoto (*91*), which envisaged the cross-links of the kinetic theory of rubber elasticity to be replaced by temporary junctions or entanglements. The simple formulation of these ideas by Lodge (40), leading to Eq. (7-F-3), has received considerable attention. The more elegant formulations deriving from Yamamoto (91) and developed upon by Wiegel (90) and Phan-Tien and Tanner (55) [see Eq. (7-F-28)] have expanded perspectives. In more recent years, the work of Doi and Edwards (19), which deviated from the entanglement junction concept and concentrated on repitating chains, has attracted the greatest attention.

Almost all of these theories are attractive. There is, however, a problem in judging them. In general, rheological probes are not good sensors of molecular structure. Different molecular theories may equally well predict rheological behavior. One needs to test the predictions of models not just with viscosity and normal stress measurements, but with their interaction with electromagnetic radiation (e.g., birefringence), neutrons, and so on.

G. FLUIDS WITH YIELD VALUES

The phenomenological rheological properties of suspensions of small particles are frequently represented by models with yield values. Three-dimensional constitutive equations for non-Newtonian fluids with yield values date to Hohenemser and Prager's (32) three-dimensional formulation of the Bingham plastic, as represented in Eq. (7-B-10b). This formulation is based upon the von Mises yield criterion and it is not a unique representation of three-dimensional behavior. Generalizations of this formulation of fluids, which respond in a viscoelastic manner above the yield value, are considered in papers by Hutton (33) and by White and his coworkers (38, 83, 84, 86, 88).

In constructing a rheological theory involving a yield surface, one must actually devise three constitutive equations representing (1) behavior at stresses above the yielding, (2) three-dimensional behavior during the yield process, that is, the yield surface, and (3) behavior at stresses below the yield value. In dealing with material undergoing large deformations, it is often convenient to represent the material to be rigid below the yield value.

Many different proposals have been made for the three-dimensional form of yielding of an elastic solid. These have generally grown out of studies of metals. As the yield surface can only depend upon stress in a manner independent of coordinate system, the functional dependence must be in terms of invariants of the stress tensor. Traditionally, the deviatoric stress tensor $\mathbf{T}$, defined through Eq. (6-B-8a), is used. As $\text{tr}\,\mathbf{T}$ is zero, this leads to

$$F(\text{II}_\text{T}, \text{III}_\text{T}, p) = 0 \qquad (7\text{-G-}1)$$

where we have noted that the yield surface could depend upon the magnitude of the applied pressure.

The simplest of the yield surfaces is the criterion of von Mises, which is generally expressed as

$$II_T = tr\, \mathbf{T}^2 = 2Y^2 \qquad (7\text{-}G\text{-}2)$$

The von Mises criterion is equivalent to a critical *distortional* strain energy:

$$E = \sum_i \sum_j T_{ij}\gamma_{ij} = \tfrac{1}{2}\, tr\,[\mathbf{T}\cdot\boldsymbol{\gamma}] = \frac{1}{2G}\, tr\, \mathbf{T}^2 \qquad (7\text{-}G\text{-}3)$$

where γ is a small-strain tensor and G is a distortional elastic modulus.

The various constitutive equations between stress and deformation for media with yield values have been almost entirely based on the von Mises yield criterion. The simplest of such constitutive equations is that usually associated with von Mises (57). We rewrite Eq. (7-G-2) as

$$\mathbf{T} = \frac{Y}{\sqrt{\tfrac{1}{2} tr\, \mathbf{T}^2}}\, \mathbf{T} \qquad (7\text{-}G\text{-}4)$$

and presume that

$$\mathbf{T} = \lambda \mathbf{d} \qquad (7\text{-}G\text{-}5)$$

This leads to the von Mises equation

$$\mathbf{T} = \frac{2Y}{\sqrt{2\, tr\, \mathbf{d}^2}}\, \mathbf{d} \qquad (7\text{-}G\text{-}6)$$

The three-dimensional Bingham plastic, that is, the one devised by Hohenemser and Prager (32), adopts Eq. (7-G-2) and has the form

$$\mathbf{T}\left[1 - \frac{Y}{\sqrt{\tfrac{1}{2} tr\, \mathbf{T}^2}}\right] = 2\eta_B \mathbf{d} \qquad (7\text{-}G\text{-}7)$$

This general formulation was used by White (83) for a material with memory exhibiting a yield value, that is, a plastic viscoelastic fluid. He writes

$$\mathbf{T}\left[1 - \frac{Y}{\sqrt{\tfrac{1}{2} tr\, \mathbf{T}^2}}\right] = \mathbf{H} \qquad (7\text{-}G\text{-}8)$$

here $\mathbf{H}$ is a deviatoric memory functional ($tr\, \mathbf{H} = 0$).

We may write $\mathbf{T}$ explicitly in terms of deformation if $tr\, \mathbf{T}^2$ can be eliminated from Eq. (7-G-8). We can achieve this by multiplying Eq. (7-G-8) by itself and taking the trace

$$tr\, \mathbf{T}^2\left[1 - \frac{Y}{\sqrt{\tfrac{1}{2} tr\, \mathbf{T}^2}}\right]^2 = tr\, \mathbf{H}^2 \qquad (7\text{-}G\text{-}9)$$

This is equivalent to

$$\sqrt{\tfrac{1}{2} \operatorname{tr} \mathbf{T}^2} = Y + \sqrt{\tfrac{1}{2} \operatorname{tr} \mathbf{H}^2} \qquad \text{(7-G-10)}$$

which leads to

$$\mathbf{T} = \left[\frac{Y}{\sqrt{\tfrac{1}{2} \operatorname{tr} \mathbf{H}^2}} + 1 \right] \mathbf{H} \qquad \text{(7-G-11a)}$$

Oldroyd (49) derived this for a Bingham plastic, that is, where $\mathbf{H}$ is $2\eta_B \mathbf{d}$:

$$\mathbf{T} = \left[\frac{Y}{\sqrt{2 \operatorname{tr} \mathbf{d}^2}} + \eta_B \right] 2\mathbf{d} \qquad \text{(7-G-11b)}$$

The first efforts at deriving 3-dimensional models for plastic fluids where the differential behavior above the yield value was more complex than Newtonian is contained in the work of Oldroyd (50), who envisaged viscous non-Newtonian behavior. Oldroyd obtains the form of Eq. (7-G-11b) with η dependent on $\operatorname{tr} \mathbf{d}^2$.

White and Tanaka (83, 88) have discussed the use of single integral formulations for the memory function $\mathbf{H}$. They proposed forms such as

$$\mathbf{H} = \int_{-\infty}^{t} \left\{ \mu_1(t-s)[\mathbf{c}^{-1} - \tfrac{1}{3}(\operatorname{tr} \mathbf{c}^{-1})\mathbf{I}] - \mu_2(t-s)[\mathbf{c} - \tfrac{1}{3}(\operatorname{tr} \mathbf{c})\mathbf{I}] \right\} ds \qquad \text{(7-G-12)}$$

Note that $\mathbf{H}$ is defined so as to be deviatoric. These authors suggested that $\mu_2(t)$ could be taken as zero and $\mu_1(t)$ expressed as

$$\mu_1(t) = \frac{f(\phi)G}{\tau} e^{-t/\tau} \qquad \text{(7-G-13)}$$

where $f(\phi)$ is a factor that depends upon volume loading. This theory was compared to experiments on particle filled polymer melts.

White (84) has presented an alternative approach by expanding $\mathbf{H}$ as an expansion of acceleration tensors in the manner of Eq. (7-F-43). The zero, first-, and second-order fluids have the forms

$$\mathbf{T} = \frac{2Y}{\sqrt{\tfrac{1}{2} \operatorname{tr} \mathbf{K}_1^2}} \mathbf{K}_1 = \frac{2Y}{\sqrt{2 \operatorname{tr} \mathbf{d}^2}} \mathbf{d} \qquad \text{(7-G-14a)}$$

$$\mathbf{T} = \left[\frac{Y}{\sqrt{\tfrac{1}{2} \operatorname{tr} b_1^2 \mathbf{K}_1^2}} + 1 \right] b_1 \mathbf{K}_1 = \frac{2Y}{\sqrt{2 \operatorname{tr} \mathbf{d}^2}} \mathbf{d} + 2b_1 \mathbf{d} \qquad \text{(7-G-14b)}$$

$$\mathbf{T} = \left[\frac{Y}{\sqrt{\tfrac{1}{2} \operatorname{tr} \{b_1 \mathbf{K}_1 + b_2 \mathbf{K}_2 + b_3[\mathbf{K}_1^2 - \tfrac{1}{3}(\operatorname{tr} \mathbf{K}_1^2)\mathbf{I}]\}^2}} + 1 \right]$$
$$\times \left[b_1 \mathbf{K}_1 + b_2 \mathbf{K}_2 + b_3[\mathbf{K}_1^2 - \tfrac{1}{3}(\operatorname{tr} \mathbf{K}_1^2)\mathbf{I}] \right] \qquad \text{(7-G-14c)}$$

where the b_j are constant coefficients and $\mathbf{K}_j$ are

$$\mathbf{K}_j = \mathbf{B}_j - \tfrac{1}{3}(\operatorname{tr}\mathbf{B}_j)\mathbf{I} \tag{7-G-15}$$

Equation (7-G-14a), for zero-order fluid, is von Mises theory of perfectly plastic solids (57). Equation (7-G-14b) is the Hohenemser–Prager (32) theory for Bingham plastics in the form developed by Oldroyd (49).

There is a class of phenomena in suspensions known as thixotropy which the formulations given above do not represent. These involve time-dependent viscosities. Slibar and Pasley (66) have attempted to represent this in terms of Hohenemser and Prager's Bingham plastic model with a time-dependent yield surface. Specifically, they write

$$\mathbf{T}\left[1 - \frac{Y(t)}{\sqrt{\tfrac{1}{2}\operatorname{tr}\mathbf{T}^2}}\right] = 2\eta_{\mathrm{B}}\mathbf{d} \tag{7-G-16}$$

The time dependence of $Y(t)$ was given by

$$Y(t) = Y_1 - \frac{\int_{-\infty}^{t} \mathrm{II}_{\mathrm{d}}^{1/2} e^{-\alpha(t-\xi)}\, d\xi}{\beta + \int_{-\infty}^{t} \mathrm{II}_{\mathrm{d}}^{1/2} e^{-\alpha(t-\xi)}\, d\xi}(Y_1 - Y_0) \tag{7-G-17}$$

The yield value decreases from an initial value of Y_1 to a steady-state $Y(\mathrm{II}_{\mathrm{d}}^{1/2})$ which, as indicated, depends upon deformation rate.

Suetsugu and White (69, 83) have considered thixotropic effects in plastic viscoelastic fluids. They write

$$\mathbf{T}\left[1 - \frac{Y(t)}{\sqrt{\tfrac{1}{2}\operatorname{tr}\mathbf{T}^2}}\right] = \mathbf{H} \tag{7-G-18}$$

which in a sense includes Eq. (7-F-15) as a special case. They take the form of $Y(t)$ to be

$$Y(t) = Y_i(\mathrm{II}_{\mathrm{d}}) - \left[\int_0^{\infty} \alpha \mathrm{II}_{\mathrm{d}}^{1/2} e^{-\alpha \mathrm{II}_{\mathrm{d}}^{1/2} z}\, dz\right](Y_i(\mathrm{II}_{\mathrm{d}}) - Y_{\mathrm{f}}) \tag{7-G-19a}$$

$$Y_i(\mathrm{II}_{\mathrm{d}}) = Y_{\mathrm{f}} + \beta \mathrm{II}_{\mathrm{d}}^{1/2} \tag{7-G-19b}$$

For long times in steady flows $Y(\mathrm{II}_{\mathrm{d}}, t)$ reduces to I_{f}, and the theory, to the theory of plastic viscoelastic fluids. The predictions of the model were compared to experimental studies of particle filled polymer melts.

H. MECHANISTIC MODELS OF SUSPENSIONS

Mechanical analyses of flow in suspensions date to Einstein's (20) study of dilute suspensions of spheres. Three-dimensional theories of flow-constitutive equations for suspensions have increasingly received attention since the publications of Batchelor (2) in the early 1970s. These have been extended in succeeding years by various investigators, notably Hinch and Leal (31). The constitutive equations generally treat dilute suspensions of particles of varying shape. The detailed development of such constitutive equations is based on the manner in which stresses distribute themselves on anisotropic particles and is beyond the scope of this paper. They have the form

$$\sigma = -p\mathbf{I} + 2\eta_0\mathbf{d} + 2\eta_0\phi[A\overline{\mathbf{pppp}}:\mathbf{d} + B(\overline{\mathbf{pp}\cdot\mathbf{d} + \mathbf{d}\cdot\mathbf{pp}}) + C\mathbf{d} + D\overline{\mathbf{pp}} \qquad (7\text{-H-1})$$

Here A–D are shape-dependent factors and $\bar{\mathbf{p}}$ is a unit vector representing the orientation of the major axis of the particles. The $\overline{\mathbf{pp}}$ and $\overline{\mathbf{pppp}}$ represent second and fourth moments of $\mathbf{p}$ with a probability distribution function. For spheres (Einstein):

$$C = 2.5, \qquad A = B = D = 0 \qquad (7\text{-H-2})$$

For a dilute suspension of chopped fibers (Batchelor):

$$A = \frac{4(L/D)^2}{9\ln L/D}, \qquad B = C = D = 0 \qquad (7\text{-H-3})$$

where L/D is the aspect ratio of the fibers.

Goddard (26) and Tanaka and White (71), facing the more difficult practical problem of dealing with concentrated systems of particles in non-Newtonian fluid matrices, have limited their attention to predicting viscous stress responses in the directions of flow. Goddard models the response of the elongational flow response of parallel fibers suspended in a power-law fluid and the latter authors, a concentrated suspension of attracting spheres in a similar fluid matrix.

In 1960 Ericksen (22) sought to develop a phenomenological theory of suspensions. He proposes a theory of anisotropic fluids in which he writes

$$\sigma = -p\mathbf{I} + \mathbf{F}[\mathbf{d},\mathbf{pp}] \qquad (7\text{-H-4})$$

This was expanded into the form

$$\sigma = -p\mathbf{I} + \lambda_1\mathbf{d} + (\lambda_2 + \lambda_3\mathbf{pp}:\mathbf{d})\mathbf{pp} + \lambda_4(\mathbf{d}:\mathbf{pp} + \mathbf{pp}:\mathbf{d}) \qquad (7\text{-H-5})$$

which is very similar to Eq. (7-H-1).

I. LIQUID CRYSTALS

There have been extensive investigations modeling flow characteristics of low-molecular-weight liquid crystals. In the quiescent static state, liquid crystals possess a complex birefringent structure. This is associated by Frank (23) with the constitutive equation

$$\boldsymbol{\sigma} = \frac{\partial W_{1c}}{\partial \nabla \mathbf{n}} \cdot \nabla \mathbf{n} \tag{7-I-1}$$

where W_{1c} is a strain energy function and $\mathbf{n}$ a unit vector denoting the direction inherent in the liquid crystal. Frank proposes special forms for $W_{1c}(\nabla \mathbf{n})$.

Leslie (38) has derived a theory of low-molecular-weight liquid crystals based on adding Frank's Eq. (7-I-1) and Ericksen's suspension model, Eq. (7-H-5). This has the form

$$\boldsymbol{\sigma} = \frac{\partial W_{1c}}{\partial \nabla \mathbf{n}} \cdot \nabla \mathbf{n} + \psi_1 \mathbf{d} + (\psi_2 + \psi_3 \mathbf{nn} : \mathbf{d})\mathbf{nn} + \psi_4 (\mathbf{d} : \mathbf{nn} + \mathbf{nn} : \mathbf{d}) \tag{7-I-2}$$

where ψ_1 to ψ_4 are material constants.

For polymer liquid crystals, most of the literature has generally dealt with molecular models based on rigid macromolecules (18, 41). White (85) has proposed a phenomenological model based upon the expression

$$\boldsymbol{\sigma} = \frac{\partial W_{1c}}{\partial \nabla \mathbf{n}} \cdot \nabla \mathbf{n} + \mathbf{H} \left[\underset{-\infty}{\overset{t}{\mathbf{c}^{-1}}}, \underset{-\infty}{\overset{t}{\mathbf{nn}}} \right] \tag{7-I-3}$$

Frank's theory, Eq. (7-I-1), represents a zero-order fluid and Leslie's formulation, Eq. (7-I-2), a first-order fluid in the sense of Coleman and Noll's representation of nonlinear viscoelastic fluids.

Doi (18) has sought to develop mechanistic theories of flow of rigid macromolecules behaving as polymer liquid crystals.

REFERENCES

1. R. Aris, *Vectors, Tensors, and The Basic Equations of Fluid Mechanics*, Prentice-Hall, Englewood Cliffs, NJ, 1962.

2. G. K. Batchelor, *J. Fluid Mech.*, **41**, 545 (1970); **46**, 813 (1971).

3. B. Bernstein, E. A. Kearsley, and L. J. Zapas, *Trans. Soc. Rheol.*, **7**, 391 (1963).

4. E. C. Bingham, *J. Wash. Acad. Sci.*, **6**, 177 (1916); *Fluidity and Plasticity*, McGraw-Hill, New York, 1922.

5. R. B. Bird in *Rheology*, Vol. I, G. Astarita, G. Marrucci and L. Nicolais, Eds., Plenum, New York, 1980.

6. R. B. Bird, R. C. Armstrong, and O. Hassager, *Dynamics of Polymeric Liquids*, Vol. 1, Wiley, New York, 1977.

7. R. B. Bird and P. J. Carreau, *Chem. Eng. Sci.*, **23**, 427 (1966).

8. D. C. Bogue, *IEC Fund.*, **5**, 253 (1966); D. C. Bogue and J. Doughty, *IEC Fund.*, **5**, 243 (1966).

9. D. C. Bogue and J. L. White, *Engineering Analysis of Non-Newtonian Fluids*, NATO Agardograph, 144, 1970.

10. L. Boltzmann, *Sitzungber Akad. Wiss.*, **10**, 275 (1874).

11. P. J. Carreau, *Trans. Soc. Rheol.*, **16**, 99 (1972).

12. A. L. Cauchy, *Ex. de Math.*, **3**, 160 (1828); **4**, 293 (1829).

13. I. J. Chen and D. C. Bogue, *Trans. Soc. Rheol.*, **16**, 59 (1972).

14. I. J. Chen, G. E. Hagler, L. E. Abbott, D. C. Bogue, and J. L. White, *Trans. Soc. Rheol.*, **16**, 472 (1972).

15. B. D. Coleman and W. Noll, *Arch. Rat. Mech. Anal.*, **6**, 355 (1960); B. D. Coleman, W. Noll, and H. Markovitz, *J. Appl. Phys.*, **35**, 1 (1964).

16. B. D. Coleman and W. Noll, *Rev. Mod. Phys.*, **33**, 239 (1961).

17. W. O. Criminale, J. L. Ericksen, and G. L. Filbey, *Arch. Rat. Mech. Anal.*, **1**, 410 (1958).

18. M. Doi, *J. Polym. Sci., Polym. Phys. Ed.*, **19**, 229 (1981); *Ferroelectrics*, **30**, 247 (1980).

19. M. Doi and S. F. Edwards, *J. Chem. Soc., Trans. II*, **74**, 1789, 1802, 1818 (1978).

20. A. Einstein, *Ann. Phys.*, **19**, 289 (1906); **34**, 591 (1911).

21. R. Eisenschitz, *Kolloid Z.*, **54**, 184 (1933).

22. J. L. Ericksen, *Kolloid Z.*, **173**, 117 (1960).

23. F. C. Frank, *Discuss Faraday Soc.*, **25**, 19 (1958).

24. A. G. Frederickson, *Principles and Applications of Rheology*, Prentice Hall, Englewood Cliffs, NJ, 1974.

25. H. Giesekus, *J. Non-Newt. Fluid Mech.*, **14**, 47 (1984).

26. J. D. Goddard, *J. Non-Newt. Fluid Mech.*, **1**, 1 (1976).

27. J. D. Goddard and C. Miller, *Rheol. Acta*, **5**, 177 (1966).

28. A. E. Green and R. S. Rivlin, *Arch. Rat. Mech. Anal.*, **1**, 1 (1957).

29. M. S. Green and A. V. Tobolsky, *J. Chem. Phys.*, **14**, 80 (1946).

30. H. Hencky, *Ann. Phys.*, **2**, 617 (1929).

31. E. J. Hinch and L. G. Leal, *J. Fluid Mech.*, **71**, 481 (1975); **76**, 187 (1976).

32. K. Hohenemser and W. Prager, *Z. Angew. Math. Mech.*, **12**, 216 (1932).

33. J. F. Hutton, *Rheol. Acta*, **14**, 979 (1975).

34. A. I. Isayev and R. K. Uphadyay *J. Non-Newt. Fluid Mech.*, **19**, 135 (1985).

35. M. W. Johnson and D. Segalman, *J. Non-Newt. Fluid Mech.*, **14**, 279 (1984).

36. R. Keunings and M. J. Crochet, *J. Non-Newt. Fluid Mech.*, **14**, 279 (1984).

37. A. I. Leonov, *Rheol. Acta*, **15**, 85 (1976); A. I. Leonov, E. H. Lipkina, E. D. Pashkin, and A. W. Prokunin, *Rheol. Acta*, **15**, 44 (1976).

38. F. M. Leslie, *Arch. Rat. Mech. Anal.*, **28**, 265 (1968); *Mol. Liq. Cryst.*, **7**, 407 (1969).

39. V. M. Lobe and J. L. White, *Polym. Eng. Sci.*, **19**, 617 (1979).

40. A. S. Lodge, *Trans. Faraday Soc.*, **52**, 120 (1956).

41. X. L. Luo and R. I. Tanner, *J. Non-Newt. Fluid Mech.*, **21**, 179 (1986).

42. T. Matsumoto and D. C. Bogue, *Trans. Soc. Rheol.*, **21**, 453 (1977).

43. J. C. Maxwell, *Phil. Trans. Roy. Soc.*, **157**, 49 (1867).

44. J. C. Maxwell, *Constitution of Bodies*, Encyclopedia Brittanica, 1867.

45. W. Minoshima, J. L. White, and J. E. Spruiell, *Polym. Eng. Sci.*, **20**, 1166 (1980).

46. M. Mooney, *J. Colloid Sci.*, **6**, 96 (1951).

47. L. Natanson, *Phil. Mag.*, **2**(6), 342 (1901).

48. W. Noll, *J. Rat. Mech. Anal.*, **4**, 3 (1955).

49. J. G. Oldroyd, *Proc. Camb. Phil. Soc.*, **43**, 100 (1947).

50. J. G. Oldroyd, *Proc. Camb. Phil. Soc.*, **45**, 595 (1949).

51. J. G. Oldroyd, *Proc. Roy. Soc.*, **A200**, 523 (1950).

52. J. G. Oldroyd, *Proc. Roy. Soc.*, **A245**, 278 (1958).

53. J. G. Oldroyd, *J. Non-Newt. Fluid Mech.*, **14**, 9 (1984).

54. N. Phan-Tien, *J. Non-Newt. Fluid Mech.*, **16**, 329 (1984).

55. N. Phan-Tien and R. I. Tanner, *J. Non-Newt. Fluid Mech.*, **2**, 353 (1977); N. Phan-Tien, *J. Non-Newt. Fluid Mech.*, **22**, 259 (1978).

56. W. Philippoff, *Trans. Soc. Rheol.*, **1**, 95 (1957).

57. W. Prager and P. Hodge, *The Theory of Perfectly Plastic Solids*, Wiley, New York, 1951.

58. M. Reiner, *Am. J. Math.*, **67**, 350 (1945).

59. M. Reiner, *Am. J. Math.*, **70**, 433 (1948).

60. R. S. Rivlin, *Proc. Roy. Soc.*, **A193**, 260 (1948).

61. R. S. Rivlin, *Phil. Trans. Roy. Soc.*, **A240**, 459 (1948).

62. R. S. Rivlin, *Phil. Trans. Roy. Soc.*, **A241**, 379 (1948).

63. R. S. Rivlin, *J. Rat. Mech. Anal.*, **5**, 179 (1956).

64. R. S. Rivlin and J. L. Ericksen, *J. Rat. Mech. Anal.*, **4**, 323 (1955).

65. T. Schwedoff, *J. Phys.*, **9**(2), 34 (1890).

66. A. Slibar and P. R. Pasley in *Second Order Effects in Elasticity, Plasticity and Fluid Dynamics*, M. Reiner and D. Abir, Eds., MacMillan, New York, 1964.

67. T. W. Spriggs, J. D. Huppler, and R. B. Bird, *Trans. Soc. Rheol.*, **10**(1), 191 (1966).

68. G. G. Stokes, *Trans. Camb. Phil. Soc.*, **8**, 287 (1845).

69. Y. Suetsugu and J. L. White, *J. Non-Newt. Fluid Mech.*, **14**, 121 (1984).

70. T. Takaki and D. C. Bogue, *J. Appl. Polym. Sci.*, **19**, 419 (1975).

71. H. Tanaka and J. L. White, *J. Non-Newt. Fluid Mech.*, **7**, 333 (1980).

72. M. Takahashi, T. Masuda, and S. Onogi, *Trans. Soc. Rheol.*, **21**, 337 (1977); *J. Rheol.*, **22**, 285 (1978).

73. L. R. G. Treloar, *Trans. Faraday Soc.*, **42**, 83 (1942); *Physics of Rubber Elasticity*, 2nd ed., Oxford University Press, Oxford, 1958.

74. C. Truesdell and W. Noll, "The Non-Linear Field Theories of Mechanics," *Handbuch der Physik*, Vol. III/3, Springer, Berlin, 1965.

75. C. Truesdell and R. A. Toupin, "The Classical Field Theories," *Handbuch der Physik*, Vol. III/1, Springer, Berlin, 1960.

76. R. K. Upadhyay and A. I. Isayev, *Rheol. Acta*, **22**, 557 (1983); R. K. Upadhyay, A. I. Isayev, and S. F. Shen, *J. Rheol.*, **27**, 155 (1983).

77. V. Volterra, *Theory of Functionals*, Blackie and Sons, London, 1930, Dover reprint.

78. M. H. Wagner, *Rheol. Acta*, **15**, 136 (1976).

79. M. H. Wagner, *Rheol. Acta*, **16**, 43 (1977); **17**, 138 (1978); *J. Non-Newt. Fluid Mech.*, **4**, 39 (1978).

80. K. Weissenberg, *Nature*, **159**, 310 (1947).

81. J. L. White, *J. Appl. Polym. Sci.*, **8**, 1129 (1964).

82. J. L. White, *J. Appl. Polym. Sci.*, **8**, 2339 (1964).

83. J. L. White, *J. Non-Newt. Fluid Mech.*, **5**, 179 (1979).

84. J. L. White, *J. Non-Newt. Fluid Mech.*, **8**, 195 (1981).

85. J. L. White, *Appl. Polym. Symp.*, **41**, 241 (1985).

86. J. L. White, L. Czarnecki, and H. Tanaka, *Rubber Chem. Technol.*, **53**, 823 (1980).

87. J. L. White and A. B. Metzner, *J. Appl. Polym. Sci.*, **7**, 1867 (1963).

88. J. L. White and H. Tanaka, *J. Non-Newt. Fluid Mech.*, **8**, 1 (1981).

89. J. L. White and N. Tokita, *J. Phys. Soc. Japan*, **22**, 719 (1967); **24**, 436 (1968).

90. F. W. Wiegel, *Physica*, **42**, 156 (1969); F. W. Wiegel and F. Th de Bats, *Physica*, **43**, 33 (1969).

91. M. Yamamoto, *Busseiron Kenkyu*, **59**, 1 (1953); **60**, 38 (1953); *J. Phys. Soc. Japan*, **11**, 413 (1956); **12**, 1148 (1957).

92. M. Yamamoto, *Trans. Soc. Rheol.*, **15**, 331, 783 (1971).

93. L. J. Zapas, *J. Res. Natl. Bur. Stand.*, **70A**, 525 (1966).

94. L. J. Zapas and T. Craft, *J. Res. Natl. Bur. Stand.*, **69A**, 541 (1965).

95. S. Zaremba, *Bull. Acad. Sci. Cracow*, 594 (1903).

VIII

APPLICATIONS TO POLYMER PROCESSING

A. INTRODUCTION

In this chapter we turn to the application of the principles developed in earlier chapters to polymer processing operations. It is indeed to the material of this chapter that much of the developments of the earlier chapters of this book are intended to lead. We begin with an overview in terms of the dimensionless groups from the rheological constitutive equations, force, and energy balances describing the process operations. We also discuss methods of analysis of flow in processing operations. Specific applications are then considered. We give attention to those applications that are strongly affected by rheological responses. The first application considered is flow through dies of varying cross sections, including die-entry flows and extrudate swell. Flow in extruder screws is then briefly described. We then turn to injection molding, where the fluid mechanics is more complex and the process nonisothermal. In the closing section, we turn to post-die flow, where external forces are applied to the emerging melt stream and the material is further shaped. Specific application is made to melt spinning and tubular film extrusion.

B. HISTORICAL BACKGROUND

Applications of rheological principles to interpret polymer processing operations probably date to the 1930s, with the earliest efforts being those of Dillon and Mooney on rubber processing (40, 98–100). Mooney's Goodyear Medal address (100) recalls the early period. The first paper in the open literature that may be classified in this manner is probably that by Dillon and Johnston (40) in 1933.

As is clear from the book (128) and later review of Schenkel (129) as well as the review of White et al (170), much of the technological development in polymer processing from 1930 onwards took place in Germany. More fundamental research, to develop understanding, was done predominantly in the United States. In the late 1940s and early 1950s, much broader efforts at rheological interpretations of processing operations appeared. These included not only Mooney's experiences (*98–101*) with synthetic rubber, but also broad-based efforts at interpreting plastics processing operations. The first major work in this direction was by R. S. Spencer and his colleagues (*133–136*) of Dow Chemical, who considered extrusion through dies (*133,134*) and injection molding (*135,136*). They also made the first serious study of molecular orientation and residual stresses in fabricated plastic parts. These publications all appeared in the period of 1948–1952. Spencer et al. made combinations of careful experimental and well-thought-out theoretical studies of the processing operations, including the importance of non-Newtonian shear viscosity and melt elasticity.

Two other significant research efforts in plastics processing started to yield results at this time. One was a joint research effort on calendering between the Armstrong Cork Company and Brown University (*12,13*). A second was an investigation of the screw extrusion process by DuPont researchers (*26*). At the end of the decade (1959), E. C. Bernhardt collected together these different research teams and compiled a monograph entitled *Processing of Thermoplastic Materials* (*16*). Three years later, J. M. McKelvey presented these developments in a more methodical manner in his book *Polymer Processing* (*88*). Much of the same material from a more technological viewpoint is contained in Schenkel (*128*).

The central thrust of applying rheology to polymer processing had been set by this period, but the level of sophistication of rheological applied had not kept up with the rapid progress in this area during the same period. The two areas were first brought together in the 1960s and then increasingly in the 1970s and 1980s. Nonlinear viscoelastic fluid models were applied to processing operations. Dimensional analysis considerations based on the theory of viscoelasticity and energy balances were first introduced in the mid-1960s (*90,120,156*). The concepts of the Weissenberg and Deborah numbers became accepted at this time.

In 1966 Tadmor and his colleagues (*139*) published the first integrated analysis of a complex polymer processing operation by analyzing the entire movement from hopper to die, including melting in a screw extruder. The 1960s also saw the first analyses of post-die flow processing involving elongational flow. Ziabicki and Kedzierska (*178,180*) and Kase and Matsuo (*74–76*) simulated melt spinning of fibers. Kase (*73,76*) and Pearson and Matovich (*112*) analyzed melt-spinning instabilities. Pearson and Petrie (*111,113*) modeled tubular film extrusion.

The last 15 years have seen rapid development in the analysis of processing operations, largely using numerical analysis procedures, particularly finite element methods (*34–36,151*). There has been much effort in process simulation. Computer-aided design has been developed notably for injection molding (*17*).

C. DIMENSIONLESS GROUPS

1. General Remarks

The origin of the use of dimensional analysis and dimensionless groups is associated with French and English scientists of the mid and late nineteenth century—Ferdinand Reech, William Froude, and Osborne Reynolds. Reech and Froude, working independently, were concerned with the scaling of model ships (125). They considered the interaction of momentum and gravitational forces and concluded that lengths should be scaled with the square root of the velocities. This is the basis of the dimensionless group which we call today the Froude number. Reynolds (121) was concerned with the transition from laminar (Poiseuille) flow of water in a tube to turbulent motion. He proposed a criterion for this transition based upon dimensional analysis of the Navier–Stokes equation. Reynolds argued that a disturbance would be propagated by inertia and damped out by viscous resistance. The flow would then break down when a criterion defined by the ratio

$$\frac{\text{inertial forces}}{\text{viscous forces}} = \frac{\rho U^2/L}{\eta U/L^2} = \frac{LU\rho}{\eta}$$

where U is a characteristic velocity and L a characteristic length, was reached. The quantity, $LU\rho/\eta$, which is now called the Reynolds number, was found to represent the transition from laminar flow to turbulence in a pipe when L was taken to be the diameter.

The first applications of dimensional analysis to heat transfer were carried out by the German engineer Wilhelm Nusselt (106) in 1915. He placed the energy equation in a dimensionless form. Dimensionless heat-transfer coefficients hL/k (Nusselt numbers) were found to be determined by the Reynolds number and the ratio of convective heat flux to heat conduction, $LU\rho c/k$, now called the Peclet number.

We now turn our attention to viscoelastic polymer melt systems.

2. Isothermal Viscoelastic Fluids

Using dimensional analysis for viscoelastic fluid systems was first considered in a 1928 paper of Herzog and Weissenberg (62). These authors discuss the state of flow in a fluid as corresponding to the relative values of three energies: kinetic, dissipation, and elasticity, that is, $E(\text{kin})$, $E(\text{warms})$, and $E(\text{elast})$. They took the ratio $E(\text{kin})/E(\text{warms})$ as the Reynolds number and speculated on the possible nature of other ratios. Their ratio $E(\text{elast})/E(\text{warms})$ represented the ratio of elastic to viscous effects. Weissenberg returned to this subject a generation later, after his discovery and interpretation of the normal stress effect. This time Weissenberg was more explicit. He wrote (153):

As dimensionless quantity of tensorial character we may quote here the recoverable strain. Just as the Reynolds number coordinates the rheological states with respect to similitude in the relative proportions of the forces of inertia and internal friction, so the recoverable strain does with respect to similitude in anisotropy in the sheared states.

The application of recoverable strain to represent the "elastic state" of viscoelastic fluids was made extensively by Philippoff (115) during the next decade. It was, however, only in the 1960s (90, 156), that the proper role of the Weissenberg number in viscoelastic fluid mechanics was developed. It was now expressed as $\tau V/L$, where τ is a characteristic time of the fluid.

A different perspective of dimensionless groups for viscoelastic fluids was developed by Reiner (120) during the early 1960s. Reiner argued that the ratio of the material time to a process time, that is, τ/t, played a critical role in defining the response of a substance as a fluid or solid.

We will now develop the dimensional analysis of a viscoelastic fluid starting with first principles. To proceed we will choose the convected Maxwell model, which is the simplest model that represents all the features of a viscoelastic fluid. Introducing a characteristic velocity U and length L, we write the equation of motion and constitutive equation for a convected Maxwell model of Eq. (7-F-8) as

$$\frac{LU\rho}{\tau_0 G}\left[\frac{\partial \mathbf{v}^*}{\partial t} + (\mathbf{v}^* \cdot \nabla^*)\mathbf{v}^*\right] = -\nabla^* p^* + \nabla^* \cdot \mathbf{P}^* + \left(\frac{LU\rho}{\tau_0 G}\right)\left(\frac{gL}{\rho U^2}\right)\mathbf{f}^* \qquad (8\text{-}C\text{-}1)$$

$$\mathbf{P}^* = 2\frac{\tau}{\tau_0}d^* - \frac{\tau}{\tau_0}\frac{\tau_0 U}{L}\frac{\delta \mathbf{P}^*}{\delta t^*} \qquad (8\text{-}C\text{-}2)$$

The stress tensor $\mathbf{P}$ and pressure p are made dimensionless with $\tau_0 G(U/L)$. The quantity $\tau_0 G$ plays the role of a viscosity. The dimensionless groups which arise in Eqs. (8-C-1) and (8-C-2) are

$$\frac{LU\rho}{\tau_0 G} = N_{\text{Re}} = \text{Reynolds number}$$

$$\frac{\rho U^2}{gL} = N_{\text{Fr}} = \text{Froude number} \qquad (8\text{-}C\text{-}3a\text{--}d)$$

$$\frac{\tau_0 U}{L} = N_{\text{ws}} = \text{Weissenberg number}$$

$$\frac{\tau}{\tau_0} \quad \text{or} \quad \frac{\tau_0 - \tau}{\tau} = N_{\text{Ya}} = \text{Yamamoto number}$$

where N_{Re} represents a ratio of inertial to viscous forces, N_{Fr} a ratio of inertial to gravitational forces, and N_{ws} the intensity of viscoelasticity or the ratio of viscoelastic to viscous forces. Here N_{Ya} is a measure of the influence of deformation history rate on the elastic memory.

We have called the quantity $\tau_0 U/L$ the Weissenberg number. The question arises as to how it is related to Weissenberg's recoverable strain concept. The simplest approach to show this follows by writing

$$\tau_0 \frac{U}{L} = \frac{\tau_0 G(U/L)}{G} = \frac{\sigma}{G} = J_e \sigma \sim S \tag{8-C-4a}$$

For a shearing flow from Eq. (6-F-6a)

$$\tau_0 \frac{U}{L} \sim J_e \sigma \sim \frac{N_1}{\sigma_{12}} \tag{8-C-4b}$$

where $\tau_0 G$ is the viscosity and $\tau_0 G(U/L)$ is a stress. Equation (8-C-4a) represents the Weissenberg number as the ratio of the stress to the modulus of the fluid. This is equivalent to the magnitude of the elastic recoil following flow in a viscoelastic fluid.

The characteristic length L in Eq. (8-C-4) may be in the direction of flow $L_{\parallel}$, as in an elongational flow or perpendicular to the direction of flow $L_{\perp}$, as in shear. If we use the interpretation of the preceding paragraph, we have in the former case elongational flow recovery and in the latter a shear flow recovery.

Another point of view is possible when the characteristic length is in the direction of flow. We may write

$$\tau_0 \frac{U}{L_{\parallel}} = \frac{\tau_0}{L_{\parallel}/U} = \frac{\tau_0}{t_{res}} \tag{8-C-5}$$

where t_{res} is the residence time of the melt during the flow. The expression τ_0/t_{res} is equivalent to what Reiner has referred to as the Deborah number (120).

We also note a dimensionless group in Eq. (8-C-2) which has the form $\tau/\tau_0(II_d^*)$. White and Minoshima (168) associate this with what they call the Yamamoto number, N_{Ya}. This represents the deformation rate softening behavior. These authors prefer the definition

$$\frac{\tau_0 - \tau}{\tau} = N_{Ya} \tag{8-C-6}$$

If one takes the deformation rate dependence of τ to be through the approximate equation

$$\tau = \frac{\tau_0}{1 + a\tau_0 II_d^{1/2}} \tag{8-C-7a}$$

it follows that

$$N_{Ya} = a \qquad \text{(8-C-7b)}$$

We would then write Eq. (8-C-2) as

$$P^* = \frac{1}{1 + N_{Ya}N_{Ws}}\left[2d^* - N_{Ws}\frac{\delta P^*}{\delta t^*} \right] \qquad \text{(8-C-7c)}$$

It may be seen that the velocity field v^*, determined by Eq. (8-C-1) may be expressed in terms of the dimensionless groups of the preceding paragraphs through

$$v^* = v^*[N_{Re}, N_{Fr}, N_{Ws}, N_{Ya}, x^*] \qquad \text{(8-C-8)}$$

Additional dimensionless groups can be introduced through the boundary conditions.

In an isothermal flowing poymer melt, inertial and gravitational forces may in general be neglected. This means that we may omit the Reynolds and Froude numbers. Equation (8-C-8) reduces to

$$v^* = v^*[N_{Ws}, N_{Ya}, x^*] \qquad \text{(8-C-9)}$$

and the velocity field is determined by the Weissenberg and Yamamoto numbers alone.

There are two limitations to these results. One is the simplicity of the constitutive equation used to represent the behavior of a viscoelastic fluid. This turns out not to be particularly important. With more complex models, $\tau_0 U/L$, the Weissenberg number, again rises. Additional dimensionless groups which charactrize the same behavior as represented by the Yamamoto number, other nonlinearities, and the breadth of the relaxation spectrum $H(\tau)$ also occur. Some are related to White's viscoelastic ratio number (156). This in no way complicates the formulation of Eqs. (8-C-8) and (8-C-9), as only one group in each case contains U and L, the Weissenberg number. The other groups remind us that v^* also depends upon the detailed nonlinearities of the material and $\tau_0 U/L$ is, by itself, not enough to determine the velocity field.

The second limitation is of greater importance. This is associated with the restriction of isothermal flows. In general, the flow of polymer melts, especially in thick cross sections, is highly nonisothermal, owing to heat generation by viscous shearing of the melt and the notoriously poor thermal conductivity of the material. Proper dimensional analysis requires consideration of the energy balance as well as Cauchy's law of motion.

3. Plastic Fluids

The arguments for viscoelastic fluids may be extended to fluids with yield values. If we introduce the deviatoric stress tensor T in place of P, then for the constitutive equation of a Bingham plastic [Eq. (7-G-11b)] we may write

$$\frac{LU\rho}{\eta_B}\left[\frac{\partial \mathbf{v}^*}{\partial t} + (\mathbf{v}^* \cdot \nabla^*)\mathbf{v}^*\right] = -\nabla^* p^* + \nabla^* \cdot \mathbf{T}^* + \left(\frac{LU\rho}{\eta_B}\right)\left(\frac{gL}{\rho U^2}\right)f^* \quad \text{(8-C-10)}$$

$$\mathbf{T}^* = \left[2\frac{YL}{\eta_B U}\frac{1}{\sqrt{2\,\mathrm{tr}\,\mathbf{d}^{*2}}} + 2\right]\mathbf{d}^* \quad \text{(8-C-11)}$$

where $\mathbf{T}$ is made dimensionless with $\eta_B U/L$. One new dimensionless group arises:

$$\frac{YL}{\eta_B U} = N_{\mathrm{Bi}} = \text{Bingham number} \quad \text{(8-C-12)}$$

Here $N_{\mathrm{Bi}}(YL/\eta_B U)$ represents a ratio of the yield stress to the differential viscous forces. This was introduced by Oldroyd (*108*) and later featured in the text of Prager (*117*), who first called it the Bingham number.

In the previous chapter, we described a theory of plastic viscoelastic fluids which combines a yield surface and a hereditary functional. This represents the flow behavior of a highly filled polymer melt. If the functional may be expressed through a modulus G and relaxation time τ, two dimensionless groups arise

$$\frac{YL}{\tau G U} \quad \text{and} \quad \tau\frac{U}{L}$$

These are the Bingham number and the Weissenberg number. This view is developed in papers by White (*158*) and by White and Huang (*164*). Groups equivalent to the Yamamoto number arise if τ is considered to depend upon II_d.

The dimensionless velocity field in a plastic viscoelastic fluid is determined by Bingham, Weissenberg, and Reynolds numbers. For highly viscous systems, the Reynolds number does not need to be considered and we obtain

$$\mathbf{v}^* = \mathbf{v}^*[N_{\mathrm{Bi}}, N_{\mathrm{Ws}}, x^*] \quad \text{(8-C-13)}$$

The Bingham number determines the shape of the yield surface in the flow field. When the Bingham number is below a critical value, the material behaves as a solid.

4. Surface Tension and Interfacial Effects

Dimensionless groups may also arise from the boundary conditions acting on a system. On studies of the dynamics of flow of low-viscosity fluids, the most important new contributions are surface and interfacial tensions that apply stresses to boundaries. This may also be critically important in the development of phase morphology in blends (85, 93). These give rise to expressions of the form

$$\Delta p = \frac{v}{R} \qquad (8\text{-}C\text{-}14)$$

where v is the surface or interfacial tension and R the appropriate radius of curvature. If we make p dimensionless with (U/L), we obtain

$$\Delta p^* = \frac{v}{\eta U} \frac{1}{R^*} \qquad (8\text{-}C\text{-}15)$$

Dimensionless groups of the form $(v/\eta U)$ trace to the work of Rayleigh for liquids in contact with the air and to Taylor (143) for contact of immiscible liquids. Min et al. (93) have applied this to interpret the development of phase morphology in immiscible polymer melt blends.

5. Energy Equation and Nonisothermal Effects

As mentioned earlier in this chapter, the flow of polymer melts is generally not isothermal. It is necessary to include the energy equation of Eq. (2-K-15) in any dimensional analysis study of this system. We may introduce a characteristic velocity U, length L and temperature difference θ, and rewrite this in the dimensionless form where we use Fourier's law of heat conduction ($\mathbf{q} = -k\nabla T$):

$$\frac{\rho c L^2}{k t_{ch}} \frac{\partial T^*}{\partial t^*} + \frac{LU\rho c}{k}(\nabla^* \cdot \mathbf{v}^*)T^* = \nabla^{*2} T^* + \frac{\eta_0 U^2}{k\theta}\mathbf{P}^*{:}\nabla\mathbf{v}^* \qquad (8\text{-}C\text{-}16)$$

where we have made $\mathbf{P}^*$ dimensionless with $\eta_0 U/L(\eta_0 = \tau_0 G)$ and t^* with a characteristic time t_{ch}. The two new dimensionless groups of Eq. (8-C-16) are

$$\frac{LU\rho c}{k} = N_{Pe} = \text{Peclet number} \qquad (8\text{-}C\text{-}17)$$

$$\frac{\eta_0 U^2}{k\theta} = N_{Br} = \text{Brinkman number} \qquad (8\text{-}C\text{-}18)$$

$$\frac{\rho c L^2}{k t_{ch}} = N_{Fo} = \text{Fourier number} \qquad (8\text{-}C\text{-}19)$$

Here $N_{Pe}(LU\rho c/k)$ is the ratio of convective heat to heat conduction flux. The $N_{Br}(\eta_0 U^2/k\theta)$ is the ratio of heat generated in flow to heat conduction and the $N_{Fo}(\rho c L^2/k t_{ch})$ represents the ratio of heat accumulation to heat conduction. The Brinkman group is most important for understanding the amount of viscous heating in flow through dies and derives from the studies of this effect in tube flow by Brinkman (*21*) and Bird (*18, 19*). The Fourier number is most important in considering transient heat conduction in solids.

The key effect of heat fluxes is to change the form of the force balance through the variation of the viscosity of the polymer melt with temperature. This would be via a relation such as

$$\eta = A e^{(E/RT_0)(1/T^*)} \qquad \text{(8-C-20)}$$

The quantity in the exponent represents a key dimensionless group which may be called a Griffith number (*50*). Here

$$N_{Gr} = \frac{E}{RT_0} \qquad \text{(8-C-21)}$$

It is usually expressed in a different form, which reflects both the temperature dependence of viscosity and viscous heating. The larger N_{Gr} in Eq. (8-C-21), the more effective is the Brinkman number in effecting the flow.

The boundary condition on the flowing melt is generally of the form

$$-k\frac{dT}{dy}\bigg|_{\text{surface}} = h(T_s - T_c) \qquad \text{(8-C-22a)}$$

where k is the thermal conductivity, h is a heat transfer coefficient T_s the surface temperature, and T_c the coolant or ambient temperature. This is equivalent to

$$-\frac{dT^*}{dy^*}\bigg|_{\text{surface}} = \left(\frac{hL}{k}\right)(T_s^* - T_c^*) \qquad \text{(8-C-22b)}$$

We obtain the dimensionless group

$$\frac{hL}{k} = N_{Nu} = \text{Nusselt number} \qquad \text{(8-C-23)}$$

The number characterizes the heat flux at the boundary. It has a value of zero for an adiabatic wall. It is often used to consider the Stanton number, representing the ratio of external heat transfer losses to convection, that is,

$$N_{St} = \frac{N_{Nu}}{N_{Pe}} = \frac{hL/k}{LU\rho c/k} = \frac{h}{\rho c U} \qquad \text{(8-C-24)}$$

6. Scale-Up

As we noted above, dimensional analysis originated in the process of scaling ships. Scale-up from small models to large-scale equipment is a problem of considerable concern in polymer processing. However, unlike studies of scale-up with low-viscosity Newtonian fluids, we cannot limit our concern to the dynamic equations alone. The high viscosity of polymer melt systems generates high levels of heat during flow. This necessitates consideration of the energy equation and dimensionless groups arising in it as well. However, it is useful from a pedagogical point of view to begin with isothermal flow and then proceed to include viscous heating.

a. Isothermal Flow

First, restricting ourselves to isothermal fluids we must consider in general,

$$\text{viscoelastic fluids:} \frac{LU\rho}{\tau_0 G}, \quad \tau_0\frac{U}{L}, \quad \frac{\tau}{\tau_0}$$

$$\text{plastic viscoelastic fluids:} \frac{LU\rho}{\tau_0 G}, \quad \frac{YL}{\tau_0 GU}, \quad \tau_0\frac{U}{L}, \quad \frac{\tau}{\tau_0}$$

Clearly, one cannot simultaneously scale the Reynolds number $LU\rho/\tau G$ with either the Weissenberg number $\tau U/L$ or the Bingham number $YL/\tau GU$ using the same fluid. The former demands scaling according to constant LU, the latter to scaling at constant U/L.

The problem of scaling under such conditions arises with polymer solutions where inertial, viscous, and viscoelastic forces are often of the same order. Clearly, this cannot be accomplished with the same fluid, but it should be possible if we use fluids of differing τ_0 but the same τ/τ_0 and G (4).

For highly viscous fluids such as polymer melts, the Reynolds number may be neglected. Only dimensionless groups with the kinematic parameter U/L arise, and scaling should be carried out by increasing U proportional to L. However, one must be wary in this situation because of the importance of viscous heating.

b. Nonisothermal Flow of Highly Viscous Fluids

We now turn to cases where we may neglect the Reynolds number, but must consider viscous heating. We must include the dimensionless groups

$$\tau_0\frac{U}{L}, \quad \frac{\tau}{\tau_0}, \quad \frac{\tau_0 GU^2}{k\theta}, \quad \frac{LU\rho c}{k}, \quad \frac{hL}{k}, \quad \frac{E}{R\tau_0}$$

It is clearly impossible to scale with the Weissenberg number $\tau_0 U/L$, the Brinkman number $\tau_0 GU^2/k\theta$, and the Peclet number $LU\rho c/k$. These demand scaling at constant U/L, U^2, and LU, respectively.

With polymer melts in isothermal flow, we may consider $\tau_0 U/L$ and τ/τ_0 alone, as described in the previous section. If cross-sectional variation changes and local deformation rates are large in the direction of flow, but thickness dimensions remain small, $\tau_0 U/L$ is probably dominant. Thus, flow into the entrance or from the exit of a die will be dominated by $\tau_0 U/L$ and τ/τ_0. It is to be noted that scaling at constant $\tau_0 U/L$ is equivalent to scaling at constant shear stress.

In internal flows where we must scale-up to large thick cross sections (i.e., in the metering region of an extruder or internal mixer) and control of heat transfer is poor, the Brinkman number $\tau_0 G U^2/k\theta$ is the dominant dimensionless group and scaling must seriously consider it. This leads to scaling at roughly constant U. In a machine with rotating parts (screw or a rotor) U is $L\Omega$. This means Ω should decrease with increase of rotating part dimension. If this is not done, viscous heat generation may lead to catastrophically high operating temperatures.

Often in practice one compromises in scale-up. One tries to scale in between constant shear stress on Weissenberg number on one hand and constant Brinkman number on the other. This is done by increasing U with L but at much less than a linear rate.

D. RHEOLOGICAL/HYDRODYNAMIC APPROXIMATIONS IN ANALYSIS OF PROCESSING OPERATIONS

1. Internal and External Flows

The experiences of a polymer melt during its process history may be roughly divided into what we may call *internal flows* and *external flows* (Fig. 8-D-1). In the former case, we have flow in extruder screws, dies, and injection molds. In the latter, we have melt-spinning of fibers, film casting, tubular film extrusion, and blow-molding of bottles. In internal flows, the melt moves between roughly parallel walls, one of which may be in motion. The polymer will in most cases adhere to the walls and be acted upon by pressure gradients. The flow will then be largely shearing in character. In external flows, the melt deforms under the action of applied tensions and inflation pressures in the region beyond the die exit. There are no interactions with walls, except in the final stages of blow molding, and the flows are elongational in character. There is another basic difference between internal and external flows. The former are roughly isothermal and the latter are highly nonisothermal. While temperature distributions in internal flows do exist because of cold walls and viscous dissipation, the temperature gradients, especially in the longitudinal direction, are minor compared with the air-quench situation existing in fiber spinning or film extrusion.

These comments have certain implications for the analysis of processing operations. In internal flows, one may use isothermal formulations of constitutive equations with more or less second-order corrections on rheological functions for temperature dependence. Depending upon boundary conditions and apparatus dimensions, one may be able to neglect or be required to include the energy balance. However, the isothermal rheological properties, and especially the shear viscosity function, play a dominant role. In external flows, the dominant phenomenon is the rapid change of the rheological behavior along the process direction due to cooling. Representation of the temperature dependence and detailed heat transfer mechanisms is key.

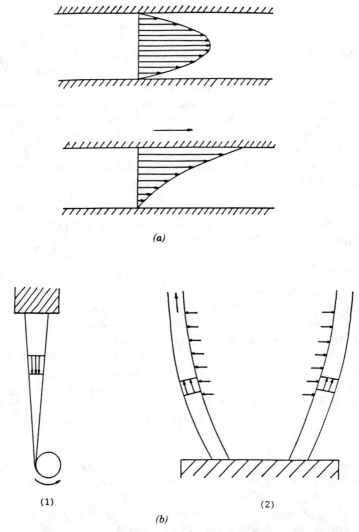

FIGURE 8-D-1. (*a*) Internal and (*b*) external flow of a polymer melt.

2. Hydrodynamic Lubrication Theory

In internal flows, the lubrication theory devised by Osborne Reynolds (*122*) to analyze the flow of lubricating oils in bearings supplies a useful approximation. Consider a narrow clearance defined by two surfaces which is roughly normal to the 2 direction, and flow is allowed in the 1–3 plane. One of the surfaces may move relative to the other with velocity components U_1 and U_3. Clearance $H(x_1, x_2)$

varies only slowly in the 1–3 plane,

$$\frac{\partial H}{\partial x_1} \ll 1, \qquad \frac{\partial H}{\partial x_3} \ll 1 \tag{8-D-1}$$

The velocity field has the form

$$\mathbf{v} = v_1\mathbf{e}_1 + 0\mathbf{e}_2 + v_3\mathbf{e}_3 \tag{8-D-2}$$

where shearing in the 2 direction is predominant. The variations in shear-flow-induced normal stresses P_{11} and P_{33} in the 1 and 3 directions will be small compared with the variation of the shear stress in the 2 direction. This allows the equations of motion in the 1 and 3 directions to be written

$$0 = -\frac{\partial p}{\partial x_1} + \frac{\partial \sigma_{12}}{\partial x_2}, \qquad 0 = -\frac{\partial p}{\partial x_3} + \frac{\partial \sigma_{32}}{\partial x_2} \tag{8-D-3}$$

These flows roughly correspond to those which are governed by the Criminale–Ericksen–Filbey theory (33), Eq. (7-F-64), of viscoelastic fluids. We may express the shear stresses in such flows in terms of the shear viscosity η and shear rates $\partial v_1/\partial x_2$ and $\partial v_3/\partial x_2$, that is,

$$\sigma_{12} = \eta\frac{\partial v_1}{\partial x_2}, \qquad \sigma_{32} = \eta\frac{\partial v_3}{\partial x_2} \tag{8-D-4}$$

Equation (8-D-3) may now be rewritten as

$$0 = -\frac{\partial p}{\partial x_1} + \frac{\partial}{\partial x_2}\eta\frac{\partial v_1}{\partial x_2}, \qquad 0 = -\frac{\partial p}{\partial x_3} + \frac{\partial}{\partial x_2}\eta\frac{\partial v_1}{\partial x_2} \tag{8-D-5}$$

One may also consider as lubrication flow cases where there is predominantly unidirectional flow v_1 in a channel whose cross-section dimensions H and W gradually change, that is,

$$\frac{\partial H}{\partial x_1} \ll 1, \qquad \frac{\partial W}{\partial x_1} \ll 1 \tag{8-D-6}$$

Shearing in both the 2 and 3 directions must be accounted for, but the tensile stress P_{11} varies slowly in the direction of flow. The equations of motion in the 2 and 3 direction may be neglected and that in the 1 direction becomes

$$0 = -\frac{\partial p}{\partial x_1} + \frac{\partial \sigma_{12}}{\partial x_2} + \frac{\partial \sigma_{13}}{\partial x_3} \tag{8-D-7}$$

The stresses σ_{12} and σ_{13} are simply related to the velocity field through expressions of the form of Eq. (8-D-4).

$$\sigma_{12} = \eta \frac{\partial v_1}{\partial x_2}, \qquad \sigma_{13} = \eta \frac{\partial v_1}{\partial x_3} \tag{8-D-8}$$

which give rise to

$$0 = -\frac{\partial p}{\partial x_1} + \frac{\partial}{\partial x_2}\left(\eta \frac{\partial v_1}{\partial x_2}\right) + \frac{\partial}{\partial x_3}\left(\eta \frac{\partial v_1}{\partial x_3}\right) \tag{8-D-9}$$

The range of application of lubrication theory is very broad. It clearly contains within its applications many, if not most, of the internal flows we have discussed. This includes flow within dies and molds where the walls are stationary and the moving melt is in a pressure-driven flow. We may also include in this formulation processing operations with moving machine members. This would include flow between screw and barrel in an extruder or possibly between calendar rolls. The applicability depends on the restrictions of Eqs. (8-D-1) and (8-D-6) and the magnitudes of the change of normal stresses in the direction of flow.

Let us consider the special case where the fluid is Newtonian and the shear viscosity η is a constant. It is then possible to easily solve Eq. (8-D-5). If we embed our coordinate system in the lower surface, the solutions of Eq. (8-D-5) for v_1 and v_3 are of the form

$$v_1 = U_1\left(\frac{x_2}{H}\right) - \frac{H^2}{3\eta}\left\{\left(\frac{x_2}{H}\right) - \left(\frac{x_2}{H}\right)^2\right\} \frac{\partial p}{\partial x_1} \tag{8-D-10a}$$

$$v_3 = U_3\left(\frac{x_2}{H}\right) - \frac{H^2}{3\eta}\left\{\left(\frac{x_2}{H}\right) - \left(\frac{x_2}{H}\right)^2\right\} \frac{\partial p}{\partial x_3} \tag{8-D-10b}$$

It is useful to determine the fluxes q_1 and q_3 per unit width. These are

$$q_1 = H\bar{v}_1 = \int_0^h v_1(x_2)\,dx_2 = \tfrac{1}{2}U_1 H - \frac{H^3}{12\eta}\frac{\partial p}{\partial x_1} \tag{8-D-11a}$$

$$q_3 = H\bar{v}_3 = \int_0^h v_3(x_2)\,dx_2 = \tfrac{1}{2}U_3 H - \frac{H^3}{12\eta}\frac{\partial p}{\partial x_3} \tag{8-D-11b}$$

Continuity of flow demands at a local level

$$\frac{\partial v_1}{\partial x_1} + \frac{\partial v_3}{\partial x_3} = 0 \tag{8-D-12a}$$

and through the cross section

$$\frac{\partial H\bar{v}_1}{\partial x_1} + \frac{\partial H\bar{v}_3}{\partial x_3} = \frac{\partial q_1}{\partial x_1} + \frac{\partial q_3}{\partial x_3} = 0 \tag{8-D-12b}$$

For the special case of a Newtonian fluid, Eq. (8-D-12b) is equivalent to

$$\frac{\partial}{\partial x_1}\left(H^3 \frac{\partial p}{\partial x_1}\right) + \frac{\partial}{\partial x_3}\left(H^3 \frac{\partial p}{\partial x_3}\right) = 6\eta\left[\frac{\partial(U_1 H)}{\partial x_1} + \frac{\partial(U_3 H)}{\partial x_3}\right] \quad (8\text{-D-}13)$$

This expression is known as the Reynolds equation. Its solution leads to the pressure field between the two surfaces. The mean velocities $\bar{v}_1$ and $\bar{v}_3$ may then be determined from Eq. (8-D-11).

There are many special cases of interest. The most important is when H is a constant and U_1 and U_3 are zero. We then obtain from Eq. (8-D-13)

$$\frac{\partial^2 p}{\partial x_1^2} + \frac{\partial^2 p}{\partial x_3^2} = 0 \quad (8\text{-D-}14)$$

This may be recognized as Laplace's equation. This has the same form as the basic equation of flow of nonviscous fluids and more generally potential theory (81). This classical area of mathematical physics may thus be used to determine pressure fields and velocity fields. This situation was first realized by Hele-Shaw (81) and is called Hele-Shaw flow.

It is possible to generalize much of this for non-Newtonian fluids, for example, those obeying the power law. This is especially the case when U_1 and U_3 are set equal to zero. Analytical generalizations of Eqs. (8-D13) and (8-D-14) are possible. However, resort to numerical procedures becomes much more necessary. This is our next topic.

The lubrication theory described in the paragraphs above may be used as the basis of numerical solutions. We shall only briefly discuss one procedure here, the simple powerful numerical technique of Tadmor et al. (140), which is known as the *flow analysis network* (FAN) method. This is based on dividing the region of interest into cells and carrying out flux balances on these cells using Eq. (8-D-11). We may set up a mesh numbered by indices (i, j) which may be used to represent cell center positions or nodes. The average plate separation at node (i, j) is $H(i, j)$. The distance between nodes is Δ. The plate-separation distribution is fed into the computer by reading in the matrix of $H(i, j)$ values. The flow regime does not extend over all nodes. The flow is confined by solid boundaries and has an entrance and an exit. It is therefore necessary to differentiative between these various kinds of nodes. This may be done on a computer by defining a fixed-point matrix $N(i, j)$ which is assigned certain values denoting the nature of a node. We distinguish (1) a field node set inside of the flow field filled with fluid, (2) gate nodes where the fluid enters the system, (3) exit nodes where the fluid exits the system, and (4) border nodes where there are walls and the fluid cannot flow.

The actual calculations are based on balances over a node. The flow rate between two nodes (i, j) and $(i + 1, j)$ is $q\Delta$. The balance principle is that the total outflow from all field nodes should be zero.

$$\Delta[q(i,j) - q(i,j+1)] + \Delta[q(i,j) - q(i,j-1)]$$
$$+ \Delta[q(i,j) - q(i+1,j)] + \Delta[q(i,j) - q(i-1,j)] = 0 \qquad (8\text{-}D\text{-}15)$$

For the special case of a Newtonian fluid in a system with stationary boundaries, the balance would take the form

$$Y(i,j)[p(i,j) - p(i,j+1)] + Y(i,j-1)[p(i,j) - p(i,j-1)]$$
$$+ X(i,j)[p(i,j) - p(i+1,j)] + X(i-1,j)[p(i,j) - p(i-1,j)] = 0 \quad (8\text{-}D\text{-}16)$$

where $X(i,j)$ and $Y(i,j)$ are flow conductivities in the 1 and 3 directions, respectively, which are defined by

$$X(i,j) = \frac{1}{12\eta}\left[\frac{H(i,j) + H(i+1,j)}{2}\right]^3, \qquad Y(i,j) = \frac{1}{12\eta}\left[\frac{H(i,j) + H(i,j+1)}{2}\right]^3$$
$$(8\text{-}D\text{-}17)$$

This approach has been used by Tadmor, Broyer, and Gutfinger (*22, 140, 141*) for dies and molds. More recently, Szydlowski et al. (*137*) have applied the technique to flow in an intermeshing corotating twin-screw extruder where there are moving boundaries. Brzoskowski et al. (*23*) have applied it to pin barrel extruders. Here one uses Eq. (8-D-11) for $q(i,j)$ in Eq. (8-D-15).

The discussion thus far has concentrated on Newtonian fluids. It is possible to extend this to non-Newtonian fluids. Aspects of this are discussed in Section 8-G-3, where U_1 and U_3 are zero. More general approaches to this problem for the case where U_3 is nonzero have appeared in recent years in analyses of corotating twin screw extruders (*138, 152*).

E. EXTRUSION THROUGH DIES

1. Introductory Remarks

Perhaps no class of polymer processing operations is more subject to rheological analysis than the flow of molten plastics through dies. Various aspects of the die-flow problem involve rheological analysis. One must be concerned with (1) the feeding of melt into a die to ensure uniformity of flow across the cross section, (2) the velocity field and pressure drop–extrusion rate characteristics within the die cross section, and (3) the relationship between extrudate shape and die cross section. We shall discuss each of these problems. We begin with the second problem because it actually forms the basis of understanding of each of the other two. It is also the simplest to analyze.

2. Flow-Through Conduits of Constant Cross Section

a. General Remarks

We now turn to the problem of flow through constant cross section conduits, that is, closed tubes of arbitrary cross sections that do not vary in the direction of flow.

The velocity field is of the form

$$\mathbf{v} = v_1(x_2, x_3)\mathbf{e}_1 + 0\mathbf{e}_2 + 0\mathbf{e}_3 \qquad (8\text{-E-1})$$

The equation of motion in the flow direction may be expressed [compare Eq. (8-D-7)] as

$$0 = -\frac{\partial p}{\partial x_1} + \frac{\partial \sigma_{12}}{\partial x_2} + \frac{\partial \sigma_{13}}{\partial x_3} \qquad (8\text{-E-2a})$$

where σ_{11} varies with x_1 only in the pressure term p. The equations of motion normal to the direction of flow are

$$0 = -\frac{\partial p}{\partial x_2} + \frac{\partial \sigma_{22}}{\partial x_2} + \frac{\partial \sigma_{23}}{\partial x_3} \qquad (8\text{-E-2b})$$

$$0 = -\frac{\partial p}{\partial x_3} + \frac{\partial \sigma_{33}}{\partial x_3} + \frac{\partial \sigma_{23}}{\partial x_2} \qquad (8\text{-E-2c})$$

b. Newtonian Fluids

It is necessary to understand the flow of Newtonian fluids (81) in various types of cross sections before we turn to viscoelastic fluids. The three-dimensional constitutive equation for a Newtonian fluid may be expressed

$$\boldsymbol{\sigma} = -p\mathbf{I} + 2\eta\mathbf{d} \qquad (8\text{-E-3})$$

For the velocity field of Eq. (8-E-1), it predicts the stress field

$$\begin{bmatrix} \sigma_{11} & \sigma_{12} & \sigma_{13} \\ \sigma_{12} & \sigma_{22} & \sigma_{23} \\ \sigma_{13} & \sigma_{23} & \sigma_{33} \end{bmatrix} = -\begin{bmatrix} p & 0 & 0 \\ 0 & p & 0 \\ 0 & 0 & p \end{bmatrix} + \begin{bmatrix} 0 & \eta(\partial v_1/\partial x_2) & \eta(\partial v_1/\partial x_3) \\ \eta(\partial v_1/\partial x_2) & 0 & 0 \\ \eta(\partial v_1/\partial x_3) & 0 & 0 \end{bmatrix}$$
$$(8\text{-E-4})$$

From Eq. (8-E-2a–c), we obtain

$$0 = \frac{\partial p}{\partial x_1} + \eta\left[\frac{\partial^2 v_1}{\partial x_2} + \frac{\partial^2 v_1}{\partial x_3^2}\right] \qquad (8\text{-E-5a})$$

$$0 = \frac{\partial p}{\partial x_2} \qquad (8\text{-E-5b})$$

$$0 = \frac{\partial p}{\partial x_3} \qquad (8\text{-E-5c})$$

The simplest cross section is a slit. Consider a slit of uniform thickness $2H$

and width W which is much greater than $2H$. We may neglect $\partial v_1/\partial x_3$ from Eq. (8-E-5a). The velocity field is found to be of parabolic character:

$$v_1(x_2) = U\left[1 - \left(\frac{x_2}{H}\right)^2\right] \tag{8-E-6a}$$

$$= \frac{H^2}{2\eta}\left(-\frac{dp}{dx_1}\right)\left[1 - \left(\frac{x_2}{H}\right)^2\right] \tag{8-E-6b}$$

where U or $H^2/2\eta(-dp/dx_1)$ is the maximum velocity. The relationship between extrusion rate and pressure gradient is given by

$$Q = 2W\int_0^H v_1\,dx_2 = \frac{2WH^3}{3\eta}\left(-\frac{dp}{dx_1}\right) \tag{8-E-7}$$

For a circular cross section, the problem is better formulated in cylindrical coordinates:

$$0 = \frac{\partial p}{\partial x_1} + \eta\frac{1}{r}\frac{\partial}{\partial r}\left(r\frac{\partial v_1}{\partial r}\right) \tag{8-E-8}$$

Consider a tube of radius R. The velocity field is again parabolic in character:

$$v_1(r) = \frac{R^2}{4\eta}\frac{dp}{dx_1}\left[1 - \left(\frac{r}{R}\right)^2\right] \tag{8-E-9}$$

and extrusion rate is related to pressure gradient through

$$Q = \int_0^R 2\pi r v_1\,dr = \frac{\pi R^4}{8\eta}\left(-\frac{dp}{dx_1}\right) \tag{8-E-10}$$

This result is known as the Hagen–Poiseuille equation.

For flow in an annulus with inner radius a and outer radius b, the equation of motion is again given by Eq. (8-E-8). The velocity profile in cylindrical coordinates is a modified parabola (81):

$$v_1(r) = \frac{a^2}{4\eta}\left(-\frac{dp}{dx_1}\right)\left[1 - \left(\frac{r}{a}\right)^2 + \frac{(b/a)^2 - 1}{\log(b/a)}\log\frac{r}{a}\right] \tag{8-E-11}$$

and the extrusion rate is given by

$$Q = \frac{\pi}{8\eta}\left(-\frac{dp}{dx_1}\right)\left[b^4 - a^4 - \frac{(b^2 - a^2)^2}{\log(b/a)}\right] \tag{8-E-12}$$

For an elliptical cross section of semiaxes a and b the velocity field proves to have the simple parabolic form (81):

$$v_1(x_2, x_3) = \frac{a^2 b^2}{a^2 + b^2} \frac{1}{2\eta} \left(-\frac{dp}{dx_1} \right) \left[1 - \left(\frac{x_2}{a} \right)^2 - \left(\frac{x_3}{b} \right)^2 \right] \qquad (8\text{-}E\text{-}13)$$

The extrusion rate is given by

$$Q = \frac{1}{4\eta} \frac{\pi a^2 b^2}{a^2 + b^2} \left(-\frac{dp}{dx_1} \right) \qquad (8\text{-}E\text{-}14)$$

c. Viscoelastic Fluids

The flow of a viscoelastic fluid through a constant cross-section channel is in general governed by the Criminale–Ericksen–Filbey theory, Eq. (7-F-64). This may be expressed as (33)

$$\boldsymbol{\sigma} = -p\mathbf{I} + \eta \mathbf{B}_1 - \tfrac{1}{2}\boldsymbol{\Psi}_1 \mathbf{B}_2 + \boldsymbol{\Psi}_2 \mathbf{B}_1^2 \qquad (8\text{-}E\text{-}15)$$

where η is the shear viscosity and $\boldsymbol{\Psi}_1$ and $\boldsymbol{\Psi}_2$ are normal stress coefficients. Note that $\mathbf{B}_1$ is (twice) the rate of the deformation tensor and $\mathbf{B}_2$ is an acceleration tensor defined by Eq. (7-F-54). The quantities η, $\boldsymbol{\Psi}_1$, and $\boldsymbol{\Psi}_2$ depend upon the second invariant of the rate of deformation tensor, that is, $\operatorname{tr} \mathbf{B}_1^2$.

The stress tensor in a viscoelastic fluid in a conduit is obtained by substituting Eq. (8-E-1) into Eq. (8-E-15). This yields

$$\begin{bmatrix} \sigma_{11} & \sigma_{12} & \sigma_{13} \\ \sigma_{12} & \sigma_{22} & \sigma_{23} \\ \sigma_{13} & \sigma_{23} & \sigma_{33} \end{bmatrix} = - \begin{bmatrix} p & 0 & 0 \\ 0 & p & 0 \\ 0 & 0 & p \end{bmatrix} + \begin{bmatrix} (\boldsymbol{\Psi}_1 + \boldsymbol{\Psi}_2)(\dot{\gamma}_2^2 + \dot{\gamma}_3^2) & \eta\dot{\gamma}_3 & \eta\dot{\gamma}_2 \\ \eta\dot{\gamma}_2 & \boldsymbol{\Psi}_2\dot{\gamma}_2^2 & \boldsymbol{\Psi}_2\dot{\gamma}_2\dot{\gamma}_3 \\ \eta\dot{\gamma}_3 & \boldsymbol{\Psi}_2\dot{\gamma}_2\dot{\gamma}_3 & \boldsymbol{\Psi}_2\dot{\gamma}_3^2 \end{bmatrix}$$

$$(8\text{-}E\text{-}16)$$

Here $\dot{\gamma}_2$ is $\partial v_1 / \partial x_2$ and $\dot{\gamma}_3$ is $\partial v_1 / \partial x_3$. The quantities η, $\boldsymbol{\Psi}_1$, and $\boldsymbol{\Psi}_2$ depend upon $\operatorname{tr} \mathbf{B}_1^2$, which is $\dot{\gamma}_2^2 + \dot{\gamma}_3^2$. Substitution of Eq. (8-E-16) into Eq. (8-E-2) yields

$$0 = -\frac{\partial p}{\partial x_1} + \frac{\partial}{\partial x_2} \eta \frac{\partial v_1}{\partial x_2} + \frac{\partial}{\partial x_3} \eta \frac{\partial v_1}{\partial x_3} \qquad (8\text{-}E\text{-}17a)$$

$$0 = -\frac{\partial p}{\partial x_2} + \frac{\partial}{\partial x_2} \boldsymbol{\Psi}_2 \left(\frac{\partial v_1}{\partial x_2} \right)^2 + \frac{\partial}{\partial x_3} \boldsymbol{\Psi}_2 \frac{\partial v_1}{\partial x_2} \frac{\partial v_1}{\partial x_3} \qquad (8\text{-}E\text{-}17b)$$

$$0 = -\frac{\partial p}{\partial x_3} + \frac{\partial}{\partial x_2} \boldsymbol{\Psi}_2 \frac{\partial v_1}{\partial x_2} \frac{\partial v_1}{\partial x_3} + \frac{\partial}{\partial x_3} \boldsymbol{\Psi}_2 \left(\frac{\partial v_1}{\partial x_3} \right)^2 \qquad (8\text{-}E\text{-}17c)$$

The primary velocity field is determined by Eq. (8-E-17a). For nonzero values of $\boldsymbol{\Psi}_2$, Eqs. (8-E-17b) and (8-E-17c) will lead to superposed secondary flows (41,

48, 83). These would seem to be of long period, compared with the length of dies, and of low amplitude. While they may have been found experimentally in polymer solutions (*41,48*), they have not been observed in melts. There is no reason not to believe that they would not occur in polymer melts under the same circumstances. It is only that their magnitude is small and the length of dies is too short to observe them.

We now consider flow in simple geometries of materials with specific non-Newtonian models. We begin with the power-law fluid model. In one dimension this is

$$\eta = K\dot{\gamma}^{n-1} \tag{8-E-18}$$

or, including velocity gradients in the 2 and 3 directions,

$$\eta = K\left[\left(\frac{\partial v_1}{\partial x_2}\right)^2 + \left(\frac{\partial v_1}{\partial x_3}\right)^2\right]^{(n-1)/2} \tag{8-E-19}$$

For flow in a slit, the velocity field equivalent to Eq. (8-E-6) has the form

$$\begin{aligned}
v_1(x_2) &= \frac{nH}{n+1}\left(\frac{H\Delta p}{KL}\right)^{1/n}\left[1 - \left(\frac{x_2}{H}\right)^{(n+1)/n}\right] \\
&= \frac{2n+1}{n+1}v\left[1 - \left(\frac{x_2}{H}\right)^{(n+1)/n}\right]
\end{aligned} \tag{8-E-20}$$

where v_1 is the average velocity. The extrusion rate is

$$Q = 2W\int_0^H v_1\,dx_2 = \frac{2n}{2n+1}WH^2\left(\frac{H\Delta p}{KL}\right)^{1/n} \tag{8-E-21}$$

We have written $(-\partial p/\partial x_1)$ as $\Delta p/L$.

The velocity profiles of Eq. (8-E-20) are plotted in Figure 8-E-1 using various power-law exponents. These are normalized to the same extrusion rate. It can be seen that the lower the power-law exponent, the flatter the velocity profile, but the higher the wall velocity gradient.

We now turn to flow through circular cross-section tubes. The velocity profile for a power-law fluid is

$$\begin{aligned}
v_1(r) &= \frac{n}{n+1}R\left(\frac{R\Delta p}{2KL}\right)^{1/n}\left[1 - \left(\frac{r}{R}\right)^{(n+1)/n}\right] \\
&= \frac{3n+1}{n+1}v\left[1 - \left(\frac{r}{R}\right)^{(n+1)/n}\right]
\end{aligned} \tag{8-E-22}$$

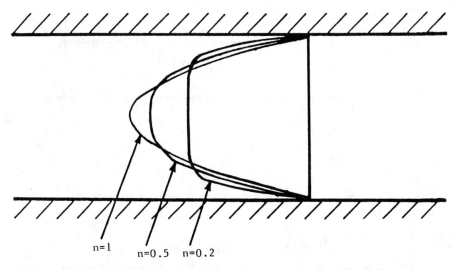

FIGURE 8-E-1. Velocity profiles of a power–law fluid in an slit.

The extrusion rate is

$$Q = \frac{n}{2(3n+1)} \pi R^3 \left(\frac{R\Delta p}{2KL} \right)^{1/n} \tag{8-E-23}$$

The velocity profiles are basically similar to those occurring in slits. Newtonian fluids where n is unity give rise to a parabolic profile, and as n decreases the velocity profile becomes flatter.

To obtain solutions for more complex cross sections, one must in general obtain numerical solutions. This is the case for annular, rectangular, and elliptical cross sections. For profiles that exhibit much greater shearing in one direction than a second, as the trapezoidal cross section of Figure 8-E-2, we may presume that the flow is locally like that in a slit. This may then be integrated throughout the entire cross section. Such a procedure was discussed by White and Huang (163). We may write

$$Q = 2 \int_0^W \int_0^{H(x_3)} v_1(x_2) \, dx_2 \, dx_3$$

$$= \frac{2n}{2n+1} \int_0^W H^2(x_3) \left[\frac{H(x_3)\Delta p}{KL} \right]^{1/n} dx_3 \tag{8-E-24}$$

As noted earlier, the flow of power-law fluids has been analyzed by various

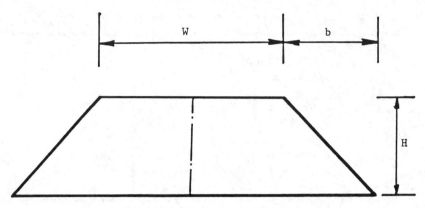

FIGURE 8-E-2. Trapezoidal die.

authors for different cross-sections. We shall summarize this here. Extrusion through an annulus was analyzed by Fredrickson and Bird (44) based on cylindrical coordinates, and was similar to that leading to Eqs. (8-E-22) and (8-E-23), except for the boundary condition that the velocity goes to zero on an inner circular circumference. The solution of this problem is not possible analytically and requires numerical computation. Rectangular ducts have been studied by Schechter (127), Wheeler and Wissler (155), Middleman et al. (91), and Rothemeyer (124). The scheme for rectangular ducts is based on substitution of Eq. (8-E-19) into Eq. (8-E-17a) to yield

$$
0 = \frac{\partial p}{\partial x_1} + K \left[\frac{\partial}{\partial x_2} \left[\left(\frac{\partial v_1}{\partial x_2} \right)^2 + \left(\frac{\partial v_1}{\partial x_3} \right)^2 \right]^{(n-1)/2} \frac{\partial v_1}{\partial x_2} \right.
$$
$$
\left. + \frac{\partial}{\partial x_3} \left[\left(\frac{\partial v_1}{\partial x_2} \right)^2 + \left(\frac{\partial v_1}{\partial x_3} \right)^2 \right]^{(n-1)/2} \frac{\partial v_1}{\partial x_3} \right] \tag{8-E-25}
$$

This equation is then solved numerically. Various schemes have been used (91, 124, 155). The results generally parallel analytical solutions for power-law fluids, for example, the flattened velocity profile and power-law pressure-drop flow-rate behavior. Typical computations are shown in Figure 8-E-3.

d. Plastic Fluids

There is a long history of analysis of the flow of plastic fluids through conduits. The simplest of models is that of the Bingham plastic where the shear stress σ is given by

$$
\sigma = Y + \eta_B \dot{\gamma} \tag{8-E-26}
$$

The earliest study was by Buckingham (24) in 1921. He showed that as no shear

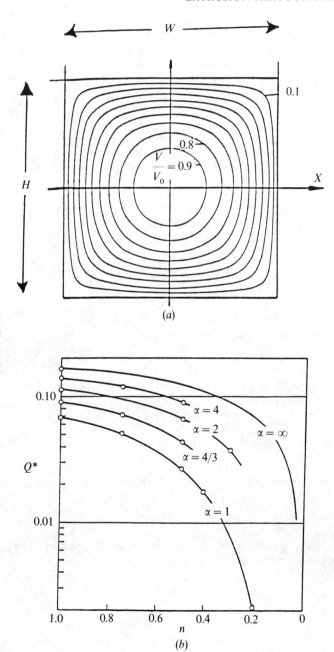

(a)

(b)

FIGURE 8-E-3. Velocity field and pressure drop of a power-law fluid in a rectangular duct. $(Q^* = Q/WH^2(2KL/H\Delta P))^{1/n}$; $\alpha = W/H$.

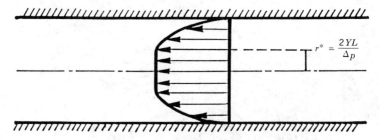

FIGURE 8-E-4. Velocity profile of a plastic fluid in a tube.

flow occurred below shear stress Y, there would be plug flow through the center of the cross-section and a shearing motion in the neighborhood of the walls where the stresses are higher (Fig. 8-E-4). The velocity field in a tube is

$$v_1(r) = \frac{1}{4\eta_B}\frac{\Delta p}{L}(R^2 - r^2) - Y(R - r) \qquad (8\text{-}E\text{-}27)$$

when

$$\frac{r\Delta p}{2L} > Y \qquad \text{or} \qquad r > \frac{2YL}{\Delta p} \qquad (8\text{-}E\text{-}28)$$

and plug flow is at lower radii. The extrusion rate is given by

$$Q = \frac{\pi R^4 \Delta p}{8\eta_B L}\left[1 - \frac{4}{3}\left(\frac{2LY}{R\Delta p}\right) + \frac{1}{3}\left(\frac{2LY}{R\Delta p}\right)^4\right] \qquad (8\text{-}E\text{-}29)$$

which is due to Buckingham (*24*).

It is possible to analyze a range of other geometries using the Bingham plastic fluid model. Various authors have considered pressure-induced flow of Bingham plastics in an annulus (*44, 80, 131*). Solid plugs occur near the center of the annular cross section.

Similar analyses are possible for plastic fluids that exhibit differentially nonlinear behavior at stresses above the yield value. These include

$$\sigma_{12} = Y + K\dot\gamma^n \qquad (8\text{-}E\text{-}30a)$$

$$\sigma_{12} = Y + \frac{A\dot\gamma}{1 + B\dot\gamma^{1-n}} \qquad (8\text{-}E\text{-}30b)$$

Equation (8-E-30a) of Herschel and Bulkley (*60*) allows analytical solutions for flow in tubes and slits. The characteristics of the solutions are similar to those of the Bingham plastic and involve solid plugs at the center of the cross section.

White et al. (*171*) present solutions for plastic fluids in a range of extrusion geometries, including elliptical and rectangular tubes. Solutions of the Herschel–Bulkley model are considered.

e. Influence of Viscous Heating

Flow of viscous fluids in tubes gives rise to heating effects due to dissipation of energy. This phenomenon was discussed in the 1930s by Hersey (61) and in the 1940s by Philippoff (114). It is, however, only with the work of Brinkman (21) and Bird (18, 19) that major attention was given to this problem. In general, we must simultaneously solve Eq. (8-E-17a) with an energy balance [Eq. (2-K-11)] which takes the form

$$\rho c v_1(x_2, x_3)\frac{\partial T}{\partial x_1} = k\left(\frac{\partial^2 T}{\partial x_2^2} + \frac{\partial^2 T}{\partial x_3^2}\right) + \eta\left[\left(\frac{\partial v_1}{\partial x_2}\right)^2 + \left(\frac{\partial v_1}{\partial x_3}\right)^2\right] \quad (8\text{-E-}31)$$

where secondary flows and axial heat conduction are neglected. Generally the temperature dependence of the viscosity must be considered. Toor (144) has argued that the decrease in density of the melt as it moves along the channel has a significant cooling effect which counterbalances viscous heating. Equation (8-E-31) should then be replaced by

$$\rho c v_1(x_2, x_3)\frac{\partial T}{\partial x_1} = k\left(\frac{\partial^2 T}{\partial x_2^2} + \frac{\partial^2 T}{\partial x_3^2}\right) + \eta\left[\left(\frac{\partial v_1}{\partial x_2}\right)^2 + \left(\frac{\partial v_1}{\partial x_3}\right)^2\right] + T\varepsilon v_1(x_2, x_3)\frac{dp}{dx_1}$$

$$(8\text{-E-}32)$$

where ε is the coefficient of volumetric expansion

$$\varepsilon = -\frac{1}{\rho}\left(\frac{\partial \rho}{\partial T}\right)_p \quad (8\text{-E-}33)$$

While in general numerical solutions are required to simultaneously solve the force balance with Eq. (8-E-31) or Eq. (8-E-32), it is possible to obtain some analytical asymptotes of interest. If heat conduction and expansion cooling are neglected, we have for one-dimensional shear

$$\rho c v_1\frac{\partial T}{\partial x_1} = \sigma_{12}\frac{dv_1}{dx_2} = x_2\frac{dv_1}{dx_2}\frac{dp}{dx_1} \quad (8\text{-E-}34)$$

Integrating across the cross section gives

$$\rho c Q\frac{dT}{dx_1} = \left[2W\int_0^H x_2\frac{dv_1}{dx_2}dx_2\right]\frac{dp}{dx_1} \quad (8\text{-E-}35)$$

and after integrating by parts,

$$\rho c Q\frac{dT}{dx_1} = Q\frac{dp}{dx_1} \quad (8\text{-E-}36)$$

or

$$\Delta T = \frac{\Delta p}{\rho c} \tag{8-E-37}$$

This gives a mean bulk temperature rise.

We will now discuss the simulation of the temperature profile across the cross section of the channel. One basic problem that arises is the nature of the boundary condition on the die walls. Should it be taken as isothermal, adiabatic, or something intermediate. For short residence times in the die, polymer melts that have low thermal conductivities would not be able to effectively conduct away heat. The heat-conduction terms in Eq. (8-E-31) mey be neglected and the equation directly integrated to give

$$T(x_1, x_2, x_3) - T_0 = \int_0^{x_1} \frac{\eta[(\partial v_1/\partial x_2)^2 + (\partial v_1/\partial x_3)^2]}{\rho c v_1(x_2, x_3)} dx_1 \tag{8-E-38}$$

If one substitutes in a reasonable velocity profile, such as for a power-law fluid in a slit channel, this becomes

$$T(x_1, x_2, x_3) - T_0 = \int_0^{x_1} KH^2 \frac{(H\Delta p/KL)^{2/n}(x_2/H)^{2/n}}{\rho c(nH/n+1)(H\Delta p/KL)^{1/n}[1 - (x_2/H)^{(n+1)/n}]} dx_1 \tag{8-E-39}$$

The temperature primarily increases in the region near the die wall with little or no change occurring in the center. Indeed, with expansion cooling the temperature is lowered in the center of the die.

We now turn to the general solution. Equations (8-E-31) and (8-E-32) may often be solved analytically using the method of Frobenius (expansions) (18, 21). The rheological properties of polymer melts, however, depend upon temperature. Thus temperature profiles influence velocity profiles. Under such conditions, it is necessary to simultaneously solve the set of equations

$$0 = -\frac{\partial p}{\partial x_1} + \frac{\partial}{\partial x_2}\left[\eta(II_d, T)\frac{\partial v_1}{\partial x_2}\right] + \frac{\partial}{\partial x_3}\left[\eta(II_d, T)\frac{\partial v_1}{\partial x_3}\right] \tag{8-E-40a}$$

$$\rho c v_1 \frac{\partial T}{\partial x_1} = k\left(\frac{\partial^2 T}{\partial x_2^2} + \frac{\partial^2 T}{\partial x_3^2}\right) + \eta(II_d, T)\left[\left(\frac{\partial v_1}{\partial x_2}\right)^2 + \left(\frac{\partial v_1}{\partial x_3}\right)^2\right] \tag{8-E-40b}$$

using a relationship for $\eta(II_d, T)$ such as

$$\eta(II_d, T) = A(II_d)e^{E/RT} \tag{8-E-40c}$$

Again, the problem is that we really do not know the wall boundary conditions. In general, one should probably use expressions of the form

$$-k\left(\frac{dT}{dr}\right)_w = h(T - T_s) \qquad (8\text{-}E\text{-}41)$$

where an adiabatic wall arises when the heat-transfer coefficient h is zero. Four dimensionless groups arise in the solution:

$$\frac{KH^2}{k\theta}\left(\frac{U}{H}\right)^{(n+1)/n} = \text{Brinkman number} = N_{Br} \qquad (8\text{-}E\text{-}42a)$$

$$\frac{HU\rho C}{k} = \text{Peclet number} = N_{Pe} \qquad (8\text{-}E\text{-}42b)$$

$$\frac{hL}{k} = \text{Nusselt number} = N_{Nu} \qquad (8\text{-}E\text{-}42c)$$

$$\frac{E}{RT_0} = \text{Griffith number} = N_{Gr} \qquad (8\text{-}E\text{-}42d)$$

The first of these relates to the ratio of viscous heating to conduction [Eq. (8-C-17)], the second to convection as opposed to conduction [Eq. (8-C-18)], the third the ratio of external heat transfer to heat conduction, and the fourth to the temperature dependence of the viscosity.

Typical numerical simulations are given by Gee and Lyon (47) Morrete and Gogos (102), and Cox and Macosko (32). Calculations even for isothermal walls lead to dog-eared temperature profiles, as suggested by Eq. (8-E-39). The larger the contribution of the convection term, the Peclet number, and the smaller and Nusselt number, the more dog-eared the temperature profiles become.

3. Converging Flow into a Capillary at a Die Entrance

a. Observations

We now consider the problem of converging flow from a reservoir into a die entrance. Such entries may involve a slowly or rapidly converging section, as shown in Figure 8-E-5. The former represents a much simpler problem to solve than the latter.

However, it is certainly best to begin with a critical review of experimental phenomena observed in die-entrance regions. Mooney and Black (101) and Bagley (5) first called attention to the occurrence of large extra-pressure losses in the extrusion of rubber compounds and molten plastics through dies. These are pressure losses beyond those expected from steady-state shearing along the die land. This was confirmed by various later investigators (8, 31, 167). It was clearly shown by Han and his coworkers (53, 54), using pressure transducers along the

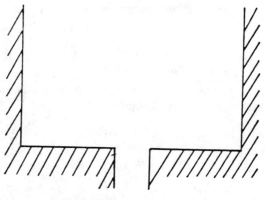

Die with 180° entrance angle

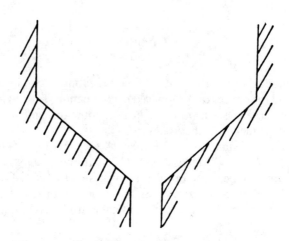

Converging entrance die

FIGURE 8-E-5. Die entrance regions.

length of a die, that there were both die-entrance and exit-pressure losses, but the former was much larger. Typical plots of these die-ends pressure losses Δp_{ends}

$$\Delta p_{\text{ends}} = \Delta p_{\text{ent}} + \Delta p_{\text{exit}} \qquad (8\text{-E-}43)$$

are shown in Figure 8-E-6 as a function of die-wall shear rate for different melts.

The pressure losses are frequently represented in terms of a multiple of the die-land shear stress, that is,

$$\Delta p_{\text{ends}} = m\sigma_{\text{w}} \qquad (8\text{-E-}44)$$

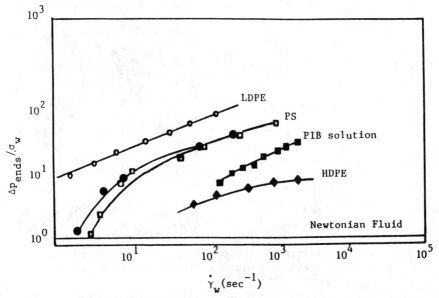

FIGURE 8-E-6. Ends pressure losses in a die entrance.

where m is called the Couette correction. It represents an effective additional number of diameters which is added to the die by this pressure loss. This can be shown by writing the total pressure loss as (compare Chapter 4)

$$\Delta p_{total} = \Delta p_{die} + \Delta p_{ends} = 4\sigma_w \frac{L}{D} + m\sigma_w \qquad \text{(8-E-45)}$$

$$= 4\sigma_w \frac{L + mD}{D} \qquad \text{(8-E-46)}$$

Values of m of order 1.2 are found for Newtonian fluids. Much larger values (i.e., up to 50) are found for molten plastics with 180° entrance angles.

Both the entrance- and exit-pressure losses have been related to melt elasticity. The arguments are largely associated with Philippoff and Gaskins (116), Han et al. (53,54), and Cogswell (31) and have been expanded upon by later investigators. Polymer melts containing large anisotropic particles such as chopped glass fibers exhibit even larger entrance-pressure losses (26). These are, however, associated with hydrodynamic particulate effects, as opposed to continuum rheological behavior.

These experiments on pressure losses induced Tordella (148), Clegg (29), Bagley, Birks, and Schreiber (6, 7), and others to make flow-visualization studies of polymer melts in the die-entrance region (see Figure 8-E-7 and Table 8-E-1). It was shown that LDPE exhibited large vortices in 180° entrances, which disappeared as the angle of convergence was increased (7). Similar vortices were

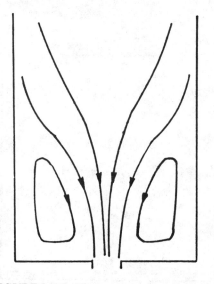

FIGURE 8-E-7. Flow patterns in a die entrance region for low-density polyethylene by Bagley and Birks (6). The right hand figure is published with permission of the American Institute of Physics.

TABLE 8-E-1   Investigations of Die Entrance Flows in Polymer Melts in 180° Entrance Angle Dies

Material	Observation	Investigator
Low-density polyethylene	Vortices in corners which grow with extrusion rate	Tordella (148) Clegg (29) Bagley and Birks (6) Ballenger and White (8) White and Kondo (167)
Polystyrene	Vortices in corners which grow with extrusion rate	Ballenger and White (8) White and Kondo (167)
High-density polyethylene	No vortices	Bagley and Birks (6) White and Kondo (167)
Polypropylene	No vortices	Ballenger and White (8) Cogswell (31)
Poly(vinyl chloride)	No vortices	Clegg (30)
Particle-filled (carbon black, TO$_2$ CaCO$_3$) low-density polyethylene	Vortices decrease in size with loading and are eventually eliminated	Ma et al. (86)

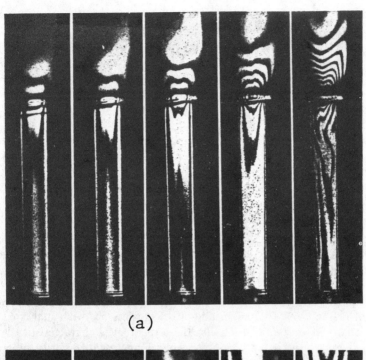

(a)

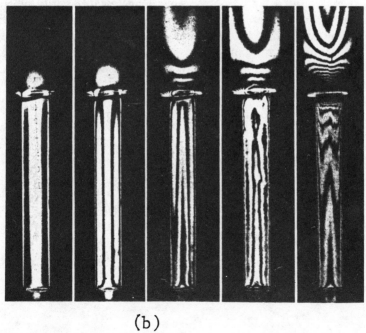

(b)

FIGURE 8-E-8. Birefringent fringe patterns for die-entrance flow after Tordella (*144*); (*a*) LDPE; (*b*) HDPE (with permission of John Wiley & Sons).

soon observed in various polymer solutions (20) and polystyrene (8), but not in HDPE (8,168), PVC (30), or PP (8,31). The vortices increase in size with increasing extrusion rate (8,167). Newtonian fluids have also been studied experimentally and they exhibit small corner vortices (20).

The influence of the addition of small particles such as carbon black, calcium carbonate, or titanium dioxide on flow patterns in LDPE was investigated by Ma et al. (86). These were found to reduce vortex size and with increasing loading to eliminate them.

Experimental studies of birefringence distributions in the entrance region of dies have dated to the 1960s. The key early investigations were by Tordella (149), who noted characteristic differences in the fringe patterns for LDPE and HDPE, with the LDPE exhibiting much larger stresses (Fig. 8-E-8).

b. Mechanisms and Modeling

The sizes of the vortices correspond with the die-entrance pressure losses (8). This would seem more of a correspondence of two aspects of a hydrodynamic phenomenon than a rheological explanation. Many researchers have, however, associated the magnitudes of Δp_{ent} with the viscoelastic properties of the melts and in particular with normal stresses (116,167) or elongational viscosity behavior (31,86,167). The magnitude of the vortices has been suggested to be thus similarly associated (8,168) with these rheological characteristics. As first suggested by Cogswell (31), an increasing elongational viscosity function is considered to create elongational stresses that disrupt the flow field. Thus, LDPE and PS exhibit vortices and HDPE and PP do not. Addition of small particles to LDPE induces a yield value and makes the elongational viscosity decreasing. Vortices are suppressed.

Quantitative measurements of flow birefringence and comparison with rheological theory have been made in the entrance region with polymer solutions by Bogue and his coworkers (1,42) and for polymer melts by Han et al. (52,55,176) and Arai et al. (2). These authors have computed stress fields in converging flow channels for planar radial flow (see discussion below). Second-order fluid and integral constitutive equations have been applied and the success is reasonably good.

There have been various efforts to analytically or numerically model flow in the entrance region and in particular to predict the large vortices and pressure losses observed. Many of these solutions are all based on upgrading earlier analyses of the same problem for Newtonian fluids, in particular Jeffrey (67) and Harrison's (58) analytical studies of radial converging flow in a cone, and continuity leads to

$$v_\theta = \frac{F(\theta)}{r^2}, \qquad v_\phi = v_r = 0 \tag{8-E-47}$$

The equations of motion for this type of flow are, in spherical coordinates,

$$0 = -\frac{\partial p}{\partial r} + \frac{\partial P_{rr}}{\partial r} + \frac{1}{r}\frac{\partial \sigma_{r\theta}}{\partial \theta} + \frac{\sigma_{r\theta}\cot\theta}{r} + \frac{2P_{rr} - P_{\theta\theta} - P_{\phi\phi}}{r}$$

$$0 = -\frac{1}{r}\frac{\partial p}{\partial \theta} + \frac{1}{r}\frac{\partial P_{\theta\theta}}{\partial \theta} + \frac{\partial \sigma_{r\theta}}{\partial r} + \frac{3\sigma_{r\theta}}{r} + \frac{P_{\theta\theta} - P_{\phi\phi}}{r}$$

(8-E-48a, b)

where all derivatives with respect to ϕ are neglected, as are shear stresses with ϕ components. Harrison found that one may satisfy Eq. (8-E-48) with the velocity field of Eq. (8-E-47) for a Newtonian fluid.

The radial flow predictions of Eq. (8-E-47) are not satisfactory for non-Newtonian and viscoelastic models. The stress components do not satisfy Eq. (8-E-48). Theoretical investigations of converging flow of viscoelastic fluids date to about 1960 (69, 82, 168). These generally involve perturbations about the solutions of Jeffrey and Harrison. We will briefly outline the approach. It is possible to achieve this only by using more complex velocity fields in place of Eq. (8-E-47). This suggests velocity fields of the form

$$v_\theta = \frac{F(\theta)}{r^2} + v'_\theta(r, \theta), \qquad v_r = v'_r(r, \theta) \tag{8-E-49}$$

where v'_θ and v'_r and small superposed velocities which generate secondary flows. Langlois and Rivlin (82) and later investigators have proceeded by using Eq. (8-E-49) in constitutive equations which have the form of Newtonian fluids plus viscoelastic terms multiplied by small parameters. If v'_θ and v'_r are of the order of the Weissenberg number,

$$v_\theta = \frac{F(\theta)}{r^2} + N_{\text{Ws}}f(\theta), \qquad v_r = N_{\text{Ws}}g(\theta) \tag{8-E-50}$$

we may obtain from perturbation arguments that

$$\begin{bmatrix} \text{Newtonian form of} \\ \text{equations of motion} \\ \text{with } v'_\theta \text{ and } v'_r \end{bmatrix} = N_{\text{Ws}} \begin{bmatrix} \text{total equations of} \\ \text{motion with Newtonian} \\ \text{terms} \end{bmatrix} \tag{8-E-51}$$

This procedure leads to a prediction of small superposed vortices. However, as pointed out by White and Kondo (167), modeling with a second-order fluid would predict them equally with the parameters of LDPE, which does exhibit them, and HDPE, which does not. This would appear to be because vortices of the magnitude that can be observed are associated with third- and fourth-order effects and not simply second-order effects.

There have been major efforts to use direct numerical solution techniques to analyze flow into a die entrance. A leading role in these efforts have been taken by Crochet and his coworkers (34, 35, 77) and others (65, 150) applying finite-element numerical methods. It appears that certain models are capable of predicting entrance vortices while others are not. The reasons for this are not as yet clear. Serious numerical stability/inaccuracy problems occur at high Weissenberg number, which increase the difficulty.

4. Flow Distribution Problem in Die Design

One of the basic problems of the design of dies is the distribution of polymer melts emerging from pumping devices such as a screw extruder into a die. This is a special problem in dies of the sort shown in Figure 8-E-9. These include a sheeting die with a high-aspect ratio and an annular die. The sheeting die is of the coathanger type, containing a manifold that distributes melt to the slit. The annular die usually contains a series of holes or a crossed channel. The proper design delivers a given polymer melt through the die opening at the same rate independent of position, that is, cross-machine uniformity. Our approach follows Pearson (110) and Tadmor and Gogos (141).

The coathanger die involves a tubular manifold of varying radius, $R(x_1)$, which delivers melt at constant rate (per unit cross-machine distance). If the die is of cross length $2W$, the total extrusion rate is $2qW$. Let ξ be the distance from the die center along the manifold. The pressure difference between the manifold entry at 0 and the entrance to the die land at ξ varies as:

$$p(0) - p(\xi) = \left(\frac{dp}{d\xi}\right)\xi \qquad (8\text{-E-}52)$$

We may also write this in terms of the pressure gradient dp/dx_1 on the die land,

$$p(0) - p(\xi) = p(0) - p(x_1) = \frac{dp}{dx_1}[L(0) - L(x_3)] \qquad (8\text{-E-}53)$$

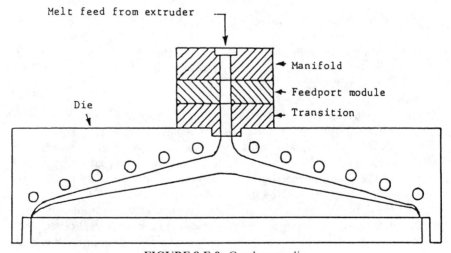

FIGURE 8-E-9. Coathanger die.

Proper die design ensures that the linear extrusion rate q given by

$$q = -\frac{dQ}{dx_3} = -\frac{dQ}{d\xi}\frac{d\xi}{dx_3} \qquad (8\text{-}E\text{-}54)$$

(where Q is the manifold extrusion rate) is independent of position (x_3) normal to flow. Thus

$$q[W - x_1] = Q(x_3) \qquad (8\text{-}E\text{-}55)$$

From Eqs. (8-E-21) and (8-E-23), we have for a power-law fluid:

$$(W - x_3)\frac{2n}{2n+1}H^2\left(\frac{H}{K}\frac{dp}{dx_1}\right)^{1/n} = \frac{n}{2(3n+1)}\eta R^3\left(\frac{R}{2K}\frac{dP}{d\xi}\right)^{1/n} \qquad (8\text{-}E\text{-}56)$$

Clearly,

$$\frac{dp}{d\xi} = \frac{dp}{dx_1}\left(\frac{dL}{d\xi}\right) \qquad (8\text{-}E\text{-}57)$$

which allows us to eliminate the pressure gradient. It follows that the manifold radius $R(x_3)$ and shape $d\xi/dL$ are related by

$$R^{(3n+1)/n}\left(\frac{dL}{d\xi}\right)^{1/n} = \frac{2(3n+1)}{(2n+1)}(W - x_3)H^{(2n+1)/n} \qquad (8\text{-}E\text{-}58)$$

Calculations with Eq. (8-E-58) are straightforward. If $dL/d\xi$ is fixed or specified, we may compute $R(x_3)$.

Manufacturing a die with a variable $R(x_1)$ may be very difficult and alternative solutions exist, such as having a curved manifold which would at least require an approximately constant $R(x_1)$.

5. Extrudate Swell

Polymer melts emerging from dies generally swell to dimensions different from the cross section of the die. In the case of simple die cross sections, such as a capillary or slit, this amounts to a simple swelling. For more complex die cross sections, the results may be a distortion of shape. Most papers in the literature deal with swell of melts from circular cross section or capillary dies and few papers deal with other cross sections (78, 163). Generally, the swell B (or d/D) of melts emerging from simple cross-section dies decreases with die length and increases with extrusion rate (Fig. 8-E-10).

The swell level appears from the view of dimensional analysis to depend upon the Weissenberg number, die geometry, and the residence time, perhaps in the

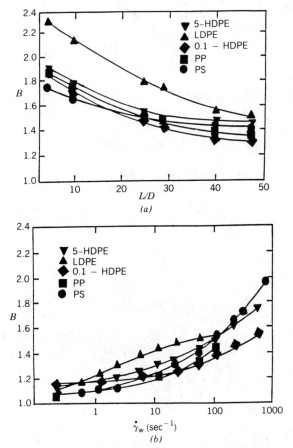

FIGURE 8-E-10. Extrudate swell $B(d/D)$ as a function of length–diameter ratio, L/D, and die-wall shear rate, $\dot{\gamma}_w$: (a) length–diameter ratio effects; (b) die-wall shear rate effects.

form of a Deborah number, that is,

$$B = B\left[\tau\frac{U}{L}, \frac{\tau}{t_{res}}, \text{entrance geometry}\right] \qquad (8\text{-}E\text{-}59)$$

where L is a characteristic length, usually the die diameter or slit thickness.

There is a long history of theoretical analyses of extrudate swell $(27, 36, 62, 63, 103, 105, 119, 142, 163, 169)$ which associates swell with elastic recovery of the melt as it exits the dies. Methods of formulating this vary considerably. Early papers usually interpreted extrudate swell in terms of elastic recovery following a fully developed Poiseuille flow. The details of the theories varied, but had this thread running through them. They are based on the idea that

$$B \sim \text{elastic recovery, } S \qquad (8\text{-}E\text{-}60a)$$

and from the theory of viscoelastic fluids [compare Eqs. (8-C-4) and (6-F-62)]

$$B \sim S \sim J_e \sigma_{12} \sim \frac{N_1}{\sigma_{12}} \tag{8-E-60b}$$

The most sophisticated of theories of this type was that formulated by Tanner (142) and developed by others (62, 63, 119, 163), which considers the swell of an integral fluid model with a Maxwell relaxation time as it emerges from flow in a slit or capillary die. The stress history is considered to consist of (1) a period of Poiseuille flow in the die, (2) a period of instantaneous recovery, and (3) a period of rest. The recovery may be expressed as

$$x_1(t) = \alpha_1 x_1(t_R) - g$$

$$x(t) = \alpha_2 x_2(t_R) \tag{8-E-61a}$$

$$x_3(t) = x_3(t_R)$$

where $\alpha_j(t)$ represents the position in the extrudate and $x_j(t_R)$, the prior positions in the die before recovery, 1 is the direction of flow, 2 again the direction of shear, and α_2 is related to swell through

$$B = \int_0^1 \alpha_2 \, dx_2 \tag{8-E-61b}$$

For flow in slit and capillary dies α_1 and α_2 are interrelated by

$$\alpha_1 = \frac{1}{\alpha_2}\text{(slit)}, \qquad \alpha_1 = \frac{1}{\alpha_2^2}\text{(capillary)} \tag{8-E-62}$$

We refer the reader to the original papers. One obtains from this theory that
capillary

$$B = [1 + \tfrac{1}{2}\tau_{\text{eff}}^2 \dot\gamma_w^2]^{1/6} = \left[1 + \tfrac{1}{2}\left[\frac{\sigma_w}{G}\right]^2\right]^{1/6} \tag{8-E-63a}$$

slit

$$B = [1 + \tfrac{1}{3}\tau_{\text{eff}}^2 \dot\gamma_w^2]^{1/4} = \left[1 + \tfrac{1}{3}\left[\frac{\sigma_w}{G}\right]^2\right]^{1/4} \tag{8-E-63b}$$

where $\tau_{\text{eff}}\dot\gamma_w$ should be recognized as a Weissenberg number.

A number of variations on this theory have been proposed. White and Roman (139) consider the influence of drawdon tension. The integral of σ_t acrossme cross section is now F. Integration of Eq. (8-E-63) for capillaries and slits leads to

Eq. (8-E-64a)

capillary

$$B^6 + \frac{4F}{\pi D^2 G} B^2 = 1 + \tfrac{1}{2}\tau_{eff}^2 \dot\gamma_w^2 = B^6 \ (0) \qquad (8\text{-E-64a})$$

slit

$$B^4 + \frac{F}{HWG} B^2 = 1 + \tfrac{1}{3}\tau_{eff}^2 \dot\gamma_w^2 = B^4 \ (0) \qquad (8\text{-E-64b})$$

increasing F, decreasing B.

White and Huang (*163*) have considered swell from dies of more complex cross-sections and present results for trapezoidal and rectangular dies. One presumes that Eq. (8-E-63a) holds locally at individual narrow-channel cross section. Huang and White (*63*) have also analyzed flow from short dies and in the particular case where $L \to 0$. Elongational flow into the die entrance is considered rather than shearing deformations.

This class of analyses have the problem that they predict no swell when the memory goes to zero. Newtonian fluids exhibit about a 10% swell at low Reynolds number, which decreases to a 12% contraction at high values (*92*). In order to correlate data on melts using Eq. (8-E-63) one must empirically add 0.1 for the capillary case and 0.2 for the slit (Fig. 8-E-11). The reason for this, as pointed out by Tanner and his coworkers (*105*), is that the velocity profile must rearrange within the die before the exit. Agreement with experiment is only good for relatively narrow molecular-weight distribution melts, which presumably may be reasonably represented by Maxwellian behavior.

Tanner and his coworkers (*105, 142*) and others (*27, 36*) have used finite-element numerical techniques to solve the problem of velocity rearrangements at the exit of the die and their implications on swell. They are able to predict the 10% expansion observed for Newtonian fluids and the qualitative characteristics of polymer melts.

The extrudate swell of highly-filled thermoplastics tends to be much lower than pure melts (*26, 85*). For materials with yield values the Bingham number would enter into correlations (*164*).

6. Extrudate Distortion

Distorted extrudates (Fig. 8-E-12) were first described by Nason (*104*) in the mid-1940s and then the later in more detail by Spencer and Dillon (*133*), who were the first to obtain an experimental ceriterion for their occurrence. They found that polystyrene exhibited helical extrudates at critical die-wall shear stress $(\sigma_{12})_w$ of order 100,000 pascals, which seemed to decrease inversely with increasing molecular weight. More extensive studies were reported by Tordella (*146–149*) in the period 1956–1963. These showed that extrudate distortion occurs in a wide

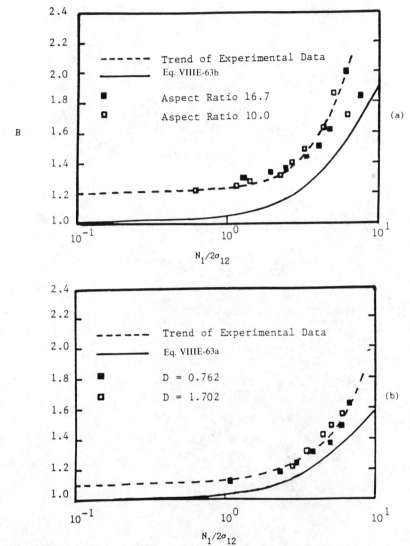

FIGURE 8-E-11. Extrudate swell as a function of $N_1/2\sigma_{12}$ and comparison to theory: (a) slit; (b) capillary.

range of polymer melts at critical shear stresses of order 100,000 pascals, independent of the diameter. The critical stress criterion is of course equivalent to a Weissenberg number, $G \cdot \tau U/L$ [see Eq. (8-C-4)].

Two classes of distorted extrudates have been observed. These are helical screw-like extrudates with varying pitch and amplitude, and gross distortions.

Tordella founds that the origins of this instability in LDPE was in the die-entrance region (148). In HDPE the origin of the instability was found to be

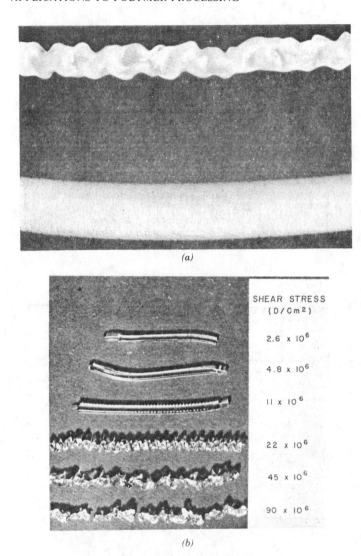

(a)

(b)

FIGURE 8-E-12. Extrudate distortion: (a) LDPE; (b) PMMA.

within the die, apparently associated with slippage at the die wall (*149*). Later studies of polystyrene and polypropylene by Ballenger and White (*8*) have indicated that the origin of the distorted extrudates are a flow instability at the die entrance. Spiralling motions are observed which resut in helically distorted extrudates. Gross distortion occurs at higher extrusion rates. There have been many theories of the mechanism of this breakdown of flow. They are ranged from "fracture" of melts (*196–198*), to wall slippage phenomena (*113*), to hydrodynamic instabilities associated with viscoelasticity (*37, 156*). The mechanism may vary from material to material. A recent paper by Kalika and Denn (*68*)

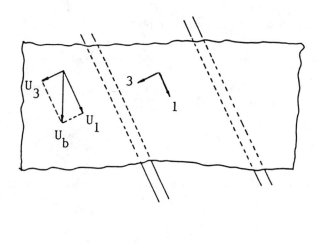

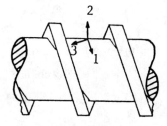

FIGURE 8-F-1. Extruder screw with coordinate system embedded in screw channel.

associates Pearson and Petrie's wall slippage mechanism (*113*) with the breakdown of melts such as LLDPE which exhibit large pressure fluctuations associated with extrudate distortion.

F. FLOW IN EXTRUDER SCREWS

Flow in a screw extruder represents one of the most important problems in polymer processing. The problem generally consists of three parts: (1) melt metering near the die, (2) melting of a particulate bed, and (3) conveying the particulate bed from the hopper to the inception of melting. In a book of this nature, we logically only consider the first of these, that is, melt metering near the die. We proceed by embedding our coordinate system in the screw root, then flattening out and unwinding the screw flights. This is the same procedure used by earlier investigators (*23, 26, 88, 111, 118, 141, 170*). The melt is dragged forward by the barrel and resisted by the screw (see Fig. 8-F-1). The velocity field may be expressed

$$\mathbf{v} = v_1(x_2, x_3)\mathbf{e}_1 + v_2(x_2, x_3)\mathbf{e}_2 + v_3(x_2, x_3)\mathbf{e}_3 \qquad (8\text{-F-}1)$$

To a good approximation, we may simplify this to [compare Eq. (8-D-2)]:

$$\mathbf{v} = v_1(x_2)\mathbf{e}_1 + 0\mathbf{e}_2 + v_3(x_2)\mathbf{e}_3 \qquad \text{(8-F-2)}$$

which is the useful approximation of Carley et al. (23). The equations of motion describing this flow are [compare Eq. (8-D-3)]:

$$0 = -\frac{\partial p}{\partial x_1} + \frac{\partial \sigma_{12}}{\partial x_2}, \qquad 0 = -\frac{\partial p}{\partial x_3} + \frac{\partial \sigma_{32}}{\partial x_2} \qquad \text{(8-F-3a, b)}$$

If we consider the flow to be Newtonian, that is, we use Eq. (8-E-3), we obtain

$$0 = -\frac{\partial p}{\partial x_1} + \eta\frac{\partial^2 v_1}{\partial x_2^2}, \qquad 0 = -\frac{\partial p}{\partial x_3} + \eta\frac{\partial^2 v_3}{\partial x_2^2} \qquad \text{(8-F-4a, b)}$$

with boundary conditions of

$$v_1(0) = v_3(0) = 0$$
$$v_1(H) = U_1 = \pi DN \cos \theta \qquad \text{(8-F-5)}$$
$$v_3(H) = U_3 = -\pi DN \sin \theta$$

where D is screw diameter, N the screw speed, and θ the screw helix angle. Equations (8-F-4) have the solution [compare Eq. (8-D-10)]:

$$v_1(x_2) = U_1\left(\frac{x_2}{H}\right) - \frac{H^2}{2\eta}\left(-\frac{\partial p}{\partial x_1}\right)\left[\left(\frac{x_2}{H}\right) - \left(\frac{x_2}{H}\right)^2\right] \qquad \text{(8-F-6a)}$$

$$v_3(x_2) = U_3\left(\frac{x_2}{H}\right) - \frac{H^2}{2\eta}\left(-\frac{\partial p}{\partial x_3}\right)\left[\left(\frac{x_2}{H}\right) - \left(\frac{x_2}{H}\right)^2\right] \qquad \text{(8-F-6a)}$$

If one neglects leakage over screw flights, the $v_3(x_2)$ component must satisfy the restriction

$$\int_0^H v_3(x_2)\,dx_2 = 0 \qquad \text{(8-F-7)}$$

This specifies $(-\partial p/\partial x_3)$ and leads to

$$v_3(x_2) = U_3\left[2\left(\frac{x_2}{H}\right) - 3\left(\frac{x_2}{H}\right)^2\right] \qquad \text{(8-F-8)}$$

This represents a superposed circulation on the down-channel flow. The throughput of the extruder screw on the other hand is only

$$Q = W \int_0^H v_1(x_2)\, dx_2 \qquad \text{(8-F-9)}$$

where W is the channel width. This leads to [compare Eq. (8-D-11)]:

$$Q = \frac{HWU_1}{2} - \frac{H^3 W}{12\eta} \frac{\partial p}{\partial x_1} \qquad \text{(8-F-10)}$$

This formulation as described above is for a Newtonian fluid. If we extend this to a viscoelastic fluid but neglect normal stresses, we retain Eqs. (8-F-2), (8-F-3), and (8-F-5). The equations of motion become

$$0 = -\frac{\partial p}{\partial x_1} + \frac{\partial}{\partial x_2}\eta \frac{\partial v_1}{\partial x_2}, \qquad 0 = -\frac{\partial p}{\partial x_3} + \frac{\partial}{\partial x_2}\eta \frac{\partial v_3}{\partial x_2} \qquad \text{(8-F-11a, b)}$$

For a power law fluid the shear viscosity has the form

$$\eta = K\left[\left(\frac{\partial v_1}{\partial x_2}\right)^2 + \left(\frac{\partial v_3}{\partial x_2}\right)^2\right]^{(n-1)/2} \qquad \text{(8-F-12)}$$

leading to

$$0 = -\frac{\partial p}{\partial x_1} + \frac{\partial}{\partial x_2} K\left[\left(\frac{\partial v_1}{\partial x_2}\right)^2 + \left(\frac{\partial v_3}{\partial x_2}\right)^2\right]^{(n-1)/2} \frac{\partial v_1}{\partial x_2}$$

$$0 = -\frac{\partial p}{\partial x_3} + \frac{\partial}{\partial x_2} K\left[\left(\frac{\partial v_1}{\partial x_2}\right)^2 + \left(\frac{\partial v_3}{\partial x_2}\right)^2\right]^{(n-1)/2} \frac{\partial v_3}{\partial x_2} \qquad \text{(8-F-13)}$$

[which is at least superficially similar to Eq. (8-E-25)]. The formulation of Eq. (8-F-13) is due to Griffith (50). We may simplify this to a first approximation by noting

$$\frac{\partial v_1}{\partial x_2} \gg \frac{\partial v_3}{\partial x_2} \qquad \text{(8-F-14)}$$

This leads to

$$0 = -\frac{\partial p}{\partial x_1} + \frac{\partial}{\partial x_2} K\left(\frac{\partial v_1}{\partial x_2}\right)^n \qquad \text{(8-F-15)}$$

with the boundary conditions of Eq. (8-F-5). Equation (8-F-15) has the solution

$$v_1(x_2) = \frac{n/(n+1)}{K^{1/n}(\partial p/\partial x_1)} \left\{\left[\left(\frac{\partial p}{\partial x_1}\right)x_2 + C_1\right]^{(n+1)/n} - C_1^{(n+1)/n}\right\} \qquad \text{(8-F-16a)}$$

with C_1 determined by

$$U_1 = \frac{n}{n+1} \frac{1}{K^{1/n}} \frac{1}{(\partial p/\partial x_1)} \left[\left(H\frac{\partial p}{\partial x_1} + C_1 \right)^{(n+1)/n} - C_1^{(n+1)/n} \right] \quad \text{(8-F-16b)}$$

for the drag flow dominated case. The throughput Q is

$$Q = W \int_0^H v_1 \, dx_2 = \frac{n}{n+1} \frac{W}{K^{1/n}(\partial p/\partial x_1)}$$

$$\times \left[\frac{[H(\partial p/\partial x_1) + C_1]^{(2n+1)/n} - C_1^{(2n+1)/n}}{(\partial p/\partial x_1)((2n+1)/n)} - C_1^{(n+1)/n} H \right] \quad \text{(8-F-17)}$$

We plot a dimensionless Q versus a dimensionless pressure gradient in Figure (8-F-2). The non-Newtonian viscosity is seen to produce nonlinear $Q - \Delta p$ screw characteristic plots. The predictions are quite different from any attempt at linearly adding forward drag flow and backward pressure flow as was sometimes suggested in the early literature. Screw pumping capability is reduced.

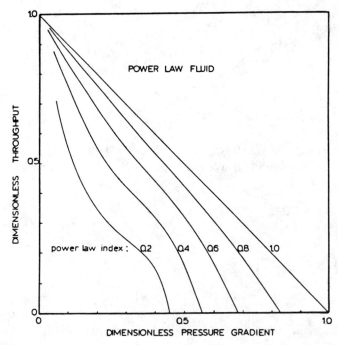

FIGURE 8-F-2. Dimensionless throughput $(2Q/U_1 WH)$ versus pressure gradient $(H^{n+1}\Delta p/6KU_1^n L)$ showing the effect of the power law exponent on flow in an extruder screw channel.

The full solution of Eqs. (8-F-13) coupled with the energy equation in the form

$$0 = k\frac{\partial^2 T}{\partial x_2^2} + \eta\left[\left(\frac{\partial v_1}{\partial x_2}\right)^2 + \left(\frac{\partial v_3}{\partial x_2}\right)^2\right] \qquad (8\text{-}F\text{-}18)$$

is described by Griffith (50). However, this formulation neglects the obviously important contribution of thermal convection. Griffith (50) also included the effect of temperature dependence of the shear viscosity. Under a wide range of conditions the Griffith solution gives similar results to Eq. (8-F-13).

It is to be emphasized that these solutions do not include the solid-conveying and melting regions of a screw extruder. They apply only to the metering region of a plasticating extruder.

Analyses of flow in complex extrusion machines generally use hydrodynamic lubrication theory. The FAN method was applied to pin barrel extruders by Brzoskowski et al. (23) and to intermeshing corotating twin screw extruders by Szydlowski et al. (137). The discussion of flow in complex extruders including multiple screw extruders is, however, beyond our scope. It may only be noted that intermeshing corotating machines and nonintermeshing counter-rotating machines resemble single screw extruders in having open screw channels. Intermeshing counter-rotating twin screw machines have closed screw flights and are rotary positive displacement pumps. Technology and flow analyses are summarized by White et al. (170).

G. INJECTION MOLDING

1. Regimes of Mold Filling

In this section we turn our attention to the filling of cold molds with molten plastics. This has been an area of active interest among rheologists since the 1950s (136, 137). Much attention has been directed towards the detailed manners in which molten plastics fill molds. As may be seen in Figure 8-G-1, there are two major regimes of flow of melts into molds. These are the movement of a simple front through the mold and jetting. The jetting of a melt into a mold is undesirable as it leaves surface blemishes on molded parts. While the jetting regime was observed by many early investigators (135, 161, 173), the literature associated with it was not clear as to the mechanisms determining the occurrence of each of the regimes.

It was only in the work of Oda et al. (107) in 1976 that the jetting phenomenon came to be understood. Jetting was found to be associated with low values of extrudate swell (Fig. 8-G-2). In isothermal injection molding processes, with hot runners and molds, jetting was associated with low injection rates. In injection through cold-runner systems, jetting is found at high injection rates. The explanation for the difference is that cold runners lower melt temperatures to greater extents when the residence time is higher or injection rate lower. Lowering melt temperatures increases viscosity and melt elasticity, or more specifically,

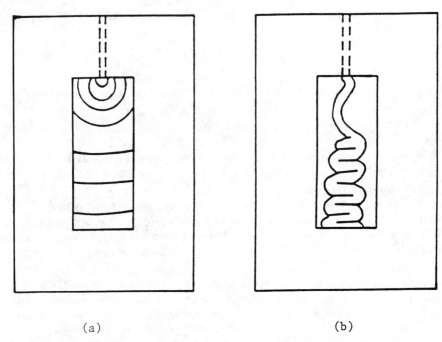

(a)

(b)

FIGURE 8-G-1. Regimes of mold filling: (*a*) simple uniform molding filling; (*b*) jetting.

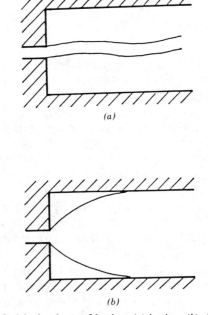

(a)

(b)

FIGURE 8-G-2. Mechanisms of jetting: (*a*) jetting; (*b*) simple flow front.

shear stresses, resulting in higher extrudate swell at the same injection rate. The viscosity increase from higher residence times in the runner will generally outweigh the extrusion rate increase in its effect on shear stresses and pressure gradients. Thus, with a cold-runner system, die-wall shear stresses increase with decreasing extrusion rate. It should not be unexpected then that extrudate swell decreases with increasing extrusion rate and that jetting is a high-extrusion rate phenomenon.

Filled polymer melts exhibit very low levels of extrudate swell (26, 85). They tend to exhibit severe jetting for this reason.

2. Injection Pressure to Fill Molds—Simple View

It is our purpose here to present a broad view of the problem of filling a cold mold with a molten plastic. If one injects melt into a hot mold, the injection pressure will increase with injection rate according to rheological relationships similar to those applying to extrusion. However, as one injects a polymer melt into a cold mold, a solid layer build up along the mold wall. The slower the injection process the more time solidification has to proceed (see Fig. 8-G-3). The available portion of the mold in which melt may flow decreases as the injection rate is lowered.

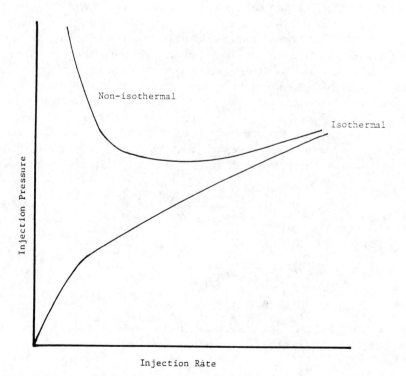

FIGURE 8-G-3. Injection pressure as a function of injection rate into a mold.

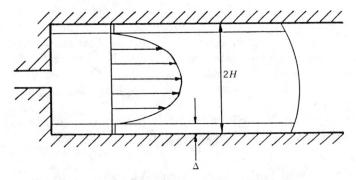

FIGURE 8-G-4. Barrie's model of injection molding.

There will be an injection rate so slow that this remaining channel will disappear. As the channel narrows the injection pressure will increase, and when it goes to zero the pressure will become infinite. This suggests a roughly hyperbolic injection rate–injection pressure relationship, as shown in Figure 8-G-3. We plot the isothermal relation for a hot mold for comparison purposes. At high injection rates, where the melt has little time to solidify, the curves come asymptotically together.

It is possible, following Barrie (*11*) [see also Dietz and White (*39, 162*)], to develop a simple one-dimensional view of the mold-filling process and the interaction of heat transfer and rheological considerations. The temperature profile in the melt filling the mold is considered to be flat near the centerline, and at a distance from the wall to drop suddenly to the mold wall temperature. This suggests that the system can be approximated as isothermal flow of a melt through a channel narrowed by an amount Δ which increases with time and varies with position (Fig. 8-G-4). We may, within the spirit of hydrodynamic lubrication theory, use the analysis of Eqs. (8-G-18) to (8-G-21) for a power-law fluid. If $2H$ is the thickness of the mold channel, the melt may be considered to move through a channel of thickness $2(H - \Delta)$. Thus, the average injection flux is

$$q = 2HV = \frac{n}{n+1} \frac{(H - \Delta)^{2n+1}}{K^{1/n}} \left[\frac{dp}{dx_1} \right]^{1/n} \qquad (8\text{-}G\text{-}1)$$

The injection pressure is then

$$\Delta p = \left[\frac{2n + 1}{n} \right]^n k q^n \int_0^L \frac{dx_1}{(H - \Delta)^{(2n + 1)/n}} \qquad (8\text{-}G\text{-}2)$$

The quantity Δ must be considered to increase with time. Barrie experimentally found that Δ was independent of position, but increases with the third power of time, that is,

$$\Delta(x_1, t) = C t^{1/3} \qquad (8\text{-}G\text{-}3)$$

The third power has been explained in terms of the characteristic of solutions of the heat convection equation (*39, 66, 162*). The heat convection equation in the absence of viscous dissipation is of the form

$$\rho c v_1 \frac{\partial T}{\partial x_1} = k \frac{\partial^2 T}{\partial x_2^2}$$

(8-G-4)

which in the neighborhood of the solid layer is

$$\rho c \dot{y} x_2' \frac{\partial T}{\partial x_1} = k \frac{\partial^2 T}{\partial x_2'^2}$$

(8-G-5)

where the origin of the coordinate system has been shifted to the solid layer surface ($x_2' = x_2 - \Delta$) and $\dot{y}$ is the local melt shear rate. This has the solution

$$\frac{T_0 - T}{T_0 - T_s} = F\left[\frac{x_2'^3 \dot{y}_w \rho c}{x_1 k}\right] = \frac{\int_u^\infty e^{-\mu^3/9}\, d\mu}{\int_u^\infty e^{-\mu^3/9}\, d\mu}$$

(8-G-6)

where μ is $x_2'^3 \dot{y}_w \rho c / x_1 k$, T_0 the initial temperature, and T_s the solid-layer surface temperature. This indicates that isotherms including the solidification temperature will vary according to

$$x_2' \sim x_1^{1/3}$$

(8-G-7)

The third power of dependence of x_2' on time is implicit.

Equations (8-G-2) and (8-G-3) are capable of semiquantitatively explaining the details of mold filling and the injection rate–injection pressure relationship of Fig. 8-G-3.

3. Quantitative Analysis of Mold Filling

In this section, we consider the simple mold-filling regime in greater detail. We conceive the flow to be in the 1–3 plane and to be sheared in the 2 direction. To analyze this problem we must simultaneously solve force and heat balance equations of the form [Compare Eq. (8-D-5)]:

$$0 = -\frac{\partial p}{\partial x_1} + \frac{\partial}{\partial x_2}\eta \frac{\partial v_1}{\partial x_2}, \qquad 0 = -\frac{\partial p}{\partial x_3} + \frac{\partial}{\partial x_2}\eta \frac{\partial v_3}{\partial x_2}$$

(8-G-8a, b)

$$\rho c\left[\frac{\partial T}{\partial t} + v_1 \frac{\partial T}{\partial x_1} + v_3 \frac{\partial T}{\partial x_3}\right] = k \frac{\partial^2 T}{\partial x_2^2} + \eta\left[\left(\frac{\partial v_1}{\partial x_2}\right)^2 + \left(\frac{\partial v_3}{\partial x_2}\right)^2\right]$$

(8-G-8c)

where heats of crystallization are neglected.

This formulation has been applied by many authors. The earliest analyses of injection mold filling were by Spencer and Gilmore (*135, 136*) and Ballman et al. (*9, 10, 145*). Since the late 1960s there has been greatly increased attention to such analysis (*15, 22, 59, 64, 69, 79, 80, 123, 138, 151, 154, 157, 161*). Hydrodynamic lubrication theory was first specifically used to analyze mold filling by Richardson (*123*). It was later explored by White (*157*). Numerical simulations involving sophisticated non-Newtonian lubrication theory began with Broyer et al. (*22*) who applied the FAN method.

To begin we refer the reader to the development of hydrodynamic lubrication theory in Section 8-D-2. The velocity field may be expressed from Eq. (8-G-8c)

$$v_1 = \int_H^{x_2} \frac{1}{\eta} \int_H^{x_2} \frac{\partial p}{\partial x_1} \, dx'_2 \, dx_2 \tag{8-G-9a}$$

$$v_3 = \int_H^{x_2} \frac{1}{\eta} \int_H^{x_2} \frac{\partial p}{\partial x_3} \, dx'_2 \, dx_2 \tag{8-G-9b}$$

The mean velocity field is

$$\bar{v}_1 = \frac{1}{H} \int_0^H v_1 \, dx_2, \qquad \bar{v}_3 = \frac{1}{H} \int_0^H v_3 \, dx_2 \tag{8-G-10a, b}$$

By continuity, we must have Eq. (8-D-12b).

For a Newtonian fluid, this continuity equation takes on an especially simple form. The quantities $\bar{v}_1$ and $\bar{v}_3$ from Eq. (8-D-11) may be written

$$\bar{v}_1 = \frac{H^2}{12\eta} \left[-\frac{\partial p}{\partial x_2} \right], \qquad \bar{v}_3 = \frac{H^2}{12\eta} \left[-\frac{\partial p}{\partial x_3} \right] \tag{8-G-11}$$

We may write Eq. (8-D-13) as

$$\frac{\partial}{\partial x_1} \left[H^3 \frac{\partial p}{\partial x_1} \right] + \frac{\partial}{\partial x_3} \left[H^3 \frac{\partial p}{\partial x_3} \right] = 0 \tag{8-G-12}$$

where we have considered the possibility of H varying in the 1 and 3 directions. These are special cases of Eqs. (8-D-11) and (8-D-13). If H is constant, we obtain Eq. (8-D-14), Laplace's equation, and Hele-Shaw flow. Mold-filling patterns are determined by potential theory. For complex molds we may use Tadmor's FAN method, described in Section 8-D-2. This is discussed by Broyer et al. (*22*).

For a non-Newtonain fluid, the situation is much more complex. We again begin with Eq. (8-G-8), and Eq. (8-D-12b) is valid. However, beyond this, the situation is much more complex. If we choose a viscosity–shear rate relationship combining low-shear-rate Newtonian and high-shear-rate power-law behavior, that is,

$$\eta = \frac{\eta_0}{1 + A\dot{\gamma}^{1-n}} \tag{8-G-13a}$$

we must begin with a velocity field involving components $v_1(x_2)$ and $v_3(x_2)$ and use the form (see Section 8-E-2c)

$$\eta = \frac{\eta_0}{1 + A[(\partial v_1/\partial x_2)^2 + (\partial v_3/\partial x_2)^2]^{(1-n)/2}} \tag{8-G-13b}$$

We may proceed as follows: from Eqs. (8-G-9) and (8-G-10), we have (151)

$$\bar{v}_1 = S\frac{\partial p}{\partial x_1}, \qquad \bar{v}_3 = S\frac{\partial p}{\partial x_3} \tag{8-G-14}$$

where

$$S = \frac{1}{H}\int_0^H \frac{z^2\, dz}{\eta}$$

From Eq. (8-D-12b),

$$\frac{\partial}{\partial x_1}(\bar{v}_1 H) + \frac{\partial}{\partial x_3}(\bar{v}_3 H) = \frac{\partial}{\partial x_1}\left[HS\frac{\partial p}{\partial x_1}\right] + \frac{\partial}{\partial x_3}\left[HS\frac{\partial p}{\partial x_3}\right]$$

or

$$HS\nabla^2 p + VHS\cdot\nabla p = 0 \tag{8-G-15}$$

The second term largely involves non-Newtonian effects. Solving Eq. (8-G-15) with Eq. (8-G-14) yields the velocity and pressure fields for an isothermal model. The problem is, however, very nonisothermal. The formulation given above is to be simultaneously solved with Eq. (8-G-4). Various research groups have carried out simulations of injection molding by solving these equations using various numerical techniques (15, 22, 64, 69, 70, 151).

The solutions of the equations have similar characteristics whether we are dealing with Newtonian or non-Newtonian formulations. Near the gate, the flow emerges in circles of ever-increasing radius until they meet with a barrier. This is basically the point-source solution of inviscid flow theory which follows readily from Eq. (8-D-14) in polar coordinates. If everything is symmetric about a point and the radial variable need only be considered,

$$\frac{1}{r}\frac{\partial}{\partial r}\left[r\frac{\partial p}{\partial r}\right] = 0 \tag{8-G-16}$$

where p has a logarithmic solution and

$$v_r = \frac{Q}{2\pi r}, \qquad v_\theta = 0 \tag{8-G-17}$$

As $r \to \infty$, a flat advancing front is found locally. However, in a real mold there are barriers and solutions are more complex.

H. STRATIFIED TWO-PHASE FLOW IN EXTRUSION AND INJECTION MOLDING

There have been extensive investigations of stratified two-phase flow of polymer melts in extrusion and in injection-molding operations. Coextrusion of polymer melts (*126*) (Fig. 8-H-1) and sandwich injection molding (Fig. 8-H-2) (*109*) are important processing operations which produce multilayer sheet/film, bicomponent fibers, and skin-core molded parts.

These process technologies exhibit a range of processing problems associated with differences in rheological properties. These are shown schematically in Figure 8-H-3 for coextrusion and Figure 8-H-4 for sandwich molding. In coextrusion one use usually seeks to achieve flat interfaces from two melts being metered side by side through a die. However, one frequently obtains a curved interface, as shown in Figure 8-H-3 with one polymer encapsulating a second (*51,84,94,132*). In sandwich injection molding, one wishes to achieve a uniform skin about a core produced by the second melt injected into the mold. This desirable result is often not what is achieved, but rather a burst of the second polymer through the first, with a very nonuniform skin, as shown in Figure 8-G-4 (*177*).

The problems of both Figures 8-H-3 and 8-H-4 have been shown to be associated with differences in viscosities of the two-melt phases. The lower-viscosity coextruded melt encapsulates the higher-viscosity melt in the coextrusion process illustrated in Figure 8-H-3 (*51, 84, 94, 132*). This has been explained (*166*) through arguments in terms of hydrodynamic lubrication theory. The same pressure gradient and flow rates must exist in each phase. This is only possible when viscosity differences exist with the lower-viscosity material being concentrated in the higher shear-rate regions, that is, at the die wall. Pressure differences across the interface will push the lower-viscosity melt in this region.

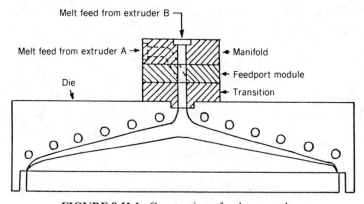

FIGURE 8-H-1. Coextrusion of polymer melts.

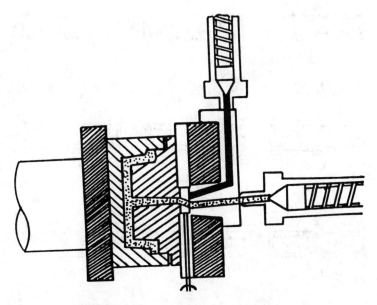

FIGURE 8-H-2. Sandwich injection molding.

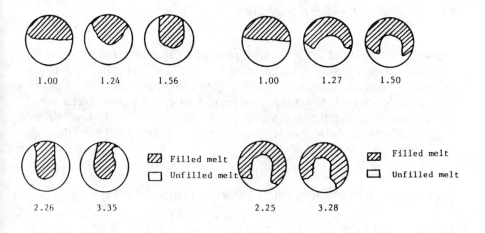

1.00 1.24 1.56   1.00 1.27 1.50

2.26 3.35   Filled melt   2.25 3.28   Filled melt
               Unfilled melt   Unfilled melt

(a)   (b)

FIGURE 8-H-3. Encapsulation during extrusion through a circular die. (a) $\eta_{\text{filled}} >$ η_{unfilled}; (b) $\eta_{\text{filled}} > \eta_{\text{unfilled}}$.

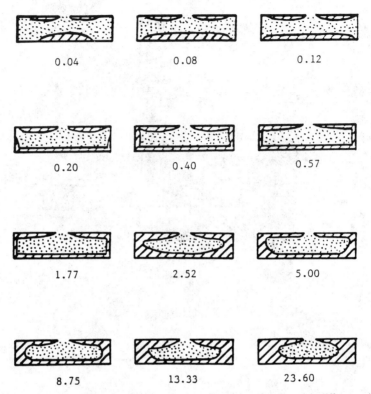

FIGURE 8-H-4. Sandwich injection molding using pairs of melts with different viscosity ratios. The number cited is the ratio of the viscosity of the second melt divided by the viscosity of the first melt.

Similar hydrodynamic lubrication theory arguments arise in sandwich molding. As the pressure gradient should be the same on both sides of the interface, lower-viscosity melts will tend to move at high mean velocities and flow rates. If the second melt injected has the lower-viscosity, it will move at a higher mean velocity through the cross section. If the difference with the first injected melt, which is along the wall, is large enough, the second melt will burst through, as has been observed.

I. POST-DIE EXTRUSION PROCESSING (FIBERS, FILM)

1. General and Force Balances

a. Melt Spinning and Cast Film

When a melt emerges from a die it is often taken up and drawn down to smaller dimensions. This is notably the case for melt spinning of fibers and tubular and

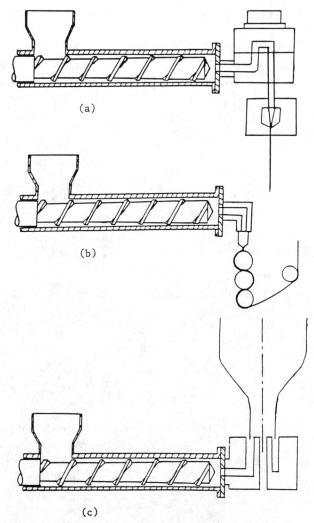

(a)

(b)

(c)

FIGURE 8-I-1. Melt-spinning and film-extrusion processes.

cast films, as shown in Figure 8-I-1. In melt-spinning and cast film the deformations and applied forces are uniaxial in character. In tubular film and when tentering frames are used in sequence with the cast-film extrusion process, the deformations are biaxial in character. In any case, the polymer is no longer encased between solid walls and the deformations are elongational and not shearing in character.

Let us first turn our attention to the melt-spinning and cast-film processes. In these basically uniaxial processes, the deformation rate tensor has the form

$$
\mathbf{d} = \begin{bmatrix} \partial v_1/\partial x_1 & 0 & 0 \\ 0 & -1/2(\partial v_1/\partial x_2) & 0 \\ 0 & 0 & -\tfrac{1}{2}(\partial v_1/\partial x_1) \end{bmatrix} \tag{8-I-1}
$$

A force balance between position x_1 in the spin line of "film line" and the take-up roll has the form (87, 159, 176, 177–180)

$$
F_L + F_{grav} - F_x - F_{drag} = G(v_L - v) \tag{8-I-2}
$$

where F_L and F_x are the forces at positions L and x. Here F_{grav} and F_{drag} are gravitational and air-drag forces given by

$$
F_{grav} = \int_{x_1}^{L} \rho g A \, dx_1, \qquad F_{drag} = \int_{x_1}^{L} \sigma_f C \, dx_1 \tag{8-I-3a, b}
$$

where A is the cross-sectional area and C the circumference. For cylindrical fibers and film, respectively,

Fibers

$$
A = \frac{\pi d^2}{4}, \qquad C = \pi d \tag{8-I-4a}
$$

Film

$$
A = WH, \qquad C = 2W \tag{8-I-4b}
$$

The term $G(v_L - v)$ is the difference in momentum flux between position L and x.

The relative magnituoe of the terms of the force balance vary with the take-up velocity and drawdown ratio v_L/v_0. At low drawdown ratios, Eq. (8-I-2) has the form

$$
F_{grav} - F_x = 0 \tag{8-I-5a}
$$

With increasing drawdown, the take-up tension F_L becomes greater than gravitation and we obtain

$$
F_L - F_x = 0 \tag{8-I-5b}
$$

This has been the most-studied asymptote. At higher take-up velocities and drawdowns, air drag and momentum fluxes become increasingly important and we have

$$
F_L - F_x - F_{drag} = G(v_L - v_0) \tag{8-I-5c}
$$

Spun-bonding processes involve fibers being drawn down by air jets rather than take-up rolls (see Fig. 8-I-1a). The dynamics of such processes have been investigated by Chen et al. (28) terms similar to those described above.

b. Tubular Film

The analysis of the dynamics of film processes where air inflates the material emerging from an annular die is more complex. The deformation rate component $\partial v_3/\partial x_3$ reflects the thinning of the film and $\partial v_2/\partial x_2$ the radial inflation. It may be seen that the deformation rate tensor will be

$$\mathbf{d} = \frac{Q\cos\theta}{2\pi RH}
\begin{bmatrix}
-\dfrac{1}{H}\dfrac{dH}{dz} - \dfrac{1}{R}\dfrac{dR}{dz} & 0 & 0 \\[2ex]
0 & \dfrac{1}{R}\dfrac{dR}{dz} & 0 \\[2ex]
0 & 0 & \dfrac{1}{H}\dfrac{dH}{dz}
\end{bmatrix}
\tag{8-I-6}$$

where z is the axial direction. The kinematics of tubular film extrusion may be represented in terms of a three-dimensional space for drawdown ratio v_L/v_0, blowup ratio R_L/R_0 and frostline height z_f.

The dynamics of the tubular film process may be analyzed using membrane theory (Section 2-I). This was first carried out by Pearson and Petrie (111, 114). We begin with Eqs. (2-I-1a–c) and neglect shear stresses. We presume symmetry in the circumferential direction (111, 114), an action which allows us to neglect Eq. (2-K-1b). Equation (2-I-1a) represents the force balance in the machine direction. We may eliminate $\sigma_{\phi\phi}$ with Eq. (2-I-1c) and integrate. This gives

$$F_L = 2\pi RH\sigma_{11}\cos\theta + \pi\Delta p(R_L^2 - R^2) \tag{8-I-7a}$$

$$\Delta p = \frac{H\sigma_{11}}{R_1} + \frac{H\sigma_{22}}{R_2} \tag{8-I-7b}$$

where θ is the angle the film makes with the vertical and R_1 and R_2 are the radii of curvature in the 1 and 2 directions.

2. Modeling

a. Melt Spinning

Two classes of modeling have been reported for the fiber spin line and for tubular-film extrusion. The simplest line of models are isothermal ones. These have been reported for both Newtonian fluids and viscoelastic models in each case. If the melts emerge and are quenched in the air, they are of little value, as the rheological properties change rapidly and continuously through solidification. They are, however, useful in systems with constant temperature chambers or in cases where there is a slow cooling followed by a rapid quench.

In these post-die highly nonisothermal processing operations, the variations of properties with temperature and rates of cooling are the key factors for understanding. The basic analysis of melt spinning is that of Kase and Matsuo (74), who combine Eq. (8-I-5b) with heat balance, presuming the polymer melt responds as a Newtonian fluid whose properties vary with temperature. Specifically, they write

$$F_L = \frac{\pi d^2}{4}\sigma_{11}(x_1) = \frac{\pi d^2}{4}\chi(T)\frac{dy_1}{dx_1} \tag{8-I-8}$$

$$\rho Q c \frac{dT}{dx_1} = -\pi d[h(T - T_s) + \varepsilon\lambda(T^4 - T_s^4)] \tag{8-I-9}$$

where

$$Q = \frac{\pi d^2}{4}v_1, \qquad G = \rho Q \tag{8-I-10}$$

Here $\chi(T)$ is a nonisothermal elongational viscosity, h is the heat transfer coefficient, ε is the emissivity, and λ is the Stefan–Boltzmann constant. Crystallization is neglected. As mentioned earlier, for a rapid-quench situation the dynamics of the problem are dominated by the temperature dependence of the properties and proper representation of heat transfer.

It is, however, of interest to discuss the solutions of Eqs. (8-I-8) and (8-I-9) when we take χ as a constant. This allows us to solve Eq. (8-I-8) to give an exponential solution for $v_1(x_1)$ and $d(x_1)$

$$v_1(x_1) = v_1(0)e^{\rho F_L x_1/G\chi}, \qquad d(x_1) = d(0)e^{-\rho F_L x_1/2G\chi} \tag{8-I-11a, b}$$

The temperature profile has the form

$$(T - T_s)(x_1) = (T_0 - T_s)e^{-\int_0^{x_1}(\pi\,dh/Gc)\,dx_1} \tag{8-I-12}$$

Both the diameter and temperature should fall roughly exponentially with position.

Kase and Matsuo (74) have experimentally determined the local heat-transfer coefficient h by blowing air over hot wires. They determine the form

$$\frac{hd}{k_a} = 0.42\left(\frac{dv\,\rho_a}{\eta_a}\right)^{1/3} \tag{8-I-13}$$

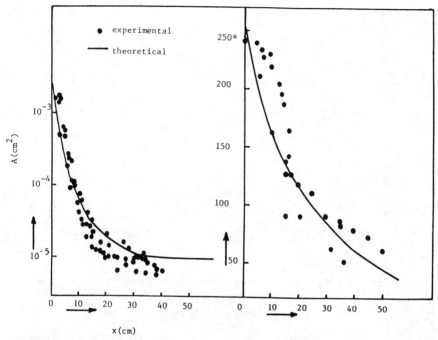

FIGURE 8-I-2. Comparison of theory and experience for the Kase–Matsuo theory of melt spinning (published with permission).

The elongational viscosity $\chi(T)$ varies with tempeature through

$$\chi = Ae^{-E/RT} \tag{8-I-14}$$

where E is an activation energy of flow. These authors describe the simultaneous solution of Eqs. (8-I-8) through (8-I-10) using this heat-transfer relationship.

The predictions of the computations are shown in Figure 8-I-2.

The effects of viscoelasticity on the behavior of polymer melts in the spinline has been explored in papers by Denn, Avenas and Petrie (*38*) and Fisher and Denn (*43*). If a convected Maxwell model such as (7-F-8) is used, the material exhibits a strain hardening behavior which causes the thread line to thin down more gradually.

b. Tubular-Film Extrusion

The dynamics of tubular-film extrusion have been analyzed by Pearson and Petrie (*113*) and various other investigators (*72, 150*). In a Newtonian fluid model,

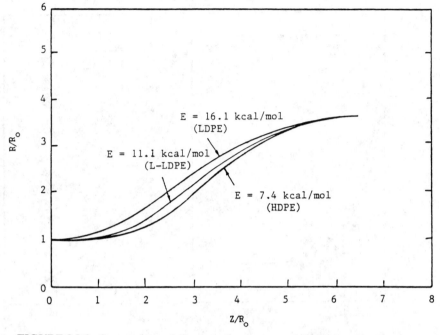

FIGURE 8-I-3. Comparison of theory and experiment for tubular-film extrusion.

one takes

$$\sigma_{11} = \sigma_{11} - \sigma_{33} = \chi \frac{Q \cos \theta}{\pi R H} \left(\frac{1}{R} \frac{dR}{dz} - \frac{2}{H} \frac{dH}{dz} \right) \qquad (8\text{-}I\text{-}15)$$

$$\sigma_{22} = \sigma_{22} - \sigma_{33} = \chi \frac{Q \cos \theta}{\pi R H} \left(\frac{1}{R} \frac{dR}{dz} - \frac{1}{H} \frac{dH}{dz} \right) \qquad (8\text{-}I\text{-}16)$$

The isothermal Newtonian fluid problem was solved by Pearson and Petrie (*113*). The nonisothermal problem is of more interest. If we neglect crystallization, the heat balance is

$$\rho c Q \cos \theta \frac{dT}{dz} = -2\pi R [h(T - T_\text{s}) + \varepsilon \lambda (T^4 - T_\text{s}^4)] \qquad (8\text{-}I\text{-}17)$$

Kanai and White (*71*) have experimentally determined the local heat-transfer coefficient, which may be expressed

$$\frac{hL}{k_\text{a}} = 0.045 \left(\frac{L U \rho_\text{a}}{\eta_\text{a}} \right)^{0.76}, \qquad z < K$$

$$h = 2.5 \, v_\text{max}^{1.6}, \qquad z > K \qquad\qquad (8\text{-}I\text{-}18)$$

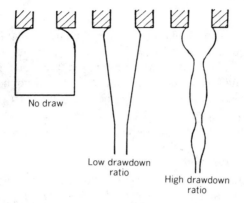

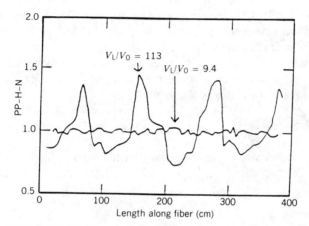

FIGURE 8-I-4. Draw resonance instability in melt spinning.

where K is the frostline height. Equation (8-I-14) was used for the temperature dependence of the viscosity.

Kanai and White (72) describe the solution of these equations using this heat-transfer coefficient. The most striking result of their calculation is that the bubble shape at fixed drawdown ratio, blowup ratio, and frostline height is primarily determined by the dimensionless group E/RT (the Griffith number). Large E/RT makes the bubble diameter increase gradually (see Fig. 8-I-3). Kanai and White also introduce the effect of crystallization, but this proves of second-order influence. Yamane and White (175) have considered this situation in more detail for a power a fluid. They find that E/RT tends to dominate non-Newtonian behavior in determining bubble shape. The wine-glass stem of HDPE bubbles as opposed to LDPE is associated with the lower value of E/RT of HDPE. The effects of melt elasticity on bubble shape are considered by Luo and Tanner (85), who also note the great importance of nonisothermal characteristics.

3. Instabilities

The fiber produced in melt spinning and cast film and tubular film are invariably nonuniform in thickness. These variations may be due to system noise, such as fluctuating extrusion rate or heat transfer. On the other hand, the system nonuniformities may be due to mechanical or hydrodynamic instabilities. Generally, the thickness fluctuations associated with the former problem are small in magnitude and random (73, 97). The latter are usually periodic in character and possess large amplitudes.

a. Melt Spinning/Cast Film

The most widely studied instability phenomenon observed in fibers and film is the draw resonance phenomenon (14, 45, 73, 76, 95, 96, 112, 174), which occurs at a critical drawdown ratio (see Fig. 8-I-4). The random noise observed in fiber or film thickness suddenly becomes a periodic high-amplitude disturbance. This phenomenon was first properly analyzed by Kase et al. (76) and by Pearson and Matovich (112) using the methods of linear stability theory [e.g., see Denn (38)]. Shah and Pearson (130) initiated studies of nonisothermal behavior and Fisher and Denn (43) studied materials with viscoelastic characteristics. One proceeds by introducing a disturbance of the form

$$v(x, t) = v(x)[1 + \beta(x, t)], \qquad d(x, t) = d(x)[1 + \alpha(x, t)] \quad \text{(8-I-19a, b)}$$

The continuity, force, and energy equations are rewritten as (for Newtonian fluid model)

$$\rho c \frac{\partial \theta}{\partial t} + \frac{\partial \theta}{\partial x} = - h\pi \, d\theta$$

$$\frac{\partial}{\partial t} \frac{\pi d^2}{4} + \frac{\partial}{\partial x_1} \frac{(\pi d^2 v_1)}{4} = 0 \qquad \text{(8-I-20a, b, c)}$$

$$F_L = \frac{\pi d^2}{4} \chi \frac{dv_1}{dx_1}$$

where θ is $T - T_s$. Equation (8-I-19) would be substituted into Eq. (8-I-20).

Quenching is found both experimentally and theoretically to stabilize the spinline. The degree of stabilization depends upon two dimensionless groups, the Stanton number, Eq. (8-C-24), of the air stream and the Griffith number, Eq. (8-C-21).

If there are quench baths or rolls close to the die, the behavior is approximately isothermal. Isothermal experimental studies have shown that Newtonian fluids have a critical drawdown ratio of about 21, which is confirmed by theoretical calculation (76, 112). Fisher and Denn (43) have found that for isothermal spinning using a convected Maxwell model, the Weissenberg number is

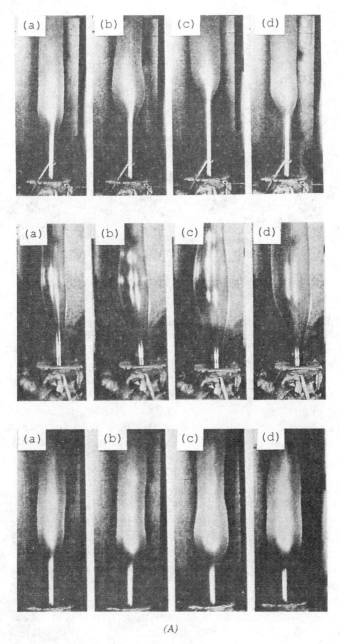

(A)

FIGURE 8-I-5. Tubular film instabilities: (*A*) Han–Park instability; (*B*) helical instability in tubular film extrusion.

(B)

FIGURE 8-I-5. (*Continued*)

stabilizing, but the deformation rate-softening parameter of what we call the Yamamoto number is destabilizing.

Linear-chain polymer melts of broad molecular-weight disturbance generally exhibit more unstable behavior than Newtonian fluids. Stability is improved by narrowing the molecular-weight distribution (*197, 173*). Long-chain branched polymer melts exhibit much higher levels of stability (*96, 172*). As noted above, stability analyses based on the convected model show melt elasticity through the Weissenberg number to be stabilizing, but the deformation rate-softening parameter of the Yamamoto number of Eq. (8-C-6) to be destabilizing (*43*). Generally, long-chain branched polymer melts such as LDPE seem to be strain hardening under extension relative to linear polymers (*96*).

b. Tubular-Film Extrusion

Under all operating conditions, the tubular-film process appears subject to low-amplitude instabilities (*46, 56, 57, 71, 96*). It is much more unstable than tubular-film extrusion. Several different instabilities have been observed and are shown in Figure 8-I-5. There is a bubble instability first described by Han and Park (*56, 57*), where periodic disturbances travel along the length of the bubble. This generally occurs at low blow-up rates. A second instability has been observed at high blow-up ratios in which a helical motion is set up (*71*). A third type of instability involves a metastable situation where two steady states exist (*96*). Figure 8-I-6 shows the occurrence of different types of instabilities for various regions of drawdown ratio v_L/v_0, blowup ratio R_L/R_0, and plane at constant frostline height.

As all experiments have been carried out under nonisothermal conditions, the relative importance of rheological and heat-transfer effects on these instabilities is not clear. Almost all studies have been for polyethylenes. The most stable polymers investigated are long-chain branched polyethylenes (*96, 172b*) (LDPE-1). Broad molecular-weight high-density polyethylenes (HDPE-1) are more

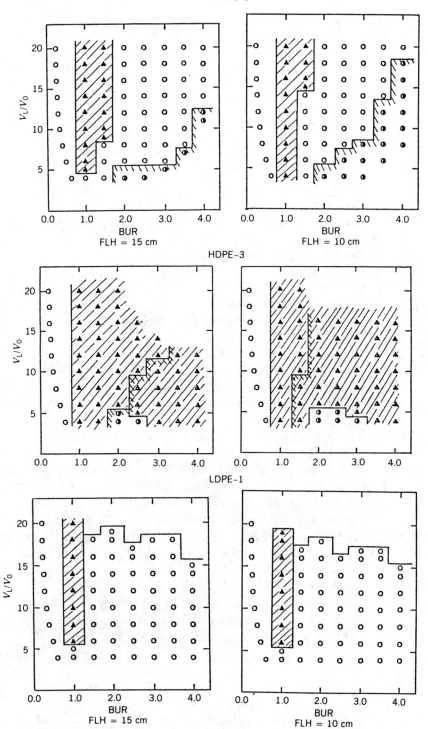

FIGURE 8-I-6. Regions of stable and unstable operating conditions: ○ stable; ▲ bubble instability (B.I.); ◐ helical instability; ▲ helical and bubble instability.

stable then narrower molecular-weight-distribution polymers (HDPE-3).

The sources of instability of tubular film bubbles has been critically considered in a recent paper by Cain and Denn (25).

REFERENCES

1. E. B. Adams, J. C. Whitehead, and D. C. Bogue, *AIChE J.*, **11**, 1026 (1965).
2. T. Arai, H. Ishikawa, and H. Hatta, *Kobunshi Ronbunshu*, **38**, 29 (1981).
3. H. Ashizawa, J. E. Spruiell, and J. L. White, *Polym. Eng. Sci.*, **24**, 1055 (1984).
4. G. Astarita, personal communication, 1976.
5. E. B. Bagley, *J. Appl. Phys.*, **28**, 624 (1957).
6. E. B. Bagley and A. M. Birks, *J. Appl. Phys.*, **31**, 556 (1960).
7. E. B. Bagley and H. P. Schreiber, *Trans. Soc. Rheol.*, **5**, 341 (1961).
8. T. F. Ballenger and J. L. White, *J. Appl. Polym. Sci.*, **15**, 1949 (1971).
9. R. L. Ballman, T. Shusman, and H. L. Toor, *Ind. Eng. Chem.*, **51**, 847 (1959).
10. R. L. Ballman, T. Shusman, and H. L. Toor, *Mod. Plast.*, **37**, (Sept.), 105 (1959); **37**, (Oct.), 115 (1959).
11. I. T. Barrie, *Plast. Polym.*, **37**, 463 (1969).
12. J. T. Bergen in *Processing of Thermoplastic Materials*, E. C. Bernhardt, Ed., Reinhold, New York, 1959.
13. J. T. Bergen and G. N, Scott, *J. Appl. Mech.*, **18**, 101 (1951).
14. A. Bergenzoni and A. J. DiCresce, *Polym. Eng. Sci.*, **6**, 45 (1966).
15. J. L. Berger and C. G. Gogos, *Polym. Eng. Sci.*, **13**, 102 (1972).
16. E. C. Bernhardt, Ed., *Processing of Thermoplastic Materials*, Reinhold, New York, 1959.
17. E. C. Bernhardt, Ed., *CAE for Injection Molding*, Hanser Munich, 1983.
18. R. B. Bird, *SPE J.*, Sept., 35 (1955).
19. R. B. Bird, W. E. Stewart, and E. N. Lightfoot, *Transport Phenomena*, Wiley, New York, 1960.
20. D. Boger and H. Nguyen, *J. Non-Newt. Fluid Mech.*, **5**, 353 (1978).
21. H. C. Brinkman, *Appl. Sci. Res.*, **A2**, 120 (1951).
22. E. Broyer, C. Gutfinger, and Z. Tadmor, *Trans. Soc. Rheol.*, **19**, 423 (1975).
23. R. Brzoskowski, J. L. White, W. Szydlowski, N. Nakajima, and J. L. White, *Int. Polym. Proc.*, **3**, 134 (1988).
24. E. Buckingham, *Proc. ASTM*, **21**, 1154 (1921).
25. J. J. Cain and M. M. Denn, *Polym. Eng. Sci.*, **28**, 1527 (1988).
26. J. F. Carley, R. S. Mallouk, and J. M. McKelvey, *Ind. Eng. Chem.*, **45**, 974 (1953).
27. Y. Chan, J. L. White, and Y. Oyanagi, *Polym. Eng. Sci.*, **18**, 268 (1978).
28. C. H. Chen, J. L. White, J. L. Spruiell, and B. C. Goswami, *Textile Res. J.*, **53**, 44 (1983).
29. P. L. Clegg in *Rheology of Elastomers*, Pergamon, London, 1958, p. 174.
30. P. L. Clegg, *Trans. Plast. Inst.*, **28**, 245 (1960).
31. F. N. Cogswell, *Polym. Eng. Sci.*, **12**, 64 (1972).
32. H. W. Cox and C. W. Macosko, *AIChE J.*, **20**, 785 (1984).

33. W. O. Criminale, J. L. Erickson, and G. L. Filbey, *Arch. Rat. Mech. Anal.*, **1**, 410 (1958).

34. M. J. Crochet, *Rubber Chem. Technol.*, **62**, 426 (1989).

35. M. J. Crochet, A. R. Davies, and K. Walters, *Numerical Simulation of Non-Newtonian Flow*, Elsevier, Amsterdam, 1984.

36. M. J. Crochet and R. Keunings, *J. Non-Newt. Fluid Mech.*, **10**, 85, 339 (1982).

37. M. M. Denn, *Stability of Reaction and Transport Processes*, Prentice-Hall, Englewood Cliffs, NJ, 1975.

38. M. M. Denn, C. J. S. Petrie, and P. J. Avenas, *AIChE J.*, **21**, 791 (1975).

39. W. Dietz and J. L. White, *Rheol. Acta*, **17**, 676 (1978).

40. J. H. Dillon and N. Johnston, *Physics*, **4**, 225 (1933).

41. A. G. Dodson, P. Townsend, and K. Walters, *Comput. Fluids*, **2**, 317 (1974).

42. T. R. Fields and D. C. Bogue, *Trans. Soc. Rheol.*, **12**, 39 (1968).

43. R. J. Fisher and M. M. Denn, *AIChE J.*, **22**, 236 (1976).

44. A. G. Fredrickson and R. B. Bird, *Ind. Eng. Chem.*, **50**, 347 (1958).

45. H. I. Freeman and H. J. Coplan, *J. Appl. Polym. Sci.*, **8**, 2389 (1964).

46. M. Fleissner, *Int. Polym. Proc.*, **2**, 279 (1988).

47. R. E. Gee and J. B. Lyon, *Ind. Eng. Chem.*, **49**, 950 (1957).

48. H. Giesekus, *Rheol. Acta*, **4**, 229 (1965).

49. H. Giesekus, *Rheol. Acta*, **7**, 127 (1968).

50. R. M. Griffith, *IEC Fund.*, **1**, 180 (1962).

51. C. D. Han, *J. Appl. Polym. Sci.*, **17**, 1289 (1973).

52. C. D. Han, *J. Appl. Polym. Sci.*, **19**, 2403 (1975); *Rheol. Acta*, **14**, 173 (1975).

53. C. D. Han, *Rheology in Polymer Processing*, Academic, New York, 1976.

54. C. D. Han, M. Charles, and W. Philippoff, *Trans. Soc. Rheol.* **13**, 455 (1969).

55. C. D. Han and L. H. Drexler, *J. Appl. Polym. Sci.*, **17**, 2329 (1973).

56. C. D. Han and J. Y. Park, *J. Appl. Polym. Sci.*, **19**, 3291 (1975).

57. C. D. Han and R. Shetty, *IEC Fund.*, **16**, 49 (1977).

58. W. J. Harrison, *Proc. Camb. Phil. Soc.*, **19**, 307 (1919).

59. D. H. Harry and R. G. Parrott, *Polym. Eng. Sci.*, **10**, 209 (1970).

60. M. D. Hersey, *Physics*, **7**, 403 (1936).

61. R. O. Herzog and K. Weissenberg, *Kolloid Z.*, **47**, 277 (1928).

62. D. C. Huang and J. L. White, *Polym. Eng. Sci.*, **19**, 609 (1979).

63. D. C. Huang and J. L. White, *Polym. Eng. Sci.*, **20**, 182 (1980).

64. A. I. Isayev, M. Sobhanie, and J. S. Deng, *Rubber Chem. Technol.*, **61**, 906 (1988).

65. A. I. Isayev and R. K. Uphadhyay, *J. Non-Newt. Fluid Mech.*, **19**, 135 (1985).

66. H. Janeschitz-Kriegl, *Rheol. Acta*, **16**, 327 (1977).

67. G. B. Jeffrey, *Phil. Mag.*, **29**(6), 455 (1915).

68. D. S. Kalika and M. M. Denn, *J. Rheology*, **31**, 815 (1987).

69. P. N. Kaloni, *J. Phys. Soc. Japan*, **20**, 152 (1965); **20**, 610 (1965).

70. M. R. Kamal, Y. Kuo, and P. H. Doan, *Polym. Eng. Sci.*, **12**, 294 (1972).

71. T. Kanai and J. L. White, *Polym. Eng. Sci.*, **24**, 1185 (1984).

72. T. Kanai and J. L. White, *J. Polym. Eng.*, **5**, 135 (1985); T. Kanai, J. L. White, M. Tomikawa, and J. Shimizu, *Seni Gakkaishi*, **40**, T-465 (1984).

73. S. Kase, *J. Appl. Polym. Sci.*, **18**, 3279 (1974).

74. S. Kase and T. Matsuo, *J. Polym. Sci.*, **A3**, 2541 (1965).

75. S. Kase and T. Matsuo, *J. Appl. Polym. Sci.*, **11**, 251 (1967).

76. S. Kase, T. Matsuo, and Y. Yoshimoto, *Seni Kikai Gakkaishi*, **19**, T63 (1966).

77. R. Keunings and M. J. Crochet, *J. Non-Newt. Fluid Mech.*, **14**, 279 (1984).

78. R. L. Kruse, *Polymer Lett.*, **2**, 841 (1964).

79. Y. Kuo and M. R. Kamal, *AIChE J.*, **22**, 661 (1976).

80. W. M. Laird, *Ind. Eng. Chem.*, **49**, 138 (1957).

81. H. Lamb, *Hydrodynamics*, 6th ed., Cambridge University Press, London, 1945.

82. W. E. Langlois and R. S. Rivlin, Technical Report #3, Division of Applied Mathematics, Brown University, Providence, 1959.

83. W. E. Langlois and R. S. Rivlin, *Rend Mat.*, **22**, 169 (1963).

84. B. L. Lee and J. L. White, *Trans. Soc. Rheol.*, **18**, 467 (1974).

85. X.-L. Luo and R. I. Tanner, *Polym. Eng. Sci.*, **25**, 620 (1985).

86. C. Y. Ma, J. L. White, F. C. Weissert, and K. Min, *J. Non-Newt. Fluid Mech.*, **17**, 275 (1985); *Polym. Compos.*, **6**, 215 (1985).

87. M. A. Matovich and J. R. A. Pearson, *IEC Fund.*, **8**, 512 (1969).

88. J. M. McKelvey, *Polymer Processing*, Wiley, New York, 1962.

89. A. B. Metzner, W. T. Houghton, R. A. Sailor, and J. L. White, *Trans. Soc. Rheol.*, **5**, 133 (1961).

90. A. B. Metzner, J. L. White, and M. M. Denn, *AIChE J.*, **12**, 863 (1966); J. L. White and N. Tokita, *J. Appl. Polym. Sci.*, **11**, 321 (1967).

91. S. Middleman, *Trans. Soc. Rheol.*, **9**, 83 (1965).

92. S. Middleman and J. Gavis, *Phys. Fluids*, **4**, 335 (1961).

93. K. Min, J. L. White, and J. F. Fellers, *Polym. Eng. Sci.*, **24**, 1327 (1984).

94. N. Minagawa and J. L. White, *Polym. Eng. Sci.*, **15**, 825 (1975).

95. W. Minoshima and J. L. White, *Polym. Eng. Rev.*, **2**, 212 (1983).

96. W. Minoshima and J. L. White, *J. Non-Newt. Fluid Mech.*, **19**, 275 (1986).

97. W. Minoshima, J. L. White, and J. E. Spruiell, *J. Appl. Polym. Sci.*, **25**, 287 (1980).

98. M. Mooney, *J. Colloid Sci.*, **2**, 68 (1947).

99. M. Mooney in *Rheology*, Vol. 2, F. R. Eirich, Ed., Academic, New York, 1958.

100. M. Mooney, *Rubber Chem. Technol.*, **35**(5), XVII (1962).

101. M. Mooney and S. A. Black, *J. Colloid Sci.*, **1**, 204 (1952).

102. R. A. Morrete and C. G. Gogos, *Polym. Eng. Sci.*, **8**, 272 (1968).

103. N. Nakajima and M. Shida, *Trans. Soc. Rheol.*, **10**, 299 (1966).

104. H. K. Nason, *J. Appl. Phys.*, **16**, 338 (1945).

105. R. E. Nickell, R. I. Tanner, and B. Caswell, *J. Fluid Mech.*, **65**, 189 (1974).

106. W. Nusselt, *Gesundh Ing.*, **38**, 477 (1915).

107. K. Oda, J. L. White, and E. S. Clark, *Polym. Eng. Sci.*, **16**, 585 (1976).

108. J. G. Oldroyd, *Proc. Camb. Phil. Soc.*, **43**, 383 (1947).

109. D. F. Oxley and D. J. H. Sandiford, *Plast. Polym.*, **39**, 288 (1971).

110. J. R. A. Pearson, *Trans. Plast. Inst.*, **32**, 239 (1964).

111. J. R. A. Pearson, *Mechanical Principles of Polymer Melt Processing*, Pergamon, Oxford, 1966.

112. J. R. A. Pearson and M. A. Matovich, *IEC Fund.*, **8**, 605 (1969).

113. J. R. A. Pearson and C. J. S. Petrie, *Proc. 4th Int. Rheological Congress*, **3**, 265 (1965).

114. J. R. A. Pearson and C. J. S. Petrie, *Plast. Polym.*, **38**, 85 (1970); *J. Fluid Mech.*, **40**, 1 (1970); **42**, 609 (1970).

115. W. Philippoff, *Trans. Soc. Rheol.*, **1**, 95 (1957).

116. W. Philippoff and F. H. Gaskins, *Trans. Soc. Rheol.*, **2**, 263 (1958).

117. W. Prager, *Introduction to the Mechanics of Continua*, Ginn, Boston, 1961.

118. C. Rauwendaal, *Polymer Extrusion*, Hanser, Munich, 1985.

119. K. R. Reddy and R. I. Tanner, *Trans. Soc. Rheol.*, **22**, 661 (1978).

120. M. Reiner, *Physics Today*, Jan., 62 (1964).

121. O. Reynolds, *Phil. Trans. Roy Soc.*, **A174**, 935 (1883).

122. O. Reynolds, *Phil. Trans. Roy. Soc.*, **A177**, 157 (1886).

123. S. Richardson, *J. Fluid Mech.*, **56**, 609 (1972).

124. F. Rothemeyer, *Rheol. Acta*, **9**, 259 (1970).

125. H. Rouse, *History of Hydraulics*, Iowa Institute of Hydraulic Research, 1957.

126. W. J. Schrenk, *Plast. Eng.*, **30**(3), 65 (1974).

127. R. S. Schechter, *AIChE J.*, **7**, 445 (1961).

128. G. P. M. Schenkel, *Kunststoff-Extruder Technik*, Hanser, Munich, 1963, and earlier

129. G. P. M. Schenkel, *Int. Polym. Proc.*, **4**, 3 (1988).

130. Y. T. Shah and J. R. A. Pearson, *IEC Fund*, **11**, 145 (1972).

131. A. Silbar and P. R. Pasley, *ZAMM*, **37**, 441 (1958).

132. J. H. Southern and R. L. Ballman, *Appl. Polym. Symp.*, **20**, 175 (1973).

133. R. S. Spencer and R. E. Dillon, *J. Colloid Sci.*, **4**, 241 (1949).

134. R. S. Spencer and G. D. Gilmore, *Mod. Plast.*, **28**, 97 (1950).

135. R. S. Spencer and G. D. Gilmore, *J. Colloid Sci.*, **6**, 118 (1951).

136. J. F. Stevenson, *Polym. Eng. Sci.*, **18**, 557 (1978).

137. W. Szydlowski, R. Brzoskowski, and J. L. White, *Int. Polym. Proc.*, **1**, 207 (1987); *Adv. Polym. Technol.*, **7**, 177 (1987).

138. W. Szydlowski and J. L. White, *J. Non-Newt. Fluid Mech.*, **28**, 29 (1988).

139. Z. Tadmor, *Polym. Eng. Sci.*, **6**, 185 (1966); I. Klein and D. Marshall, *Polym. Eng. Sci.*, **6**, 199 (1966); Z. Tadmor, I. J. Duvdevani, and I. Klein, *Polym. Eng. Sci.*, **7**, 198 (1967).

140. Z. Tadmor, E. Broyer, and C. Gutfinger, *Polym. Eng. Sci.*, **14**, 660 (1974).

141. Z. Tadmor and C. G. Gogos, *Principles of Polymer Processing*, Wiley, New York, 1979.

142. R. I. Tanner, *J. Polym. Sci.*, **A-2**(8), 2067 (1970).

143. G. I. Taylor, *Proc. Roy. Soc.*, **A150**, 322 (1935).

144. H. L. Toor, *Ind. Eng. Chem.*, **48**, 922 (1956); *Trans. Soc. Rheol.*, **1**, 203 (1957).

145. H. L. Toor, R. L. Ballman, and L. Cooper, *Mod. Plast.*, **38**, (Dec.), 117 (1960); Proceedings of the International Congress on Technology in Polymer, Processing, October, 1960.

146. J. P. Tordella, *J. Appl. Phys.*, **27**, 454 (1956).

147. J. P. Tordella, *Rheol. Acta*, **1**, 216 (1958).

148. J. P. Tordella, *Trans. Soc. Rheol.*, **1**, 203 (1957).

149. J. P. Tordella, *J. Appl. Polym. Sci.*, **7**, 215 (1963).

150. M. H. Wagner, *Rheol. Acta*, **15**, 40 (1976).

151. K. K. Wang, S. F. Shen, C. A. Hieber, and J. F. Stevenson in *Science and Technology of Polymer Processing*, N. P. Suh and N. H. Sung, Eds., MIT, Cambridge, MA, 1979.

152. Y. Wang and J. L. White, *J. Non-Newt. Fluid Mech.*, **32**, 19 (1989).

153. K. Weissenberg, *Proc. 1st Int. Rheol. Cong.*, I-29 (1948).

154. G. Williams and H. A. Lord, *Polym. Eng. Sci.*, **15**, 533 (1975).

155. J. A. Wheeler and E. H. Wissler, *AIChE J.*, **11**, 206 (1965).

156. J. L. White, *J. Appl. Polym. Sci.*, **8**, 2339 (1964).

157. J. L. White, *Polym. Eng. Sci.*, **15**, 44 (1975).

158. J. L. White, *J. Non-Newt. Fluid Mech.*, **8**, 195 (1981).

159. J. L. White, *Polym. Eng. Rev.*, **1**, 297 (1981).

160. J. L. White in *Rheometry: Industrial Applications*, K. Walter, Ed., Wiley, New York, 1980.

161. J. L. White and H. B. Dee, *Polym. Eng. Sci.*, **14**, 212 (1979).

162. J. L. White and W. Dietz, *Polym. Eng. Sci.*, **19**, 1081 (1979).

163. J. L. White and D. C. Huang, *Polym. Eng. Sci.*, **21**, 1101 (1981).

164. J. L. White and D. C. Huang, *J. Non-Newt. Fluid Mech.*, **9**, 223 (1981).

165. J. L. White and Y. Ide, *J. Polym. Sci.*, **22**, 3057 (1978).

166. J. L. White and B. L. Lee, *Trans. Soc. Rheol.*, **19**, 457 (1975).

167. J. L. White and A. Kondo, *J. Non-Newt. Fluid Mech.*, **3**, 41 (1977).

168. J. L. White and W. Minoshima, *Polym. Eng. Sci.*, **21**, 1113 (1981).

169. J. L. White and J. F. Roman, *J. Appl. Polym. Sci.*, **20**, 1005 (1976).

170. J. L. White, W. Szydlowski, K. Min, and M. H. Kim, *Adv. Polym. Technol.*, **7**, 295 (1987).

171. J. L. White, Y. Wang, A. I. Isayev, N. Nakajima, F. C. Weissert, and K. Min, *Rubber Chem. Technol.*, **60**, 337 (1987); also Y. Wang and J. L. White, to be published; Y. Wang Ph.D. Dissertation, Department of Polymer Engineering, University of Akron, 1988.

172. J. L. White and H. Yamane, *Pure Appl. Chem.*, **57**, 1442 (1985); *Pure Appl. Chem.*, **59**, 193 (1987).

173. Y. Yamaguchi, Y. Oyanagi, Y. Sugai, and M. Kunitake, *Res. Rept. Kogakuin Univ.*, **24**, 71 (1968).

174. H. Yamane and J. L. White, *Polym. Eng. Sci.*, **23**, 516 (1983).

175. H. Yamane and J. L. White, *Int. Polym. Proc.*, **2**, 107 (1987).

176. H. Yasuda and H. Sugiyama, *Seni Gakkaishi*, **37**, T-497 (1981).

177. H. J. Yoo and C. D. Han, *J. Rheol.*, **25**, 115 (1981).

178. S. S. Young, J. L. White, E. S. Clark, and Y. Oyanagi, *Polym. Eng. Sci.*, **20**, 798 (1980).

179. A. Ziabicki, *Kolloid Z.*, **175**, 14 (1961).

180. A. Ziabicki, *Fundamentals of Fiber Formation*, Wiley, New York, 1976.

181. A. Ziabicki and K. Kedzierska, *Kolloid Z.*, **171**, 51 (1961).

INDEX